Technology, Power and Society

Technology, Power & Society

VOLUME 1

The titles published in this series are listed at *brill.com/tpso*

Technology, Power and Society

Critical Perspectives on the Global Digital Transformation

Edited by

Dennis Nguyen, Jing Zeng and Bruce Mutsvairo

BRILL

LEIDEN | BOSTON

The Library of Congress Cataloging-in-Publication Data is available online at https://catalog.loc.gov
LC record available at https://lccn.loc.gov/2025006386

Typeface for the Latin, Greek, and Cyrillic scripts: "Brill". See and download: brill.com/brill-typeface.

ISSN 2950-1598
ISBN 978-90-04-71138-9 (hardback)
ISBN 978-90-04-71139-6 (e-book)
DOI 10.1163/9789004711396

Contents

Acknowledgements

The editors thank each contributor for their chapters compiled in this volume. It was their hard work, dedication, and passion for critical research that made this book possible. Additionally, the editors would like to express their gratitude to Utrecht University. This concerns specifically the Department for Media and Culture Studies and the Institute for Cultural Inquiry (ICON) at the Faculty of Humanities as well as the Focus Area Governing the Digital Society (GDS).

Finally, we would like to thank Brill Publishers for enabling the publication of this inaugural volume and hosting the book series *Technology, Power & Society*, which aims to offer diverse perspectives and voices a platform to share their research with a broad audience interested in important questions and challenges related to the global digital transformation.

Dennis Nguyen — Utrecht, The Netherlands
Jing Zeng — Zurich, Switzerland
Bruce Mutsvairo — Utrecht, The Netherlands

Figures and Tables

Figures

Tables

Notes on Contributors

Crystal Abidin
(professor) is an anthropologist and ethnographer of internet cultures, focusing especially on influencer cultures, internet celebrities, online visibility, and social media pop cultures, especially in the Asia Pacific region. Crystal has published extensively in academic and popular media outlets, with over 80 articles and chapters on various aspects of vernacular internet cultures, and her sixth book is *TikTok and Youth Cultures* (2023, Emerald Publishing). Crystal is Professor and ARC DECRA Fellow in Internet Studies at Curtin University, Director of the Influencer Ethnography Research Lab (IERLab), and Founder of the TikTok Cultures Research Network. Reach her at wishcrys.com.

Afonso de Albuquerque
(PhD) is professor of Communication Program at Federal Fluminense University. Coordinator of the National Institute of Science and Technology in Disputes and Informational Sovereignty. Researcher of national council for scientific and technological development (CNPq).

Rhied Al-Othmani
is a PhD student at the University of Applied Sciences Utrecht and Utrecht University in the Netherlands. Currently, she teaches at the master's programme in Data-driven Design, where she develops courses and coaches' students on Human-AI interaction. Rhied is inspired by the role of "new" technologies in people's and organisations' lives, which is why she is pursuing her PhD on Conversational Agents. Her research explores chatbots and voice assistants, as well as their impact on organisational trust and public perception. This interdisciplinary study critically examines how citizens' interactions with automated communication shape trust in public organisations.

Stefan Bächtold
(PhD) is a research fellow at Monash University Malaysia interested in the intersections of technologies, political conflict, and power in Southeast Asia. A former research fellow at the Technical University of Munich and an associate researcher at the swisspeace foundation, his research draws on international political sociology, science & technology studies, but equally on

years of experience with practice and research in international development and peacebuilding in Africa, Asia, Europe, and the Middle East.

Bjorn Beijnon
is a PhD candidate at the Amsterdam School for Cultural Analysis, University of Amsterdam, and a lecturer at the Institute for Communication, HU University of Applied Sciences Utrecht. His research focusses on subjectification through datafication practices by big tech companies in contemporary surveillance cultures.

Christophe Declercq
(PhD) is an assistant professor at the University of Utrecht and Senior Honorary Research Fellow at University College London. After his PhD (Imperial College London) on the linguistic and socio-cultural identity of a temporary displacement, Christophe has moved into the domain of cross-cultural communication at times of crisis and conflict, either in a historic setting (including three books on Languages and the First World War, co-edited with Julian Walker, Palgrave-MacMillan 2016, Bloomsbury 2021) or in a contemporary setting (including *Intercultural Crisis Communication: Translating, Interpreting and Languages in Local Crises*, co-edited with Federico Federici, Bloomsbury 2019).

Daria Delavar-Kasmai
is a master's student in the New Media and Digital Culture program at the University of Amsterdam. She initiated the Empatia Program for addressing the mental health issues of displaced Ukrainian people and works as a journalist for Euromaidan Press.

Sarper Durmuş
previously held positions in journalism and publishing prior to embarking on his postgraduate studies. He now has a Ph.D. in communications from Istanbul Bilgi University, where he teaches undergraduate media studies and journalism courses. His general interests are digital platforms and global media.

Gys-Walt van Egdom
(PhD) holds a doctor's degree in Linguistics and Literary Studies and a master's degree in Translation Studies. He lectures on Translation and Translation Studies. His research interests include translation didactics, translation evaluation, translation ethics, translation processes and human-computer interaction. In 2022, he held the Émile Verhaeren Chair

at the Vrije Universiteit Brussel. He also recently served as a guest-editor for three "dossiers" on literary translation and technology: 'Literair vertalen en Technologie' (Filter [2021], co-editor: Joke Daems), 'Vertaling en technologie' (Kennisbank ELV [2022]), 'Computer-aided Literary Translation' (Tradumàtica Revista [2023], co-editor: Christophe Declercq)

Marloes Geboers
(PhD) is a postdoctoral researcher Platform (Sub)Cultures in the Humane AI and Global Digital Cultures research areas at the University of Amsterdam. She is also affiliated to the Digital Methods Initiative, UvA. Her research blends traditional visual methods with digital methods and automated analyses, to arrive at novel ways to study social media images. Her dissertation centred around social platforms and how they shape socio-technically constructed regimes of visibility.

Michelle H. S. Ho
(PhD) is an assistant professor of Feminist and Queer Cultural Studies in the Department of Communications and New Media (CNM) at the National University of Singapore (NUS). Her research interests lie broadly in issues of gender, sexuality, race and ethnicity, affect and emotion, and media and popular cultures in contemporary Asia. Her recent works on gender and media in Japan are published in *Feminist Media Studies* and *Social Media + Society*.

Julia Luteijn
is currently teaching in the Communication and Multimedia Design department at the University of Applied Sciences in Rotterdam. Her primary focus is on the field of creative technologies. Through her teaching, she aims to bridge the gap between practical technological skills and ethical considerations in our rapidly evolving digital society. In addition to her teaching, she works as an artist. Her art practice combines technology and data to explore digital transformations and the evolving zeitgeist, with a particular emphasis on themes like immersive technologies and artificial intelligence.

Yishu Mao
is a PhD candidate at the Department of Communication and Media Research, University of Zurich, and a Predoctoral Fellow at the Research Group "China in the Global System of Science" at Max Planck Institute for the History of Science. Her PhD project examines the emerging AI governance

regimes in China, Germany and the US and current research interests include the co-production of science and social order, sociotechnical imaginaries and communicative construction of AI, and global governance of AI.

Tatiane Mendes Pinto
(PhD) is a postdoctoral fellow of the Communication Program at Federal Fluminense University.

Bruce Mutsvairo
is a professor in the Department of Media and Culture Studies at Utrecht University. He studies the development of journalism and media in non-Western contexts.

Dennis Nguyen
is an assistant professor for Digital Literacy and Digital Methods at Utrecht University in the Netherlands. He holds a PhD in Media, Culture & Society from the University of Hull (UK). His main research interests are critical data studies, public discourses on datafication, difference between digital cultures with focus on the Global South, and computational methods for media research.

Elena Pilipets
(PhD) is a postdoctoral researcher in Digital Media and Methods at the Department of Media Studies, University of Siegen (Media Studies, University of Klagenfurt). Her work focuses on the processes of social media circulation, with particular attention to the study of platforms through combinations of visual and digital methods. This includes online-grounded empirical investigations of Internet memes, platform subcultures, and social bots, the latter being studied in the context of a working group funded by the Center of Advanced Internet Studies (CAIS) in Bochum.

Thaiane Oliveira
(PhD) is professor of Communication Program at Federal Fluminense University. Researcher of the National Institute of Science and Technology in Disputes and Informational Sovereignty, in Comparative Studies in Conflict Manager and in Public Communication of Science and Technology. Member of the Brazilian Academy of Science.

Sijun Shen
received her PhD in Media & Communication Studies from Monash University. She currently works as a Research Fellow at the Influencer Ethnography Research Lab at Curtin University. Her projects study online popular cultures and digital platforms to understand contemporary society in terms of social media's political, socio-economic, and psycho-sexual implications. She specialises in digital ethnography research (psychoanalytic) feminism and takes a special interest in China's governmentality of its internet industry and hyper-consumerism.

Hiromi Tanaka
(PhD) is an associate professor of Gender, Media and Culture in the School of Information and Communication at Meiji University, Japan. She was a visiting scholar at the University of Amsterdam (2022–2023) and Utrecht University (2023–2024). Her research interests include gender and media, popular culture, postfeminist digital culture, posthumanism, and new materialism. She has written on Japanese manga, women's magazines, digital media and technologies among others.

Ozlem Demirkol Tønnesen
(PhD) is a Postdoctoral Researcher at University of Bergen where she is working as a part of the ERC funded PREPARE project that explores citizens connections to the public sphere via social media in their datafied everyday lives. She has collaborated with academics on several research projects that explored personalisation, health communication and community building on a variety of social media platforms. She has also been serving as the elected graduate student representative for the Association of Internet Researchers since 2021.

Yusuf Yüksekdağ
is an assistant professor at Istanbul Bilgi University's Faculty of Communication, specializing in applied ethics, media studies, and political philosophy. His research focuses on the ethics of digital technologies, smart cities, urban space ethics, and migration ethics. Yüksekdağ received his PhD in 2019 from the Department of Culture and Communication, Linköping University, Sweden. He then spent a year as a researcher at the University of Bern's Institute of Philosophy before joining Istanbul Bilgi University. His work has been published in journals like *Journal of Urban Affairs*, *Public Health Ethics*, *Developing World Bioethics*, and *Health Care Analysis*.

Jing Zeng
(PhD) is an assistant professor of Computational Communication Science at the Department of Media and Communication Research at the University of Zurich and a previous Assistant Professor of Digital Methods and Critical Data Studies at Utrecht University. Her research concerns the sociocultural implications of digital communication technologies.

Introduction

Dennis Nguyen, Bruce Mutsvairo and Jing Zeng

The present book, *Technology, Power and Society: Critical Perspectives on the Global Digital Transformation,* is intended to provide a platform for a diverse spectrum of academic perspectives aiming to critically unpack the complex implications of emerging technologies, with an emphasis on the socio-cultural and political configurations within which they come into being and through which their impact unfolds. As such, this inaugural edited volume is the first publication in our new book series of the same name hosted by Brill publishers. The series encourages researchers to take a decidedly global perspective on digital transformation that is sensitive to cultural differences and technologies' ambivalent effects on societies worldwide. The use of digital technologies is constantly expanding in scope and depth throughout societal domains and cultural contexts, marking profound shifts in social interaction that entail various beneficial and harmful potentials. Digital technologies' promises of progress, economic growth, and individual empowerment stand here in stark contrast to complex challenges related to the configuration of power hierarchies, the unequal distribution of wealth and opportunities, and exploitation. There is hardly a place in the world where these transformations and frictions do not matter, with billions around the globe using some form of digital media that create, collect, and process data for increasingly automated processes in all spheres of daily life. Large-scale data processing and automation through algorithmic systems may appear ephemeral but are arguably omnipresent.

Relatedly, suggestions and fears that technology has become a major source of isolation among humans have long been raised (Ellul, 1963; Postman, 1992; Amundsen, 2021) and profound critiques of how technology, capitalism, and politics intersect drive much of the current academic debate on digitalization, datafication, and automation (Zuboff, 2019; Couldry & Meijas, 2019; Taylor, 2017; Hepp et al., 2024). However, critical views sometimes tend to overemphasise the downsides of technological developments and underemphasise opportunities for positive change and empowerment (Arora, 2024). For example, while in some contexts the dominance of social media platforms may complicate democratic discourse and come with harmful consequences such as disinformation, in other settings the very same digital technologies offer a space for dissent and alternative political views. We thus view technology as neither inherently "good"

 | DOI:10.1163/9789004711396_002

nor “bad.” However, technology is never just a neutral tool somehow “agnostic by default” to the ends to which it is deployed, as it always comes with some sort of change in how people communicate with each other and “do things.” Like others before us, we posit that social practice, cultural norms, technology conception, adoption, and use are all interconnected and mutually influential (Baym, 2015).

More specifically, our overall goal is to shed light on contexts where different forms of digital media, data, and artificial intelligence (hereafter A.I.) play an important role but that are not always in immediate focus of public discussions about technology, generally leaving considerable room for further exploration. This often concerns the blind spots and margins across the globe. Who is excluded from shaping technology and determining its desired place in digital societies? How do different communities adopt and utilize digital media? What are the mutual affectivities between global transformations and local situatedness of people, technologies, and the systems that they form? What is the role of technology in social and cultural conflicts? These are just a few pertinent questions that challenge us to advance critical thinking, theorization, methodology development, and empirical research with an eye on the underrepresented and underexplored in digital transformation.

The wide reach of emerging technology trends gives rise to various social challenges, resonating differently across cultural contexts and necessitating a critical lens that considers local specifics and their complex relationships with global developments. Around the world, societies constantly explore and negotiate a balance between the benefits and potential downsides of technological trends, from big data to A.I. Challenges and risks come in diverse – and often unpredictable – forms that are highly context-dependent, making it mandatory to develop a deeper understanding of specific cultural settings and their histories of technology adoption and experience.

Objectives of This Inaugural Volume and the Book Series

The present volume, and the book series that it kicks off, aim to diversify and advance scholarly debates around technologies by focusing on cross-national and intercultural perspectives. By doing so, this book decentralizes critical research on digital technology, expanding its focus to underrepresented sociocultural contexts across the globe. The rise of digital technology is global in scope but takes different shapes, determined by a host of social, cultural, political, and economic factors. To understand how digital media transform social relations, interactions, productivity, and creativity, it is imperative to take a critical angle attuned to the specificities of local cultural settings in which people live. The stakes are high, as current technology trends bear the risk of deepening existing inequalities and creating new ones, serving forms of control and subjugation that maintain power imbalances, economic exploitation, and overall unfairness. Simultaneously, it is assumed that these same technological developments have the capacity to contribute to emancipation and empowerment. The question is how to translate this potential into concrete beneficial change for different social groups and to "democratize" the steering of current technology trends.

Empirically, the pursuit of critical exploration of what has not yet been researched (much) about the intricate interplay between technology, society, and power can quickly point us to the vast and diverse spaces of the so-called "Global South", i.e. Africa, the Middle East, Asia, and Latin America. Indeed, there is still much work to be done here to better understand how technologies unfold their impacts under what circumstances, to which ends. This also challenges us to reconsider still widely held misconceptions about contexts that have been the focus of what could be described as a digital-Orientalist gaze that misreads how non-Western societies engage with technology in practice. What is needed is a fresh look at the digital transformation in non-Western cultural spaces that drops its preconceptions (Arora, 2019). However, this issue is not limited to external views on distant places as implied in a global North-South or East-West divide. Questions of inclusion/exclusion, power, and influence also matter across "the West" in various situations where some groups have more of a share in influencing the future than others (Dencik et al., 2019). Put differently, unexplored territories are not only to be found outside of the affluent societies of e.g., Europe and North America but also exist within societies that are considered "developed".

To illuminate knowledge on these and various other related issues, we deliberately chose a thematic rather than a regional approach. Taking this route encourages us to collectively and creatively critique technology from a wide range of pathways and positions. For instance, when discussing data transparency (Tong, 2022), we need to embrace divergent meanings and approaches to the phenomenon, given the multiple cultural practices and interpretations that shape our ideas on what constitutes transparency and who has the power to decide what is transparent or not.

In pursuing this broader research agenda, we consider an open-minded approach that embraces interdisciplinary cooperation and the pragmatic yet sensible as well as critical integration of different methodologies as essential for achieving progress (Nguyen, 2020). Technologies and societies shape one another through complex discursive processes that – at least temporarily – determine the benefits, risks, and uses of new material means for communicating, networking, and creating (Nguyen, 2023; Jasanoff, 2015; Verbeek 2005). Cultural factors are shaping expectations and ethical considerations pertaining to specific technological developments. Critically investigating who dominates the "making of meaning" of technology and who can harness its perceived potentials inevitably raises questions of power.

To make sense of this complexity, it takes a holistic view that is not confined to a single academic discipline. We need to connect diverse fields and facilitate synergies across relevant research traditions. This includes but is by no means limited to media and communication studies, cultural studies, social science, political science, regional studies, critical data studies, sciences and technology studies but also seemingly more distant subjects such as computer and data science. We are strongly in favour of building interdisciplinary bridges in addressing pressing questions and issues that concern the intersection of technology, power, and society. Accordingly, it is imperative to approach relevant empirical cases with valid yet flexible methodologies for generating knowledge, spanning across and effectively integrating qualitative, quantitative, manual, and computational approaches. Readers of this book and potential contributors to our book series will notice our openness to all sorts of ideas that advance knowledge on the growing relevance of digital technology. These ideas may be philosophical, sociological, media-centric, or political. They may address power dynamics ranging from the increasing influence of neoliberalism or government regulation, the ability or inability of groups and individuals to exploit technological changes, ethical quagmires shaping technological use, threats to democratic values, and ubiquitous calls for resistance to technology, including expanding advocacy opposed to surveillance capitalism and market capitalism.

We seek to debate and decentre longstanding appeals and conversations about technology to champion fresh perspectives. This approach will help us understand the long-term consequences of technological transformations, the neglect of which could lead to adverse implications. We appeal to the academic community to deliver multiple viewpoints that can shape a unique understanding of technology. This includes accentuating historical insights, probing divergent theoretical influences beyond digital divides or tech pessimism and optimism, and highlighting empirical and critical accounts that provide a comprehensive appreciation of technology and its expanding role in different societies and cultures.

Despite the importance of understanding diverse local experiences when discussing the sociopolitical implications of technologies, existing literature remains dominated by a Euro-American-centric perspective. This leads to the second aspect of this book series' approach to global perspectives: endorsing inclusiveness and diversity in scholarship. In the context of academic discussions of digital technologies, this entails efforts to de-westernize and decolonize academic voices. Prior critical discussions of decolonizing research challenge the Western-centric global knowledge system wherein methods and intellectual approaches from the West are regarded as sacred, while those originating from the Global South are considered to have lesser ontological significance (Connell, 2020; Dutta et al., 2021; Schoon et al., 2020). This book series aims to challenge such a hierarchical order of knowledge by amplifying the visibility and impacts of empirical and theoretical work from and about the Global South.

Decolonizing scholarship about technology, power, and society involves more than applying a dominant theoretical framework to replicate "case studies" from non-Western contexts. Moving beyond Western-centrism requires us to move beyond "using the Global South as a place for data collection and theory testing" (Zeng & Chan, 2023, p. 610). To counteract neocolonial knowledge production structures (Dutta et al., 2021), the academic community needs to critically reflect upon inequality and prejudice in both epistemological and academic practices while promoting indigenous conceptualizations and approaches (Baum & Zhukov, 2019; Tully et al., 2022). As Connell (2020) points out, social theories originating from the so-called peripheral societies possess intellectual rigour comparable to those from more dominant regions and frequently hold greater sociopolitical relevance in diverse contexts. The book series' pursuit of a nuanced understanding of local dynamics in the Global South does not promote separatism or exceptionalism. Rather, it aims to problematize universalism in discussing technological trends while demonstrating productivity in two-way dialogues between the experiences from different social and cultural contexts. The aim is, as Schoon et al. (2020, p. 7) put it, "not to simply "reverse the gaze" but rather "gaze into the world" from where we stand."

With these overarching goals in mind, the present volume aims to inspire other researchers by presenting the diversity of topics that our book series aspires to further explore over the coming years. The volume's chapters offer a selection of critical analyses, methodological innovations, and normative reflections on the role of digital technologies in the formation of social relationships and the power hierarchies they entail. Its scope is decidedly global, with a keen eye for contexts where these developments concern marginalized groups within and across societies, as well as world regions often overlooked in mainstream academic research and public discourses in the Global North. Put differently, the book offers a "taste" of the varied range of topics and issues what we hope to cover in greater detail in dedicated future publications, whether they are monographs or edited volumes.

To this end, we invited a diverse group of contributors to engage with the challenging questions sketched out above and to share their insights throughout 13 chapters. Our contributors have diverse academic roots and backgrounds, representing Europe, Asia, Africa, the Middle East, and Latin America. Importantly, we also invited authors at different (academic) career stages and aimed to give junior researchers an outlet for sharing their ideas. All of this signals that our book series is intended to be interdisciplinary in nature and open to researchers with diverse academic roots, perspectives, and methodological approaches.

Structure of the Book

The volume is structured around three broader topics that thematically group the 13 individual chapters: 1) *Concepts*, 2) *Digital Cultures and Digital Politics*, and 3) *Ineqalities, Resistance and Alternatives*. Together they offer important theoretical, methodological, and empirical insights into critical research on the varied implications and challenges that digital technologies give rise to in specific contexts around the globe.

The first part, *Concepts*, sets the stage by offering critical reflections on methodological approaches that allow us to embrace global transformations from different angles that acknowledge the complexity of communicative flows and their impact from local to global trajectories and vice versa. Contributions in this part introduce a range of important concepts that help with critically researching technology issues in their complex social, cultural, and political contexts. These concern methodological considerations for critical cosmopolitan research with a focus on technology (Dennis Nguyen), the politics of digital sovereignty (Thaiane Oliveira, Afonso Albuquerque, and Tatiane Mendes Pinto), practices of data colonialism (Stefan Bächtold), and the relationship between expanding datafication and exploitation (Bjorn Beijnon). Drawing from diverse academic disciplines in critical sociology, media studies, and critical data studies, these chapters all take different (empirical) routes to make their argument but are united in their focus on important terminology and concepts that can facilitate and stimulate critical research in the context of technology, power, and society.

The part starts with the chapter "Cosmopolitan Critical Data Studies" by Dennis Nguyen, in which the author provides a reflection on how concepts, methodological practices, and insights from critical cosmopolitanism could be more firmly incorporated in critical data studies. The chapter argues that while critical data studies in general are not ignorant of non-Western contexts and the varied effects of technology therein, existing scholarship is still largely centred on how datafication and automation unfold in the developed societies of the global North. Nguyen posits that "[i]ntegrating a cosmopolitan view can enhance CDS" analytical scope while safeguarding them from the fallacies of methodological nationalism, i.e., the assumption that societies and nation-states are somehow congruent as analytical units for critical sociological inquiry" (p. 14). Moreover, he observes that "datafication and automation, and especially the risks that they carry, are similar in scope and complexity to the fundamental challenges that proponents of cosmopolitan sociology identified in the early 2000s, such as the globalization of finance and economies, terrorism, climate change, and migration" (p. 17). Nguyen argues that any sort of cosmopolitan critical data studies would benefit from expanding the methodological toolset by including more quantitative and especially computational methods for more empirically grounded insights into transnational formations that shape datafication and automation effects in concrete contexts.

Chapter 2, "Digital and Epistemic Sovereignty in the Science Ecosystem in Latin America" by Thaiane Oliveira, Afonso Albuquerque, and Tatiane Mendes Pinto, takes a historical view on the concept of digital sovereignty as a subject of regional resistance to imperialistic control and domination in the concrete context of academia. Oliveira et al. show how the implementation of open and multilingual publication systems and development of digital access to research metadata are some approaches adopted to disseminate knowledge produced in the Latin American region. Despite these efforts, privatization and deregulation in the region allowed oligopolistic academic publishing companies to enter, threatening scientific autonomy. The authors observe "a neoliberal shift in power dynamics, with non-state actors gaining significant influence. This power shift has paved the way for the rise of economic models that prioritize efficiency, competitiveness, and privatization over social well-being and the production of diverse knowledge" (p. 32). Achieving epistemic and digital sovereignty requires resisting knowledge control and privatization while promoting open and inclusive science. It faces challenges related to digital governance and exclusion. Oliveira et al. conclude that: "[r]eclaiming digital and epistemic sovereignty requires taking back control of production and circulation domains and placing the state in a fundamental role to create contexts that foster digital technological sovereignty through interventions guided by principles of human rights, autonomy, and independence" (p. 41).

In the third chapter "Experimenting on the Frontier: Imperial Laboratories and Facebook's Political Effects in Myanmar", Stefan Bächtold offers a critical investigation into how big technology companies from Silicon Valley (USA) make sustained attempts to expand their influence over internet access in the Global South through the example of Myanmar. The country underwent a series of political upheavals over the past decade that pitted a military junta, a democratization movement, rebel groups, and different ethnic minorities against each other. Digital media play an essential role in Myanmar's society, but the country's experience in the 2010s drastically revealed the inherent perils of an internet dominated by a single algorithm-governed, feed-based platform – and also how tech companies' practices of datafication, extraction, and expansion become increasingly entangled with (bio)political ones. Facebook's techno-optimistic discourse constituted Myanmar as digital *terra nullius*: an isolated frontier awaiting connection for progress. Bächtold argues that, in essence, Facebook turned the country into an "*imperial* laboratory" (p. 48). This allowed the company to test different features and strategies camouflaged as apolitical forms of technology development in an allegedly undeveloped area of the world. This apolitical image of technology not only obscured the platform's political effects in Myanmar. It also entailed an approach of experimentation and bricolage with techn(olog)ical "solutions" to political problems – even when the platform overflowed with hate speech and calls to violence against Muslims.

The first part of the volume concludes with Chapter 4, "From Datafication to Interpellation: Becoming a Data Subject in Contemporary Surveillance Cultures", by Bjorn Beijnon. Taking a critical perspective on data capitalism and practices of data surveillance, the chapter offers a conceptual framework for how datafication processes turn people into "data subjects." Beijnon posits that when interacting with platforms, people become users who generate a modulated media ecology in which they become a data subject by continually recognizing themselves in stratified data profiles. This interpellation happens because users constantly listen to the hail of personalized attentional triggers by data profiles. With their platforms, big tech companies have become able to construct a dispositif in which users conduct their own conduct and which makes them consequentially consider data profiles as an accurate reflection of their lived lives. Beijnon observes that "platforms contribute to a governance over their users from a distance in which users govern themselves in generating their modulated media ecology" (p. 64). Consequently, "[d]atafication in current surveillance cultures has made platform users govern their own interpellation by acknowledging that data profiles indexically reflect their past, present, and future" (p. 73).

The second part of the present volume focuses on *Digital Cultures and Digital Politics*. The respective chapters provide a variety of empirical studies that zone in on specific contexts in which perceptions and imaginations of technology shape social, cultural, and political practices. Each case highlights the role of technology in a different corner of the world, within distinct cultural and political contexts with their own histories. Collectively, however, the chapters illustrate the ambiguities between benefits and risks of technology that may take different shapes in specific places but are as such fundamental to technological developments across the globe. They also showcase different methodological approaches ranging from qualitative to computational.

Specifically, the contributions discuss the complex relationships between cultural norms, social media, and consumerism in China (Sijun Shen and Crystal Abidin), the trade-offs political activists face between empowerment and privacy risks of digital media in Turkey (Sarper Durmuş and Yusuf Yüksekdağ), the weaponisation of social media in Ukraine (Elena Pilipets, Marloes Geboers, and Daria Delavar-Kasmai), the role of social media during political turmoil in Myanmar (Paula Romero Jiménez, Ana Melchor Pérez, Siebe Dekker, Miguel Oliveira Royo, and Jing Zeng), and how imaginaries and power shape AI governance in China (Yishu Mao). The different studies exemplify how to conduct empirical research on views, ideas, and practices shaping technology in specific local contexts.

In Chapter 5, "'BM Girl' Influencers on Xiaohongshu: Tracing Beauty Discourse, Social Media Challenges, and Consumption Practices in Chinese Society", Sijun Shen and Crystal Abidin critically explore how digital platforms and consumer cultures shape one another in the Chinese context. They provide an in-depth analysis of how the social media and e-commerce platform Xiaohongshu (小红书) makes use of community-generated content to promote commercial products. Focusing on young female users, a recent trend popularised by the platform is *BM style* (Brandy Melville, a slim-sized Italian fashion brand with the slogan "one-size-fits-most"). The authors draw on content analysis of BM's ecology on Xiaohongshu to trace how discourses about standards, values, and beliefs are shaped on the platform and travel to the broader digital media ecology in Chinese society. Shen and Abidin observe a complex situation where "[c]ollectively, BM girls on Xiaohongshu – whether they are traditional celebrities paid for their sponsored journal entries, BM employee influencers, BM influencers at large, or ordinary users who are fans of BM – constitute an ecology of women who leverage fashion and consumption trends as a way of policing their own and each other's body image, and who push up the barometer for the ideal "thin body" in Chinese society" (p. 79). Their critical investigation raises important questions about how commercially oriented platforms facilitate influencer promotional cultures and perpetuate fixations on thinness as a body ideal in Chinese consumerism.

Chapter 6, "Privacy Expectations and Norms: Perceptions of the Individual Digital Activists in Turkey's Xsphere (Twittersphere)" by Yusuf Yüksekdağ and Sarper Durmuş, shifts our attention to political activism and social media. This contribution surveys the perceptions of digital activists in Turkey regarding digital privacy in general, privacy regulations on Twitter, and how they balance their privacy concerns with other values and interests pertinent to being an activist in the Twittersphere. Following a postphenomenological approach, this study, in particular, seeks to examine how perceptions and values regarding privacy are formed in conjunction with the daily interactions of digital activists on Twitter, interspatial concerns of the activists attributed to their safety and other concerns between the digital and physical worlds, and how digital spaces themselves are perceived given their prospects and benefits for digital activism. By conducting semi-structured interviews in Turkey with 15 digital activists who each have different agendas, the authors provide a useful framework for future ethnographic research that goes beyond perceptions and problematizes the nature of interaction itself. Yüksekdağ and Durmuş find that "activists" privacy-related decisions seem rooted in their pursuits and are often influenced by external factors such as cyber-harassment" (p. 103). They conclude, "digital activists seem to maintain an internal compass guided mainly by contextual privacy concerns. Recognising that the digital realm can be both a tool for empowerment and a potential threat, they adopt privacy norms for their safety and as a shield for the vulnerable communities and individuals they advocate for" (p. 103).

Moving on to the role of digital media in conflicts and wars, Chapter 7, "Detouring, Rerouting, Weaponization: The Secondary Orality of WarTok", by Marloes Geboers, Elena Pilipets and Daria Delawar-Kasmai provides a critical analysis of war-related TikTok performances. Their corpus consists of Ukrainian and Russian video posts amalgamating multiple media forms – *sounds, sticker texts, effects, hashtags, emojis* – through remix. The authors' main research objective is to understand how these multimodal and layered posts channel affective engagement, heavily relying on TikTok practices of hash-tagging and song-appropriation. To achieve this, Pilipets and Geboers adopt a platform-grounded approach, analysing posts and video metadata that are networked by either of the two TikTok songs: 1) "Доброго вечора Where Are You From?," commonly known by its incipit "Good evening, we are from Ukraine" (Ukrainian: Доброго вечора, ми з України) or 2) "Good morning, we are from Russia" (Russian: Доброе утро, мы из России). Their innovative research design unpacks the complex, media-specific ways through which social media platforms and digital content are exploited for propagandistic purposes. The authors argue that "[d] riven by the referential dynamics of memetic riffing and practices of attention hijacking, war propaganda on TikTok is formed through counter-imitative acts of sharing through combinations of hashtags, effects, duets, and sounds. Sound and its secondary orality on TikTok works as an affective amplifier" (p. 122).

Chapter 8, "Making Sense of Post-coup Myanmar through Facebook" by Paula Romero Jiménez, Ana Melchor Pérez, Siebe Dekker, Miguel Oliveira Royo, and Jing Zeng, also focuses on the role of social media in conflicts. Specifically, the authors present the case of the 2021 coup d'état in Myanmar that disrupted the country's democratization process. They offer a critical analysis of political Facebook discussions in the country utilizing computational methods, namely topic modeling and sentiment analysis. Despite attempts by the military junta to ban the platform, Facebook remains the central arena where information is circulated, and collective actions are organized by activists at home or in exile. Their analysis of 42,596 posts on public Facebook pages centred on three dimensions: content, content creators, and audience engagement. Through their empirical investigation, Jiménez et al. demonstrate that "while the unfolding political situation in the region was a salient topic, our analysis also revealed that spam and other commercial posts hijacked attention on Myanmar from international audiences. At the same time, the findings of this study demonstrate that social media analytics can provide insights into the unfolding of major events and associated public responses" (p. 139). Importantly, throughout the political upheaval the social media platform "remained an important medium through which an international audience was informed about the unfolding political unrest and showed support for civilians and activists" (p. 140).

The second part of the volume concludes with Chapter 9, "Artificial Intelligence Governance Made in China: Negotiating Imaginaries and Power" by Yishu Mao. Governance of A.I. for domestic purposes and international influence is an increasingly important item on most governments' agendas. China is one of the most ambitious and influential countries in this domain. Mao traces the public imaginaries around A.I. in China and how they shape specific regulations for recommendation algorithms, one of the fundamental components of A.I. systems and among the most widespread types of A.I. in use today. She argues that A.I. "is both the result of and inspiration for rich sociotechnical imaginaries embedded with cultural and historical diversity around the world" (p. 146). In her study, Mao argues that China's AI governance approach is characterized by three interpretations of the "Made in China" trademark: first, the participation of various actors in steering the country's sociotechnical trajectories, despite their different levels of power; second, the imagery of China's past economic success achieved through a labour-intensive development model; and lastly, China's future aspirations that rely on advanced technological capabilities. She observes "a clear discrepancy [...] in terms of power in negotiating the development trajectory of AI. The state still commands the highest power in defining the purposes of scientific and technological advancement and what constitutes the public good, that is, upgrading China's economy and global position to achieve the national "rejuvenation" and techno developmentalist goals" (p. 148).

The third and final part of the volume comprises four chapters that share a focus on *Inequalities, Resistance and Alternatives*. Processes of conceiving, developing, and applying technology have always been marked by forms of exclusion and unfairness, creating divides across various social dimensions. The promises of individual liberation, optimization, and general advancement prominent in many tech narratives often tend to cover up how only a relative few are in a position to determine the directions in which societies should head with emerging technologies. In addition, digital media and A.I. specifically can reproduce age-old stereotypes and discourses of marginalization. Each of the chapters highlights a different manifestation of inequality in the context of technology, especially regarding A.I. and social media, respectively. However, they also point to potentials for resistance and alternatives to current practices. The concrete cases covered in the third part of the volume focus on how microcelebrities engage in anti-governmental protests on social media (Ozlem Demirkol Tønnesen), the implications of gendered, racialized, and sexualized A.I. discourses (Hiromi Tanaka and Michelle Ho), the challenges of reaching gender equality in the tech industry (Julia Luteijn and Rhied al-Othmani), and how technology in the form of machine translation may offer ways to democratize global access to literature but also how its history reveals persistent patterns of inequality (Christophe Declercq and Gys-Walt Egdom).

Chapter 10, "'With great power comes great responsibility': Lending Visibility to Risky Political Content" by Özlem Demirkol Tønnesen, explores how X (formerly Twitter) microcelebrities justify and negotiate their participation in the anti-government political discourse during Turkish elections. While there is a growing interest in political microcelebrities in recent years, their expressive political participation in repressive settings remains largely unexplored. The chapter presents an analysis of 97 microcelebrities during the 2018 Turkish elections. Ozlem Demirkol Tønnesen sees social media as a double-edged sword: "while having high visibility and popularity on social media is necessary to mitigate the erasure of dissent from other media platforms, it also increases vulnerability to hostile surveillance" (p. 164). Her findings show the delicate dynamics of maintaining attention and status while grappling with backlash from sponsors, audiences, and authorities, which culminates in the framing of political participation as a duty for all microcelebrity accounts. She observes that "[m]oralisation of political participation is of particular importance within and beyond the Turkish case, as it creates pressure that all microcelebrities are expected to channel all their efforts to amplify the reach of political tweets. These pressures arise from the idea that visibility is fundamental for the dissemination of information online which naturally puts the onus on those that command the most attention" (p. 174).

Chapter 11, "Artificial Intelligence's Sexual Politics: Three Modes and the Case of Japan" by Hiromi Tanaka and Michelle Ho, examines entanglements of technology, society, and power from a critical feminist perspective. Tanaka and Ho posit that AI constitutes an important part of contemporary sexual politics characterized by "gendered entanglements of human bodies and non-human entities such as computers, technologies, cultural ideas, human emotions and desires, whereby they have the capacity to affect each other and form new relations in a new human-machine ecology" (p. 190). The authors review critical discourses about gendered (and racialised) AI from the mid-2010s to date and identify three modes of its sexual politics: 1) algorithmic discrimination, 2) AI-facilitated sexual violence, and 3) gendered embodiment of intelligent machines. Using this framework, Tanaka and Ho analyse the case of Japan to demonstrate how contemporary sexual politics of AI is played out in a specific cultural context. They assert that gendered and racialised AI such as algorithmic biases have been framed in Japanese media discourses mostly as problems that are happening abroad. This, however, does not mean that there is no gender/race issue in Japan. The authors eventually argue for the importance of cultural analysis of sexual politics of A.I. for a more just society.

Touching upon a related topic, Chapter 12, "The Tech Gender Gap: A Closer Look at Women's Experiences in the Technology Industry" by Julia Lutijn and Rhied Al-Othmani, provides a critical assessment of the limitations of seemingly progressive policies for more gender equality in the tech industry. The authors discuss the case of the Netherlands, which aims to foster gender parity across society on paper but falls short of its ambitions in practice: "despite these efforts, the workplace culture in the technology sector appears still marred by a lack of gender diversity, ultimately resulting in an imbalanced work environment" (p. 199). Lutijn and Al-Othmani show that women tend not to work in male-dominated sectors due to subtly and subconsciously unequal treatment based on stereotypes and associations, also referred to as implicit gender bias. To gain deeper insights into how individuals experience these biases, they interviewed 35 women working in Dutch technology jobs such as development, architecture, engineering, and consulting. The results show that the majority of women encounter various forms of implicit gender bias. Lutijn and Al-Othmani argue that the gender gap and implicit gender bias are intertwined and affect women at various stages of their careers, from choosing to work in technology to encountering bias in the workplace and working in non-diverse teams. Their study underlines the urgency to address these issues, as "[m]ale-dominated workplaces can produce inequalities and power imbalances that affect women's experiences in various ways" (p. 203).

The final chapter, "Beyond the Strictest Computation of the General Proportion: Language Automation, Machine Translation and Their Complicated Possibilities for Cultural Exchange" by Christophe Declercq and Gys-Walt Egdom, takes a broader historical view on how emerging technologies in the context of translations of literature promise democratization but how existing political, cultural, and economic inequalities put clear confines on such potentials. The authors base their critical analysis on the argument that intercultural relations are influenced by hierarchy and proximity, with political, economic, and cultural factors traditionally driving these dynamics. Literature plays a pivotal role in this exchange, but liberalization in the 20th century reduced political influence, leading to a dominance of English-language literature. To address this imbalance and promote cultural diversity, some translation scholars propose using language technology, particularly machine translation, in opening literary markets of lesser-resourced and less-diffused languages. While data-driven language technology may indeed offer here possible routes for more diverse cultural offerings in the field of literature, Declercq and Egdom also identify clear risks: "[u]nequal representation in terms of cultural, gender, or racial biases in translated texts are not easily remediated technically; it requires measures throughout the development process: pre-training and fine-tuning techniques can be employed to mitigate bias, including careful curation of training data, i.e.

conscious diversification data sources instead of "blind" harvesting of data" (p. 226).

Taken together, the present volume's chapters over three thematically organised parts aim for 1) extending the analytical lens on technology, power, and society by considering interdisciplinary perspectives sensitive to the complex interplay between local, regional, national, transnational, and global layers of socio-cultural and political interaction; 2) introducing key concepts for articulating concrete research objectives that attempt to critically unpack imbalances in power but also misconceptions about agency (individual and collective); 3) illustrating the different methodological and theorisation-related challenges in describing and researching complex socio-technical configurations but also productive ways forward; and 4) identifying potentials for challenging, resisting, and reversing current socio-political configurations that foster forms of inequality and unfairness. In this way, the chapters illustrate the ambitions and goals of the book series and showcase different research directions for approaching the complex interplay between technology, power, and society.

References

Amundsen, D. (2021). Digital technologies as a panacea for social isolation and loneliness among older adults: An intervention model for flourishing and wellbeing: Visual technologies as a panacea for social isolation. *Video Journal of Education and Pedagogy*, 5(1), 1–14.

Arora, P. (2024). *From pessimism to promise: Lessons from the global south on designing inclusive tech.* MIT Press.

Arora, P. (2019). *The next billion users: Digital life beyond the West.* Harvard University Press.

Baum, M. A., & Zhukov, Y. M. (2019). Media ownership and news coverage of international conflict. *Political Communication*, 36(1), 36–63.

Baym, N. K. (2015). *Personal connections in the digital age* (2nd ed.). Polity.

Connell, R. (2020). *Southern theory: The global dynamics of knowledge in social science.* Abingdon, UK: Routledge.

Couldry, N., & Mejias, U.A. (2020). *The costs of connection: How data are colonizing human life and appropriating it for capitalism.* Oxford University Press.

Dencik, L., Hintz, A., Redden, J., & Treré, E. (2019). Exploring Data Justice: Conceptions, Applications and Directions. *Information, Communication & Society*, 22(7), 873–881. https://doi.org/10.1080/1369118X.2019.1606268

Dutta, M., Ramasubramanian, S., Barrett, M., Elers, C., Sarwatay, D., Raghunath, P., Kaur, S., Dutta, D., Jayan, P., Rahman, M., Tallam, E., Roy, S., Falnikar, A., Johnson, G. M., Mandal, I., Dutta, U., Basnyat, I., Soriano, C., Pavarala, V., Sreekumar, T. T., Ganesh, S., Rathina Pandi, A., & Zapata, D. (2021). Decolonizing open science: Southern interventions. *Journal of Communication*, 71(5), 803–826. https://doi.org/10.1093/joc/jqab027

Ellul, J. (1963). The technological order. In Stover, C.F (ed). *The Technological Order* (pp. 10–37). Detroit: Wayne State University Press.

Hepp, A., Loosen, W., Dreyer, S., Jarke, J., Kannengießer, S., Katzenbach, C., Malaka, R., Pfadenhauer, M., Puschmann, C., & Schulz, W. (2023). ChatGPT, LaMDA, and the hype around communicative AI: The automation of communication as a field of research in media and communication studies. *Human-Machine Communication*, 6 (2023), 41–63. https://doi.org/10.30658/hmc.6.4

Jasanoff, S. (2015). Future imperfect: Science, technology, and the imaginations of modernity. In S. Jasanoff & S.-H. Kim (Eds.), *Dreamscapes of modernity: Sociotechnical imaginaries and the fabrication of power* (pp. 1–33). University of Chicago Press.

Nguyen, D. (2023). How news media frame data risks in their coverage of big data and AI. *Internet Policy Review*, 12(2). https://doi.org/10.14763/2023.2.1708

Nguyen, D. (2020). Media and communication studies in the age of digitalization and datafication: How practical factors and research interests determine methodological choices. In: Nguyen, D., Dekker, I., & Nguyen, S. (Eds.), *Understanding Media and Society in the Age of Digitalisation* (pp. 13–34). Palgrave Macmillan, Cham. https://doi.org/10.1007/978-3-030-38577-4_2

Postman, N. (1992). *Technopoly: The surrender of culture to technology.* New York: Alfred A. Knopf.

Schoon, A., Mabweazara, H. M., Bosch, T., & Dugmore, H. (2020). Decolonising digital media research methods: Positioning African digital experiences as epistemic sites of knowledge production. *African Journalism Studies*, 41(4), 1–15.

Taylor, L. (2017). What is data justice? The case for connecting digital rights and freedoms globally. *Big Data & Society*, 4(2), 1–14.

Tong, J. (2022) *Data for journalism: Between transparency and accountability.* London: Routledge.

Tully, M., Madrid-Morales, D., Wasserman, H., Gondwe, G., & Ireri, K. (2022). Who is responsible for stopping the spread of misinformation? Examining audience perceptions of responsibilities and responses in six Sub-Saharan African countries. *Digital Journalism*, 10(5), 679–697.

Verbeek, P.P. (2005). *What things do – Philosophical reflections on technology, agency, and aesign.* Penn State: Penn State University Press, ISBN 0-271-02539-5 (264 pp.)

Zeng, J., & Chan, C. H. (2023). Envisioning a more inclusive future for digital journalism: A diversity audit of journalism studies (2013–2021). *Digital Journalism*, 11(4), 609–629.

Zuboff, S. (2019). *The age of surveillance capitalism: The fight for a human future at the new frontier of power*. Profile Books.

PART 1

Concepts

CHAPTER 1

Cosmopolitan Critical Data Studies

A Research Agenda and Its Methodological Challenges

Dennis Nguyen

1 Introduction

In this chapter, I propose that Critical Data Studies (CDS) can benefit from integrating a cosmopolitan-empirical methodological approach that explicitly focuses on the transnational and global dimensions of datafication and automation. Arguably, the concrete technological developments through which these trends become manifest resonate differently across socio-cultural contexts due to various localisation processes. For example, while applications of artificial intelligence (AI) are based on largely the same technological-scientific foundations, exact use cases, perceptions, evaluations, and regulations differ between cultural contexts (Nguyen & Hekman, 2022). However, to understand local contexts of technology adoption and its effects, it is imperative to critically consider transnational and/or global formations from which political, economic, and socio-cultural influences emerge. The local, regional, national, international, transnational, and global spheres of social interaction form a continuous spectrum and are mutually influential (Nguyen, 2017). Indeed, methodological propositions in the wider field of CDS underline that the situatedness of datafication and automation connects the global with the local (Loukissas, 2019). Decolonial approaches suggest a "de-Westernization" (Milan & Trere, 2019) of CDS towards more research on the Global South, while critical feminist frameworks shift focus to marginalised groups with emphasis on intersectionality, including transnational social formations (D'Ignazio & Klein, 2020). On a conceptual level, CDS can be explicitly linked to critical discussions of cosmopolitanism as an analytical perspective further unpacking the border-transcending nature of contemporary socio-technical transformations and their socio-political implications (Beck, 2003; Delanty, 2006).

Critical cosmopolitanism provides a possible venue for empirical research and subsequent normative critique on current societal challenges related to data, Artificial Intelligence (AI), and technology. It is sensitive to

 | DOI:10.1163/9789004711396_003

the complex interrelations along the local-to-global spectrum of fundamental societal challenges (Delanty, 2006). Integrating a cosmopolitan view can enhance CDS' analytical scope while safeguarding them from the fallacies of methodological nationalism, i.e., the assumption that societies and nation-states are somehow congruent as analytical units for critical sociological inquiry (Chernilo, 2011). Rather, we should critically consider the border-transcending dimensions of the origins and effects of AI and datafication trends in specific socio-cultural configurations. This approach helps with explaining where local factors matter, determining where and how challenges arise from beyond immediate localities, and identifying potential for resistance based on cosmopolitan-transnational cooperation (Beck, 2003). Such a lens is not blind to the continued relevance of nation-states as focal points of individual and collective imagination – including self-perception – as well as frameworks for social interaction. However, it emphasises the necessity of integrating multiple levels of analysis and diverse perspectives to fully grasp the complex dynamics at play (Nguyen, 2017).

Pertinent questions in CDS place emphasis on inherently global transformations such as emerging power hierarchies in digital societies, political economies of technology, representation through data, and discursive practices involving large amounts of data and algorithmic systems that enact forms of governance (Hepp et al., 2023). There are numerous important issues that warrant critical inquiries from a humanistic-sociological perspective (Lindgren, 2023) as well as governance and regulation-centred approaches (Taylor, 2017). Accordingly, CDS are best considered as an interdisciplinary field located at the intersection of, among others, media- and communication studies, policy research, and critical social theory.

Generally, there is room to more thoroughly outline what a cosmopolitan perspective in CDS should entail. Additionally, methodologies are needed that enable CDS to contextualize and explain the various transnational and global layers that shape societal challenges accompanying expanding technological trends. Respectively, CDS may benefit from connecting a decidedly cosmopolitan lens to flexible empirical research methods by connecting dominant qualitative approaches in the field with more quantitative, and especially computational, methods. The point is: aside from more clearly conceptualizing how cosmopolitan research subjects need a cosmopolitan perspective in CDS, it is equally important to explore how we can utilise empirical methods that are still underrepresented in "traditional" CDS research. This is a tricky challenge as 1) quantitative-computational approaches are not commonly used in the core disciplines of CDS that are often at home in the humanities; and 2) there is a lack of both computational methods and access to data for research within a critical cosmopolitan framework on CDS' research subjects.

In what follows, I outline a proposal for a critical cosmopolitan outlook in CDS along with a general research agenda regarding pertinent societal challenges. The starting point is to acknowledge that current trends subsumed under catchphrases such as big data and AI are inherently global. This does not only concern the seemingly obvious point of their global proliferation, which happens at different speeds and at different depths of penetration across social-cultural contexts around the world. More pressingly, it concerns the different social, political, economic, and increasingly environmental challenges, inequalities, and risks that come with the rise of data-driven technologies on a global scale (Crawford, 2021). To be clear, it is not my intention to falsely discredit CDS as generally ignorant of the complex interplay between local, transnational, and global transformations: that would be factually wrong. The goal is rather to explore how a cosmopolitan view can be more firmly positioned within CDS' ontological and epistemological outlook. I then continue with a critical reflection on where to start with developing empirical methods that go beyond qualitative investigations while remaining critically conscious of the various practical challenges for pursuing such research endeavours.

2 Datafication and Automation as Cosmopolitan Challenges in CDS Research

CDS is an interdisciplinary field that emerged as a response to the hype around big data across business and science in the early 2010s (boyd & Crawford, 2012). It draws from a diverse range of academic subjects mainly located in the humanities, sociology, and law, such as media studies, communication studies, cultural studies, human geography, and policy and governance studies. Broadly speaking, research in CDS examines socio-technical configurations surrounding data collection, processing, and utilisation with a focus on power dynamics, ethics, and social impacts. CDS critically interrogate the assumptions, values, and contexts shaping data practices and technologies, aiming to uncover how these data regimes influence and are influenced by social, cultural, political, and economic structures (Iliadis & Russo, 2016). Related disciplines closely connected to CDS are critical algorithm studies, internet studies, and platform studies.

The starting point for CDS is the critique of datafication – the translation of social action into quantifiable digital data – and its socio-political implications. According to Meijas & Couldry (2019 p. 1), datafication refers to "the quantification of human life through digital information", i.e., how everyday human activities and interactions are converted into data points that can be tracked, analysed, and exploited for commercial purposes and/or forms of control. The rise of AI, intrinsically linked to datafication, has naturally become another central research subject in CDS. Automation, as described by Katzenbach and Ulbricht (2019), involves the increasing use of algorithmic systems that automate decision-making processes to varying degrees. These systems typically perform classification and ranking tasks within complex data and information pipelines employing various forms of machine learning. In contemporary digital societies both datafication and automation come together in the form of "data assemblages" (Kitchin, 2022), which are often the focus of CDS inquiries.

What is important is that datafication and automation are not considered from only – not even primarily – a technical perspective but are rather approached as complex assemblages that involve a variety of social, economic, political, and cultural factors. As Illiadis and Russo (2016) state, CDS centre on big data "to explore [its] unique cultural, ethical, and critical challenges." These include forms of unfairness, exclusion, suppression, and exploitation through data and technology. The field's expanding research agenda often concentrates on issues such as data governance, justice, and abuse, particularly focusing on the intersection between private and public sectors. Concretely, CDS critically explore how data and AI have become embedded in the practices of public organizations and governments, examining disruptions caused by tech platforms, and the commercial exploitation risks posed by big tech companies (Rieder & Simon, 2016; Nguyen & Beijnon, 2023; Couldry & Meijas, 2020).

This body of work often indicates that global and transnational layers matter in understanding how data assemblages come into being, as illustrated in propositions such as data colonialism (Couldry & Meijas, 2019), surveillance capitalism (Zuboff, 2019), or data feminism (D'Ignazio & Klein, 2020). Furthermore, CDS are not delimited to critical reflections on data practices and their socio-political implications in the Global North (Milan & Trere, 2019). Research is increasing on cultural contexts across the Global South, such as East Asia, Africa, and Latin America (e.g, Zhu, 2024; Morales & Reilly, 2023; Cinnamon, 2020). CDS as a wider field are global in scope. Indeed, progress in and dissemination of AI and big data technologies shaping digital transformation are mainly driven by private enterprises that operate on a global level and enable social interactions that transcend legal and geographical borders. They promote their own techno-capitalist notions of (market-liberal) cosmopolitanism encapsulated in quasi-utopian visions of a borderless future for tech business. Simultaneously, initially local or regional developments expand beyond their context of origin. A case in point are digital culture trends that emerge locally and then scale globally, such as TikTok (Wagner, 2023). Moreover, seen from a geopolitical angle, similar aspirations and motivations seem to spur the data- and AI-strategies of most governments that aim to position themselves as contenders for influence in the global digital economy, exporters of technology policies, and leaders in AI weaponisation (Nguyen & Hekman, 2022).

Arguably, datafication and automation, and especially the risks that they carry, are similar in scope and complexity to the fundamental challenges that proponents of cosmopolitan sociology identified in the early 2000s, such as the globalization of finance and economies, terrorism, climate change, and migration (Beck, 2003). Ulrich Beck's *World Risk Society* (2008) outlines how in an interconnected world such challenges transcend national spaces and thus demand a post-national, cosmopolitan approach to finding solutions. According to Beck, the "semantics of risk" (2008, p. 19) emerging in modernity leads societies to focus on "future threats discussed in the present that are often an outcome of the successes of civilisation" (Beck, 2008, p. 19, *my translation*). Risk always has a dual character of "chance and threat" (Beck, 2008, p. 19) and the perceptions of the balance of the two as well as the imaginaries about possible risk consequences are at the centre of contemporary public discourses that steer concrete action. What complicates the matter is that risk definitions, causes, effects, and entailed visions for action are not confined to singular social contexts. Risks that are "ecological, financial, military, terrorist, biochemical, and informational" (Beck, 2012) are virtually omnipresent, inherently global, but their perception is also culturally contingent.

Datafication and automation play an important role in all these global risk contexts. Arguably, they too have a dual character: they are critical factors in global risks but are at the same time posing their own global risks, such as mass surveillance, data exploitation, misinformation, global inequality, tech colonialism, depletion of natural resources, and environmental harms. These risks often emerge in the contexts where data and AI solutions are implemented, possibly exacerbating problems that they are supposed to overcome. However, seen from a bird's eye view, risks stemming from current technology trends have their own distinct qualities that put them next to other fundamental challenges that the world is facing. Curiously, however, CDS rarely seem to explicitly connect to cosmopolitanism. So far, one does not find many – or any – results on search platforms such as Google

Scholar for keywords including "transnational data studies" or "cosmopolitan data studies," with or without the adjective "critical" interjected. This seems like a missed opportunity to contextualise datafication and automation more concretely as profound cosmopolitan transformations that become manifest through interdependences across geographical and legal boundaries.

As I will explain below, cosmopolitanism as a theoretical framework and critical methodology shares ontological and epistemological similarities with CDS, but it offers a more concise conceptualisation of how global, transnational, national, and sub-national dimensions of social interactions interconnect and affect one another. Linking cosmopolitanism to CDS can provide the latter with a more comprehensive perspective on the global dynamics of data and technology, advocating for a more integrated approach to understanding these complex phenomena.

3 Critical Cosmopolitanism and Its Potential for CDS

While critical reflections on cosmopolitanism and the closely related concept of transnationalism have been discussed in critical sociology since at least the 1970s, the academic discussion produced some of its most influential theoretical-methodological frameworks in the early 2000s, popularised by sociologist Ulrich Beck (2002; 2003). His concept of cosmopolitanism builds on a transnational-global understanding of the social world, highlighting how global interconnectedness reshapes individual and collective identities, politics, and ethics (Beck, 2003). He perceived the technological, social, economic, and political transformations at the beginning 21st century as a sign for the emergence of "a new system in which everyday practices involve an exceptional level of cosmopolitan interdependencies" (Beck, 2003, p. 455).

Cosmopolitanism extends beyond acknowledging global diversity, addressing the realities of an interconnected world where issues such as climate change, turbulent financial markets, and terrorism necessitate transnational cooperation and solutions that surpass national borders and interests (Beck, 2008). Beck departs from a critique of the predominantly national angle of much of critical sociological research in the 20th and early 21st century: "[m]ethodological nationalism takes the following premises for granted: it equates societies with nation-state societies and sees states and their governments as the cornerstones of social-scientific analysis. It assumes that humanity is naturally divided into a limited number of nations, which internally organize themselves as nation-states and externally set boundaries to distinguish themselves from other nation-states" (Beck, 2003, p. 453). This assumption is highly problematic, as it confines the analytical lens from the outset and ignores how external influences and context-dependent networks of social actors across different spaces play pivotal roles in the materialisation of challenges and their consequences.

As an alternative, Beck (2003; 2006) advocates for a cosmopolitan outlook in which individuals and societies recognize their interconnectedness and interdependence. This fosters a new kind of politics and ethics capable of addressing global challenges, marking a shift from methodological nationalism to methodological cosmopolitanism that considers the global condition as its starting point: "[c]osmopolitan social science entails the systematic breaking up of the process through which the national perspective of politics and society as well as the methodological nationalism of political science, sociology, history, and law confirm and strengthen each other in their definitions of reality" (Beck, 2003, p. 456). Beck's vision of cosmopolitanism goes beyond classical ideals like universal moral obligations and world citizenship to focus on the practical and empirical realities of global interdependence. He builds his approach on "cosmopolitical realism" (Beck, 2004): the cosmopolitanisation of societies is not the result of some guided political programme towards a unified world society but rather "a side effect of actions" that have transnational origins and effects without any explicit intent by societal actors. That does not change the fact that national policies and practices should reflect the transnational realities of global risks. This realism advocates for approaches grounded in the practical and empirical realities of global interdependence, necessitating political and ethical frameworks capable of navigating and mitigating global challenges. That also means that cosmopolitanism needs to consider multiple perspectives and narratives from diverse cultural contexts (Beck & Sznaider, 2006). The point is not to disregard the relevance of local and even national frameworks for social interaction but to critically consider the transnational contexts in which they are embedded. Yet cosmopolitanism is decidedly post-Western and focuses on the "interface of the local and the global" (Delanty & Harris, 2018, p. 95).

On the one hand, cosmopolitanism is a theoretical proposition for describing how globalisation of human activity – mediated through communication technology – led to the emergence of social contexts that transcend geographical and socio-political boundaries from the national dimension to a transnational one (Nguyen, 2017). Flows of communication, social interactions, political developments, economic transactions, and cultural exchange are not bound by geographical locations and the confinements of nation-states. The second half of the 20th century saw a rapid dissolution of various borders, questioning the integrity and separability of countries as somehow closed containers or autopoietic systems that can be understood and described by themselves and through themselves (Beck, 2007; Nguyen 2017). Rather than seeing societies as grounded in and delimited by historically contingent nation-states, it is more accurate to perceive them as complex networks that may cluster around certain political-cultural focal points but that are inevitably linked to global social configurations.

On the other hand, methodological cosmopolitanism intends to guide research practice accordingly. Instead of stopping the analysis of a social phenomenon and especially crises at the borders of a country, it takes a cosmopolitan research framework to critically question such potentially artificial borders and to engage with the complex interconnectedness of issues. One concrete example is the COVID-19 pandemic, which affected societies on a global scale, shaped transnational and international relations, and steered national governance all the way down to its regional and local sub-levels, which in turn could affect similarly all other dimensions, cascading upwards in a similar fashion. A local super-spreader event had the potential to change national policies, which in turn could trigger changes to transnational openness in e.g., the European Union (EU). Similarly, the very platforms that build and maintain technological systems for the commercial exploitation of digital data operate on a global scale, yet regulatory interventions are developed on a state-level with the notable exception of the transnational frameworks devised by the European Union.

However, it is important to recognise the conceptual differences between cosmopolitanism and transnationalism. A strong link exists between the two, yet the former has an explicitly normative function, while the latter is best considered as an empirical phenomenon representing the existence of non-national but also not necessarily fully global spheres of social interaction. Previously, I conceptualised transnationalism as both a discursive process and a condition where social, political, and economic activities and discourses transcend national borders (Nguyen, 2017). This leads to the formation of networks and communities that are not strictly defined by geographical boundaries or national identities. What makes transnational scopes different from global scopes is that they still present concrete, discernible socio-cultural spaces that are interlinked through specific societal actors, concrete issues, and their relationships.

Transnational contexts do not simply concern the "whole world" as such, which would be implied by the broader term *global*, or, put differently:

"[f]orms of transnationalism are a consequence of globalisation, but not all transnational relations are necessarily global in scope" (Nguyen, 2017, p. 41). While globalism refers to the idea of a fully integrated world where global connectivity and cooperation lead to the erosion of national borders and identities, often focusing on the economic and corporate dimension of globalisation, transnationalism offers a distinct perspective. It does not necessarily imply a homogenization of culture or governance but highlights the existence of multiple, overlapping communities and networks that operate across and beyond national borders. Transnationalism focuses on the lived experiences of individuals and communities navigating these cross-border connections, showcasing the complexities of modern identities and social formations in a globally interconnected landscape. Examples of transnationalism are political-legal entities transcending nation-states, such as the EU, migrant networks spanning different host countries, or social, cultural, and political discourse configurations centred on specific issues such as climate change. The formation of transnational social contexts is greatly facilitated by digital technologies and the internet, which enable the rapid exchange of information, ideas, and resources across borders (Hall, 2018). This perspective underscores how identities, cultures, and political actions are determined by context-dependent flows of communication that cannot be considered in any way confined to nation-states as somehow closed containers.

Cosmopolitanism is not the same as transnationalism; a critical take on cosmopolitanism always includes a strong normative component and "as a mode of critique, [it] is not simply manifest in the fact of hybridity or transnationalism" (Delanty & Harris, 2018, p. 99). Delanty and Harris (2018) emphasise that cosmopolitanism acts as a normative framework for understanding social and political contexts. Although there are differing opinions on the extent and type of its normativity, adopting a cosmopolitan perspective generally leads "towards a critical as opposed to an affirmative view of current realities" (Delanty & Harris, 2018: 91). From this angle, cosmopolitanism is about critical analysis and potentials for action that can initiate change. Gilabert (2006: 1) locates here "three coordinates: a search for universal rights, sensitivity to contextual specificities, and autonomous empowerment of all individuals." Critical cosmopolitanism aims for both critique and emancipation from existing forms of domination based on empirical research, with the goal of "exploring new ways of seeing the world" (Delanty & Harris, 2018, p. 94).

This normative mission and its goals strongly resonate with CDS. For example, Heeks and Shekhar (2019) show how context sensitivity can help with better understanding and assessing concrete local initiatives that try to make use of data for urban development projects in the global South. Concerning empowerment and civic resistance to oppressive data regimes emerging in the private-public intersection, normative cosmopolitanism strongly connects to "data justice" (Taylor, 2017), which promotes critical work to protect and empower especially marginalised social groups and the powerless in datafication trends.

However, it can appear that the need to consider the locality of data (Loukissas, 2019) comes at the expense of more in-depth analyses of the border-transcending dimensions shaping socio-technical data assemblages. And while the analytical lens is expanding beyond the Global North, critical views on data practices, even rooted in frameworks centred on marginalised social groups, still tend to take on a predominantly "Western" perspective that is potentially blind to non-Western experiences (Tacheva, 2022). Despite the various conceptual and normative common ground concerning their underlying ontological and epistemological outlooks, CDS do not always concretely state when and how their research subjects are inherently transnational, potentially warranting a critical cosmopolitan lens. They tend to describe how global transformations surrounding technology explain local manifestation of impact, rightly arguing that it is in specific and immediate contexts where datafication and automation need to be understood. This implies that CDS may risk over-focusing on these two extremes, the global and the local. Transnational moments are different and need a fitting methodological approach for which critical cosmopolitanism offers a framework that easily connects to the core mission of CDS. For example, to explain datafication trends within specific European countries' national governance systems, it would be equally important to consider the transnational influences, conflict constellations, and negotiations on the European level.

Furthermore, the transnational layers of datafication and automation challenges could be positioned as a CDS research subject in their own right, approached from a cosmopolitan perspective. While research on the datafication of particular municipalities or data activism movements in a given country from a local perspective is essential, critically assessing them also from a cosmopolitan viewpoint can enrich critical understanding of the underlying societal dynamics (e.g., where do technology narratives shaping local data imaginaries originate from? Who are the stakeholders outside national borders and what is their relevance?). It is against this background that I argue there is considerable potential for CDS to integrate a cosmopolitan lens more systematically in its research practices.

4 A Cosmopolitan Research Agenda for CDS

There are various ontological and epistemological similarities between CDS and critical cosmopolitanism: both approach social reality through a critical analysis of communication processes and discursive practices in which media technologies, socio-political configurations, political economies, cultural contexts, and societal actors' – individual and collective, human, and non-human – positionalities are the main research focus. For both, it is all about how people and the techno-human systems that they form organise social life and what ethical challenges this entails. Thus, CDS and critical cosmopolitanism have a similar outlook on what constitutes research questions worth pursuing: how to identify, explain, and address the harms and risks posed by fundamental societal transformations. They are similarly broad in scope and diverse concerning their empirical foci. Cosmopolitan configurations matter across economic, political, social, and cultural contexts. A deep understanding of the context specifics is crucial, as the cosmopolitan dimensions of, e.g., ecological challenges are different from conflicts or financial collapses, though the intricate interdependencies between these need to be considered as well (Nguyen, 2017). Empirical cases and their contexts are rather diverse within CDS as well. There are shared technological developments and social practices in how big data and AI transform, e.g., education, politics, and culture, but the constellations of relevant societal actors, the stakes involved, and ethical challenges vary.

Where critical cosmopolitanism and CDS differ but also have potential for supplementing one another concerns their respective conceptual outlook and empirical focus. As outlined above, cosmopolitanism provides a more elaborate conceptualization of the relevance of transnational contexts for social interaction and how to critically research them from a normative perspective. Media technology plays an important role here (Beck, 2008). Previous critical cosmopolitan discussions of digital media focused on "whether political spaces, over the Internet, can be democratising or politically polarising?" (Hall, 2018, p. 412). However, the urgency of addressing datafication and automation specifically was not yet widely acknowledged at the time when core tenets of critical cosmopolitanism were articulated. Over the past decade, CDS has developed a comprehensive theorisation of data and AI as essentially mediating technologies that reconfigure how social relationships are formed, maintained, and governed. Simply put, cosmopolitanism offers detailed theoretical-methodological frameworks for critically analysing the transcendence of social configurations, while CDS focuses on the role of data-driven technologies and the (ideological) discourses around them.

The question is what conceptual and methodological considerations CDS could productively integrate from critical cosmopolitanism. Firstly, the concept of data/technology assemblages already highlights interconnectedness across societal domains and sectors, often pointing to global trends driving their formation (Kitchin, 2022). However, the transnational dimension of these formation processes should be more explicitly addressed and unpacked where possible. What are the concrete configurations of stakeholders in the complex spectrum from local to global? How can we differentiate concrete transnational moments that matter for the materialisation of a given assemblage? These are important questions that could be approached from a cosmopolitan angle. Secondly, CDS may expand on cross-cultural comparative research to identify commonalities and differences in perceptions of datafication and automation, especially pertaining to risks. Charting culturally dependent views on technology is a starting point to understanding and bridging differences, allowing us to underline shared experiences. Connected to this, CDS would benefit from paying more attention to transnational technology discourses and how they factor into politicisation of big data and AI. Thirdly, CDS may draw from critical cosmopolitanism to locate transnational potentials for the formation of civic resistance to data regimes as well as the appropriation and re-imagination of data assemblages. Many people are part of transnational networks (Boccagni, 2012) that are maintained with the help of digital media, no matter the cost in data and how large the sacrifice of personal privacy.

Over a decade ago, Boczkoswki and Siles (2014) outlined four research areas for a cosmopolitan view on media technologies that still largely hold up in the age of big data and AI: "the production of content, the consumption of content, the production of materiality, and the consumption of materiality." In the following, I outline different research areas that would fall within the scope of CDS with a cosmopolitan-transnational outlook. This overview is not intended to be complete and merely suggests general directions for critical inquiry into current societal challenges and transformations that emerge around datafication and automation trends on a global scale.

1. *Researching transnational AI discourses and their role in the politicisation of technology.* One important area of research that requires a cosmopolitan lens is the critical analysis of AI discourses with a focus on the exploration of commonalities, mutual affectivities, and differences between cultural contexts (Nguyen & Hekman, 2022). Discourses about AI concern questions about socio-technical imageries,

technology framing, and entailed questions of power (who determines the agenda? What are the dominant narratives and to what extent do they scale beyond national contexts?). Cosmopolitan research in media and communication studies has a tradition of describing and assessing potentials for and manifestations of transnational discursivity in concrete contexts, such as the EU (Nguyen, 2017) or around ecological challenges, conflicts, and human rights issues (Dieter, 2010). Data and AI are increasingly relevant factors in diverse societal domains that transcend national borders, raising various political, social, economic, and cultural questions. This connects in the European context to transnational attempts at regulation, as exemplified in the 2024 AI Act. AI development, deployment, and integration include transnational social formations that create meanings of AI in discursive settings, factoring into visions, imaginations, and eventually decision-making across borders. Here a critical cosmopolitan lens offers frameworks and theories that can help describe prevailing discourse dynamics and normative implications.

2. *The human labour behind and material costs of AI.* The massive material and human costs of AI are often left out of dominant technology discourses (Crawford, 2022; Lindgren, 2023), since acknowledging them easily breaks the illusion of digital technologies' allegedly transcendental and immaterial nature, which is part of their recurring framing as universal. Crucial infrastructures in AI assemblages rely heavily on human labour that is barely mentioned. Often outsourced to cheaper labour markets across the Global South, tech companies developing cutting-edge AI inventions rely on a constant supply of hand-labelled training data. In addition, natural resources are extracted under questionable conditions from world regions that have been historically subject to colonial value-extraction. These supply chains and the exploitation of human labour and ecological ramifications they entail form complex global networks. The critical analysis of specific sections of such networks requires an understanding of transnational dynamics between companies, governments, labourers, and activists. Maps revealing these transnational contexts are one analytical tool that CDS can apply here (D'Ignazio & Klein, 2020; Crawford, 2022).

3. *State-control and tech ideologies.* As Tang argues, "transnational cloud platforms pose challenges to cross-border data governance" (2022: 2397). How to rein in the expanding influence of technology companies is a persistent challenge for regulators around the world, often raising questions about possibilities for transnational collaboration

that need to consider cosmopolitan reflection on normative propositions and their potential consequences. State-control takes shape under global-transnational transformations, often trying to counter these developments. Furthermore, it is not only that technologies are exported as material matters; they come with narratives and imaginaries for how they should be used. Geopolitically, different self-proclaimed AI powers attempt to dominate technology development as well as the policies that should steer their implementation in global arenas selected for expanding political influence, often located in the Global South (Nguyen, Wang & Mutsvairo 2024). Another major concern is how authoritarian repression utilising digital technology and data is not confined to nation-states but can take a transnational dimension (Dukalskis et al., 2022). While this research area overlaps with critically analysing AI discourses, the focus here is more specifically on governmental practices and how frameworks are negotiated, challenged, and implemented from a cosmopolitan perspective.

4. *Cosmopolitan data activism.* Grass-roots movements challenging existing data-centric power hierarchies and raising critical questions about "exploitative forces inherent in processes of datafication" (Lehtiniemi & Ruckenstein, 2019: 2) are growing in number across cultural, social, and political contexts. Data activism can take different forms and aim for different goals. On the one hand, data can become a tool to promote a specific agenda within a particular context, such as the use of data in climate activism or to address social injustice (D'Ignazio & Klein, 2020). Data can serve here as empirical evidence to point to issues affecting social groups and to instrumentalise them as a means of political persuasion that supports a given movement's set of objectives. On the other hand, data power can be the very target of political activism itself. In such contexts, civil society groups tend to criticise and politicise the very use of data against their interests (e.g., Cinnamon, 2020). Transnational social movements are a staple research subject in critical cosmopolitanism that is mainly interested in factors that enable their formation; the communicative dynamics that shape agency, action, and impact in such contexts; as well as the negotiation of normative agendas steering them. It is thus easy to apply a similar lens on empirical cases where civic political engagement and activism either utilise data and/or aim for overcoming forms of oppression, inequality, dispossession, subjugation, and exploitation that are enacted through emerging data assemblages.

5. *Datafication of migration.* Critical research into digital migration is a well-established research domain (Leurs, 2023) that often engages with the datafication of the governance of migration but also the transformative role of data in migrant experiences, their transitions to host countries, and the formation of transnational networks in the form of digital diasporas spanning across borders (Nguyen & Nguyen, 2022). Digital migration studies establish a strong link between CDS and critical cosmopolitanism within a specific empirical context. For example, Ponzanesi (2020) proposes digital cosmopolitanism based on a critical postcolonial framework to research digital connectivity among migrants. CDS concepts matter greatly in describing, understanding, and critiquing the role of data and AI in migration processes and the formation of transnational networks, while cosmopolitanism provides methodological frameworks for critical reflection on border-transcending dynamics in both migration movements and nation-states' attempts to control them.
6. *Transnational conflict and AI.* Contemporary conflicts almost always have a transnational dimension through their extension into digital networks where different state- and especially non-state actors clash (Karatzogianni, 2009). Extremist political streams can form networks across different countries using digital media platforms to engage in different forms of "cyberconflict" (Karatzogianni, 2009), such as coordination of activities, hacking, data theft, and information warfare, including the spread of misinformation. It is important to acknowledge that forms of misinformation are culturally contingent, and more research is needed into how they emerge – and are instrumentalised for various political goals – especially in the Global South (Zeng & Brennen, 2023). The role of AI as a weaponised tool to undermine opponents in various ways is growing in these different areas of digital conflict. On a nation-state level, governments around the world actively pursue policies of datafying and automating warfare, partially aiming to export their technologies in the global digital arms market. From a cosmopolitan perspective, critical questions concern the dynamics of conflict materialisation and the constellations of actors involved that engage in political struggles with digital means.

This is only a selection of pertinent research domains in which CDS and cosmopolitanism could be productively integrated to enhance analytical approaches for addressing specific research questions. There is a plethora of empirical cases where insights from both are needed to critically unpack socio-technical configurations and their ethical challenges from a decidedly normative perspective.

5 Cosmopolitan CDS and Empirical Research with Computational Methods

Both CDS and cosmopolitanism as methodological approaches often leave it up to the researcher to define for themselves what exact methods should be chosen for collecting and analysing relevant data. In this respect, CDS and cosmopolitanism share a general tendency towards qualitative approaches. As I will argue in the final section of this chapter, there is potential to introduce more mixed-methods within a computational- and digital methods framework. This is not yet common in either field but could enhance empirical research practice in their intersection.

One of the reasons for the skewedness of the existing set of common methods in CDS is their aforementioned background in the humanities and critical sociology; they share certain ontological-epistemological paradigms favouring qualitative methods. This concerns a variety of approaches, including e.g., critical discourse analyses, ethnographic methods, policy analyses, affordance analyses, app walkthroughs, data walks, data logbooks, interviews, and participatory research designs such as workshops and hackathons. There are also quantitative study designs using social science-oriented methods, such as surveys and experiments. To a lesser extent, CDS studies deploy computational- and digital methods or combine different methods in mixed research designs. Despite certain tendencies, CDS appear agnostic about what empirical methods fit into their research agenda and there is no formalised framework for data collection and analyses. This reflects their interdisciplinary roots and offers researchers considerable freedom to choose approaches fitting the research problem at hand. The situation is similar for cosmopolitan research practice, which is to a considerable extent informed by critical sociology (Beck, 2003). However, research investigating transnational discourses, transnational networks, and cosmopolitan identity has applied e.g., quantitative content analyses, surveys, and network analyses in the past (Nguyen, 2017). The application of computational-/digital methods is still in a nascent stage though.

There are various cases in which these methods could be applied productively to gain more empirically grounded insights and to develop and answer new research questions that expand the interdisciplinary scope of cosmopolitan CDS. For example, transnational AI discourses can be analysed on scale with text analytical methods to chart technology narratives and the socio-technical imaginaries that they carry (Nguyen, 2023). Methods such as topic modeling, word embeddings, and automated text classification can provide an overview of topical foci and trends in large corpora about CDS relevant issues. News text, social media posting, governmental communication, and corporate communication can be scanned and summarised relatively quickly and then be combined with in-depth discussions of representative texts for different categories identified during the quantitative-computational analysis. Methods such a knowledge graphs can provide further context and help charting what societal actors get to speak about what datafication- and automation-centric topics. In this way, the politicisation and transnationalisation of technology issues can be traced empirically. Similarly, terms and conditions and data policies could be analysed on scale to assess their level of accessibility and degree of transparency as well as variations across countries. Network analysis is another important method to critically analyse social relationships transcending borders and what their implications are with quantitative data. Interaction data can be scanned for how exactly users engage with digital media in contexts that matter for cosmopolitan CDS, such as community formation and data activism beyond national borders. In short, charting the scope of transnational dimensionality and its social dynamics in datafication and automation contexts can be empirically approached with diverse data-driven, computational methods and the observations can then be linked to critical reflection.

It depends on the exact research interest, but there are plenty of opportunities for the useful application of computational methods. Studies taking a cosmopolitan outlook within CDS may benefit from further exploring how novel research approaches bridging data science with critical investigations from a humanities and/or sociological approach can enrich their empirical work. While there have been attempts to introduce critical qualitative methods to data- and computer sciences (Moats & Seaver, 2019), efforts to integrate large-scale data analyses using algorithms are limited in CDS. A general lack of hands-on training in such methods in CDS's core disciplines is one important factor here (Nguyen, 2020) but training in computational methods tailored to the humanities and social sciences is increasing.

Maybe more importantly, there are several very practical challenges even if researchers have the relevant expertise. Access to data is a primary challenge regardless of the chosen methodological approach. Web scraping has become more difficult, especially with the rise of generative AI services such as ChatGPT that collect large volumes of data from the Internet without asking for explicit consent from websites. Researchers who want to collect e.g., news articles, press releases, or website texts need to assess the possibilities of retrieving data for each potential source. Alternatively, they need to consult intermediaries that keep text material in their vaults (e.g., Nexis), which often comes with considerable financial costs. Most social media platforms either can no longer be accessed to retrieve data freely or their access is heavily restricted. Infamously, Twitter's rebranding to X came with a cancellation of free academic data access. Furthermore, user interaction data is notoriously difficult to acquire, due to proprietary and privacy-related reasons. While it would be quite insightful to understand user practices on specific platforms across different markets on scale, it is often not easily possible to pursue such research interests with quantitative-computational methods. However, data donation frameworks through which users can share their personal data with researchers in a secure and privacy-savvy way offer here a possible alternative (Data Donations, n.d.) – but it is so far limited to the European Union and to what extent it could be replicated elsewhere remains an open question. Other challenges pertain to limitations of different computational methods. For instance, many tools for natural language processing (NLP) needed for text analysis are optimised for English. There is work on other languages, but considerable imbalances remain, and the quality can be comparatively low due to a lack of available training data (especially concerning the Global South). Any research design focusing on transnational contexts that include different languages needs to find fitting strategies to deal with this challenge.

Nevertheless, exploring how computational methods can enhance cosmopolitan CDS offers various opportunities to move beyond the limitations of anecdotal data. Societies engage with digital media on a massive scale, warranting a data-driven and automated approach to understanding and criticising data practices across local, transnational, and global dimensions. Put differently, critically applying big data and AI to important questions about big data and AI can reveal trends and configurations that may remain invisible with qualitative methods. This further enables the testing of existing CDS theories for their robustness and explanatory value. In addition, integrating computational methods within mixed research designs can strengthen interdisciplinary collaboration and thus expand CDS to data science and computer science. Most cosmopolitan CDS research problems need diverse academic expertise to devise methodologies and to find comprehensive answers to complex questions. Admittedly, there are various logistical as well as resource-related challenges, not to mention persistent ontological-epistemological divides between disciplines. However, collaborations have become more common, going beyond mere dialogue and centering on joint research ventures. It is in these interdisciplinary moments where innovation in methodological development and theorisation has the potential to happen. This also concerns the articulation of shared research ethics using a novel approach relying on data and AI. The task here for researchers focusing on cosmopolitan CDS is to cross borders themselves and to embrace the challenge to develop flexible, creative, integrative, and reproducible research strategies that fit their research problems. Universities and research institutions should try their best to support such projects where they can.

6 Conclusion

In this chapter, I tried to conceptually link CDS with insights from critical cosmopolitanism before arguing why such interdisciplinary research on border-defying socio-technical transformations would benefit from expanding the empirical toolset with computational methods. It is important to again underline that CDS as practiced over the past decade have never been ignorant of the global and transnational dimensions of datafication and automation. However, what I hope to have offered is a more explicit integration of key ideas, propositions, and concepts from one field to the other. In sum, I see two general benefits of a cosmopolitan outlook in CDS: on the one hand, it may inspire scholars to pursue more research into concrete transnational moments of datafication and automation. On the other, it can help with explaining how transnational and global influences factor into local data practices. In addition, a cosmopolitan research attitude can further expand CDS to engage with more empirical cases from non-Western world regions. It seems only prudent to consider these different aspects in critically understanding current technology trends of global scale and their risks, not least to identify potentials for change. As Delanty and Harris point out: "Cosmopolitanism is thus both a normative theory (which makes cognitive claims) and also a particular kind of social phenomenon" (2018, p. 95). Arguably, CDS share this core tenet.

Methodologically, any integration of both frameworks needs to eventually address the question of how to pursue empirical research. Here a shift from prevalent qualitative methods towards method mergers that also include quantitative-computational methods seems equally prudent given the nature and scope of the research subjects in focus. Conceptual clarity is an important starting point, necessitating coherent definitions for what constitutes local, transnational, and global factors. In addition, interdisciplinary collaboration may support methodological innovation and help CDS to further mature as an academic field. Transcending borders matters for CDS not only with respect to its research subjects but also the very methodologies and the underlying assumptions of how the social world can be critically analysed in reliable and valid ways.

References

Beck, U. (2002). The cosmopolitan society and its enemies. *Theory, Culture & Society*, 19(1–2), 17–44.

Beck, U. (2003). Toward a New Critical Theory with a Cosmopolitan Intent. Constellations, 10(4), 453–468.

Beck, U. (2004), Cosmopolitical Realism: On the Distinction between Cosmopolitanism in Philosophy and the Social Sciences. Global Networks, 4: 131–156. https://doi.org/10.1111/j.1471-0374.2004.00084.x

Beck, U. (2006). *Cosmopolitan Vision*. Polity.

Beck, U. (2008). *Weltrisikogesellschaft: Auf der Suche nach der verlorenen Sicherheit*. Suhrkamp.

Beck, U. (2012). Global Risk Society. In G. Ritzer (Ed.), *The Wiley-Blackwell Encyclopedia of Globalization*. https://doi.org/10.1002/9780470670590.wbeog242

Beck, U., & Sznaider, N. (2006). Unpacking cosmopolitanism for the social sciences: A research agenda. *British Journal of Sociology*, 57(1), 1–23.

Boccagni, P. (2012). Rethinking transnational studies: Transnational ties and the transnationalism of everyday life. European Journal of Social Theory, 15(1), 117–132. https://doi.org/10.1177/1368431011423600

Boczkowski, P. J., & Siles, I. (2014). Steps toward cosmopolitanism in the study of media technologies. *Information, Communication & Society*, 17(5), 560–571. https://doi.org/10.1080/1369118X.2013.808358

boyd, danah, & Crawford, K. (2012). CRITICAL QUESTIONS FOR BIG DATA: Provocations for a cultural, technological, and scholarly phenomenon. *Information, Communication & Society*, 15(5), 662–679. https://doi.org/10.1080/1369118X.2012.678878

Chernilo, D. (2011). The critique of methodological nationalism: Theory and history. *Thesis Eleven*, 106(1), 98–117. https://doi-org.proxy.library.uu.nl/10.1177/0725513611415789

Cinnamon, J. (2020). Attack the Data: Agency, Power, and Technopolitics in South African Data Activism. *Annals of the American Association of Geographers*, 110(3), 623–639. https://doi.org/10.1080/24694452.2019.1644991

Crawford, K. (2021). *Atlas of AI: Power, Politics, and the Planetary Costs of Artificial Intelligence.* Yale University Press.

Couldry, N., & Mejias, U. A. (2020). *The costs of connection: How data are colonizing human life and appropriating it for capitalism.* Oxford University Press.

Data Donations, Welcome at Europe's Data Donation Infrastructure. (n.d.). Data Donation EU. Retrieved April 24, 2024, from https://datadonation.eu

Delanty, G. (2006). The cosmopolitan imagination: Critical cosmopolitanism and social theory. *The British Journal of Sociology*, 57(1), 25–47.

Delanty, G., & Harris, N. (2018). The idea of critical cosmopolitanism. In G. Delanty (Ed.), *Routledge International Handbook of Cosmopolitanism Studies* (2nd ed., pp. 10–25). Routledge. https://doi.org/10.4324/9781351028905

Dieter (2010): The making of a comprehensive transnational discourse community, In M.L. Djelic, Marie-Laure & S. Quack (Eds.), *Transnational Communities: Shaping Global Economic Governance* (pp. 305–326). Cambridge University Press.

D'Ignazio, C., & Klein, L. F. (2023). *Data Feminism.* The MIT Press.

Dukalskis, A., Furstenberg, S., Gorokhovskaia, Y., Heathershaw, J., Lemon, E., & Schenkkan, N. (2022). Transnational repression: data advances, comparisons, and challenges. *Political Research Exchange*, 4(1).

Gilabert, P. (2006). Cosmopolitanism and Discourse Ethics: A Critical Survey*. *New Political Science*, 28(1), 1–21. https://doi-org.proxy.library.uu.nl/10.1080/07393140500518125

Hall, Oliver 2018. "The Internet and Cosmopolitanism." In Routledge International Handbook of Cosmopolitanism Studies, 406–18. Abingdon: Routledge. https://doi.org/10.4324/9781351028905-36.

Heeks, R., & Shekhar, S. (2019). Datafication, development and marginalised urban communities: an applied data justice framework. *Information, Communication & Society*, 22(7), 992–1011. https://doi-org.proxy.library.uu.nl/10.1080/1369118X.2019.1599039

Hepp, A., Loosen, W., Dreyer, S., Jarke, J., Kannengießer, S., Katzenbach, C., Malaka, R., Pfadenhauer, M., Puschmann, C., & Schulz, W. (2023). ChatGPT, LaMDA, and the hype around communicative AI: The automation of communication as a field of research in media and communication studies. Human-Machine Communication, 6, 41–63. https://doi.org/10.30658/hmc.6.4

Iliadis, A., & Russo, F. (2016). Critical data studies: An introduction. *Big Data & Society*, 3(2), 1–7. https://doi.org/10.1177/2053951716674238.

Karatzogianni, A. (2009). *The Politics of Cyberconflict.* Routledge.

Katzenbach, C. & Ulbricht, L. (2019). Algorithmic governance. *Internet Policy Review*, 8(4). https://doi.org/10.14763/2019.4.1424

Kitchin, R. (2022). The Data Revolution: Big Data, Open Data, Data Infrastructures and Their Consequences. SAGE.

Leurs, K. (2024). *Digital Migration*. Sage.

Lehtiniemi, T., & Ruckenstein, M. (2019). The social imaginaries of data activism. *Big Data & Society*, 6(1). https://doi.org/10.1177/2053951718821146

Lindgren, S. (2024). *Critical Theory of AI*. Wiley & Sons.

Loukissas, Y. A. (2022). *All Data Are Local: Thinking Critically in a Data-Driven Society*. The MIT Press.

Mejias, U. A. & Couldry, N. (2019). Datafication. *Internet Policy Review*, 8(4). https://doi.org/10.14763/2019.4.1428

Milan, S., & Treré, E. (2019). Big Data from the South(s): Beyond Data Universalism. Television & New Media, 20(4), 319–335. https://doi.org/10.1177/1527476419837739

Morales, E., & Reilly, K. (2023). The unhomed data subject: negotiating datafication in Latin America. *Information, Communication & Society*, 26(12), 2457–2471. https://doi.org/10.1080/1369118X.2023.2250436

Moats, D., & Seaver, N. (2019). "You Social Scientists Love Mind Games": Experimenting in the "divide" between data science and critical algorithm studies. *Big Data & Society*, 6(1). https://doi-org.proxy.library.uu.nl/10.1177/2053951719833404

Nguyen, D. (2017): *Europe, the Crisis, and the Internet. A Web Sphere Analysis*. London: Palgrave MacMillan.

Nguyen, D. (2020). Media and Communication Studies in the Age of Digitalization and Datafication: How Practical Factors and Research Interests Determine Methodological Choices. In *Understanding Media and Society in the Age of Digitalisation* Palgrave Macmillan. https://doi.org/10.1007/978-3-030-38577-4_2

Nguyen, D., & Hekman, E. (2022). A 'New Arms Race'? Framing China and the U.S.A. in A.I. News Reporting: A Comparative Analysis of the Washington Post and South China Morning Post. *Global Media and China*, 7(1), 58–77. https://doi.org/10.1177/20594364221078626

Nguyen, D., & Nguyen, S. (2022). Data Literacy as an Emerging Challenge in the Migration/Refugee Context: A Critical Exploration of Communication Efforts Around "Refugee Apps". *International Journal of Communication*, 16, 5553–5557. https://ijoc.org/index.php/ijoc/article/view/17406

Nguyen, D., Wang, B., & Mutsvairo, B. (2024). (Pragmatic) collaboration for progress or threat to autonomy? African news discourses about Chinese technology in Nigeria and Ghana. Global Media and China, 0(0). https://doi.org/10.1177/20594364241232757

Ponzanesi, S. (2020). "Digital Cosmopolitanism. Notes from the Underground". *Global Perspectives*, 1(1). https://doi.org/10.1525/gp.2020.12548.

Rieder, G., & Simon, J. (2016). Datatrust: Or, the political quest for numerical evidence and the epistemologies of Big Data. *Big Data & Society*, 3(1). https://doi.org/10.1177/2053951716649398

Tacheva, Z. (2022). Taking a critical look at the critical turn in data science: From "data feminism" to transnational feminist data science. *Big Data & Society*, 9(2). https://doi.org/10.1177/20539517221112901

Tang, M. (2022). The challenge of the cloud: between transnational capitalism and data sovereignty. *Information, Communication & Society*, 25(16), 2397–2411. https://doi-org.proxy.library.uu.nl/10.1080/1369118X.2022.2128598

Taylor, L. (2017). What is data justice? The case for connecting digital rights and freedoms globally. *Big Data & Society*, 4(2). https://doi.org/10.1177/2053951717736335

Wagner, K. B. (2023). TikTok and its mediatic split: the promotion of ecumenical user-generated content alongside Sinocentric media globalization. *Media, Culture & Society*, 45(2), 323–337. https://doi.org/10.1177/01634437221136006

Zhu, Y. (2024). Privacy cynicism and diminishing utility of state surveillance: A natural experiment of mandatory location disclosure on China's Weibo. *Big Data & Society*, 11(2). https://doi.org/10.1177/20539517241242450

Zeng, J. & Brennen, S. B. (2023). Misinformation. *Internet Policy Review*, 12(4). https://doi.org/10.14763/2023.4.1725

Zuboff, S. (2019). *The Age of Surveillance Capitalism: The Fight for a Human Future at the New Frontier of Power*. Profile Books.

CHAPTER 2

Digital and Epistemic Sovereignty in the Science Ecosystem in Latin America

Thaiane Oliveira, Afonso de Albuquerque and Tatiane Mendes

1 Introduction

Digital media and technology play a crucial role in Latin America, particularly in the context of promoting digital sovereignty. The region faces challenges such as a lack of digital training and limited access to technological resources in higher education (Antón-Sancho et al., 2023). The transition from nationalistic media regulations to pragmatism in digital policies in Latin America provides valuable insights into the conditions that foster nationalism and sovereignty in digital policies (Becerra & Waisbord, 2021). In the context of universities, the role of the university in Latin America has been redefined in a growing context of academic capitalism, where universities could become important actors of development in the region (Riveros et al., 2022; Arocena & Sutz, 2005).

As a growing body of literature has demonstrated, structural inequalities have persistently characterized the academic system around the world. A limited number of countries, most of them former colonial powers located in the West, have exerted firm control over the agenda of scholarly research (Alatas, 2013; Albuquerque, 2021; Demeter, 2020). The United States stands out as the dominant player in this system, which adopted English as the lingua franca (Suzina, 2021). Some authors have used the label "intellectual imperialism" to describe this asymmetric system (Alatas, 1969, 2000; Bourdieu & Wacquant, 1999). This system allows scholars working in countries of the 'center' to exert influence on the manner in which scholars elsewhere perceive the world and the place their own societies occupy in it. This enables the countries that occupy the center of this system to influence the public policies of those located on its periphery (Dezalay & Garth, 2002). However, the exercise of this asymmetric power does not happen without resistance.

This chapter intends to discuss how Latin American scholarship locates itself in this globally unbalanced academic system. It explores the manner

 | DOI:10.1163/9789004711396_004

in which such scholars resist western domination, seeking mechanisms to build spaces of scientific production and sovereign knowledge with the aim of strengthening their epistemic identity. Within academic circuits, one of the approaches that has been adopted is the implementation of open and multilingual publishing systems, allowing the dissemination of knowledge produced in the region to a broader audience, including peripheral countries (Vessuri et al., 2014). On the other hand, digital systems for accessing research metadata, ensuring accountability and transparent evaluation, and avoiding external control and manipulation of academic results by foreign forces, are not yet fully developed in Latin American universities. These actions are ways to promote Latin America's epistemic and digital sovereignty. The relevance of the Latin American case arises from the fact that Ibero-America (which includes Latin America and two European countries: Spain and Portugal) constitutes the largest alternative environment to the hegemonic Anglophone academic circuit. This is for two reasons. First, most inhabitants of Ibero-America share two languages that are at least mutually understandable (Spanish and Portuguese). The second reason is the growing presence of Ibero-American journals in Scopus, which provides better conditions for the scholars working in this region to achieve better visibility.

During the 1990s, Latin American efforts to achieve digital and epistemic sovereignty experienced a backlash. In the wake of the process of neoliberal globalization, a broad process of deregulation and privatization created the conditions leading to the entry of companies from the academic, scientific, and technological publishing oligopoly into a digitalization process of universities. In spite of this, digitalization, understood as a growing form of epistemic imperialism, is intrinsically linked to a historical process in which power holders acted as disruptors of the state's role in supporting universities. The weakening of the role of the state in providing digital systems opened the door for transnational technology companies to influence academic production and practices with their own commercial interests, jeopardizing autonomy and epistemic sovereignty in research. This phenomenon, combined with a neoliberal model of precarization, has contributed to the dismantling of the information, communication, and technological infrastructures of academic institutions.

Over time, we have observed a neoliberal shift in power dynamics, with non-state actors gaining significant influence. This power shift has paved the way for the rise of economic models that prioritize efficiency, competitiveness, and privatization over social well-being and the production of diverse knowledge. This context creates an increasing dependence on private digital platforms, massive data collection by transnational companies, and the centralization of control over access to information, which contributes to the marginalization of voices and perspectives not aligned with dominant interests.

In the context of Latin America, universities, which have historically played a crucial role in the production and dissemination of knowledge, now face fundamental challenges in preserving their autonomy and social mission in the face of the growing influence of external forces and neoliberal logics. Given this context, this chapter aims to provide a critical overview of academic imperialism and Latin America's resistance to the development of its digital sovereignty. The goal of this chapter is to draw some lessons from Latin America's successful resistance to the encroachment of the epistemic domain within universities, even amidst neoliberal policy advances in academia. We aim to identify strategies employed by the region to withstand the impacts of neoliberal digital initiatives on the local academic landscape.

2 US Imperialism after World War II and the Latin American Reaction to It

The concept of imperialism has been mostly associated with the colonization perpetrated by Western European countries on societies located in other parts of the world, such as the Americas, Africa, Asia, and Oceania, since the 16th century. Since the end of World War II, however, the United States has replaced these countries as the most important reference for imperialism (Steinmetz, 2005; Kettell & Sutton, 2013; Riani, 2017). At that time, most Western European countries were exhausted as a consequence of the destruction provoked by the war, which led them to progressively lose control over their colonies. On the other hand, the United States emerged from the war as the dominant power in the world, both economically and militarily, and as the leader of the Western bloc in competition with the Soviet Union.

As part of its effort to legitimize itself in that role, the United States made a huge investment to disseminate its worldviews and sensibilities to other countries. In particular, the United States promoted Modernization Theory as a model to be followed by the countries belonging to the Majority World – then called "underdeveloped" or "developing" countries (Alam, 2008). According to this model, these countries could become "developed" by following the model of modernization established by the United States (Lerner, 1958).

To achieve this objective, the US has used various means: it has used mass media for conquering the "hearts and minds" of other countries (Schiller, 1999; Dorfman & Mattelart, 1971); assistance policies (Hayler, 1971; Lugo-Ocando, 2018) in promoting the ideology of development as the only correct form of "revolution," opposing communist perspectives; and massive funding for academic research in other countries to exert intellectual influence over the elites of other countries. Some studies have underscored the role of philanthropic foundations such as Ford, Rockefeller, and others (Parmar, 2012), as well as the dominance of the academic system by the West, especially the United States (Alatas, 2000; Dezalay & Garth, 2002), as a cultural imperialistic strategy.

Much of the United States' initial efforts to exert cultural influence abroad were directed at Latin America. This happened in part because, according to the Monroe Doctrine, Latin America was a "natural" zone of influence for the United States, its "backyard." Latin America provided a testing ground for the United States to experiment with techniques it would later apply elsewhere. The Cuban Revolution, in 1959, provided the United States with an additional stimulus for exerting cultural influence in Latin America to contain the expansion of communism in this region (Lugo-Ocando, 2018). Added to these "soft" means, the United States actively supported military coups in the region, which in most cases led to repressive political regimes led by anticommunist hardliners.

Latin American intellectuals were among the first to react against the massive export of US media and cultural products, calling it an example of cultural imperialism. While the United States justified this initiative in the name of the free flow of communication, these intellectuals contended that it was a threat to cultural diversity (Dorfman & Mattelart, 1971; Ramiro Beltran, 1976). Many of them used Dependency Theory as a means to criticize Modernization Theory (Simonson et al., 2022). They argued that the main consequence of the model of development championed by the United States was to enhance the economic and cultural dependence of the countries adopting it vis-à-vis the United States. Their approach to the problem was not merely academic but also activist (Freije, 2019). Under the government of Salvador Allende (1970–1973), Chile became one of the most important sites promoting the anti-imperialist agenda. Latin American intellectuals' criticism against US cultural imperialism inspired a document published by UNESCO, calling for a New World Information and Communication Order (NWICO), in 1980. The United States reacted vehemently to this initiative, leading to their withdrawal from UNESCO. The United Kingdom followed its steps (Carlsson, 2003).

3 The Neoliberal Global Order and the Second Wave of Academic Imperialism

The NWICO document represented a victory for the intellectuals mobilized against US cultural imperialism, but it was a pyrrhic one. The following years testified to the rise of a US-centered global order in the wake of the collapse of the Soviet Union in 1991. Together with international financing institutions (IFIs) like the World Bank and the International Monetary Fund (IMF), the United States led a process of massive reorganization of the world order based on neoliberal principles. Beginning in the 1980s, the IFIs obliged countries with large financial debts to make a series of mandatory structural changes in their economies in exchange for new loans. These changes included, among others, privatization of public services, deregulation, reduction of social policies, and opening their economies to foreign capital. Different authors have referred to neoliberal globalization as an imperialistic process (Teichman, 2004; Carvalho, 2022).

Latin America paid a heavy toll for implementing these neoliberal economic policies. The period of economic stagnation and social misery that followed the adoption of these policies became known as the lost decade (Schwartzman, 1991). In fact, Chile served as a laboratory for the neoliberal experiment. In 1973, after a brutal military coup supported by the United States, General Augusto Pinochet took his place as its president, and soon began to implement reforms under the guidance of US neoliberal economists, especially those affiliated with the University of Chicago (the "Chicago Boys"). The autocratic character of Pinochet allowed him to adopt very unpopular measures, as protests were met by the government with brutal repression (Dezalay & Garth, 2002).

The changes associated with neoliberalism were not limited to the economy but also affected many other areas of social life, such as politics, law, and education. Specifically, with respect to the academic milieu, two consequences of the neoliberal logic were academic capitalism and the global ranking system. According to these principles, scholars and academic institutions must compete for resources based on their academic performance. The global rankings establish the parameters according to which scholars and academic institutions are evaluated.

This entire system orbits around the United States and, secondarily, other Western countries (Albuquerque, 2021; Oliveira et al. 2021). English became the international lingua franca (Suzina, 2021), and US (and other Anglophone) journals are overwhelmingly present in the rankings and in their higher positions. A similar logic applies to the universities, whose rankings are also dominated by US and other Anglophone institutions. Scholars working at US and other Anglophone institutions dominate the editorial boards of the most prestigious journals (Albuquerque et al., 2020; Goyanes, 2020; Goyanes & Demeter, 2020). A recent study estimated that the University of Texas had twice the number of editorial board members compared to the entire Latin America and Caribbean region combined (Albuquerque et al, 2020).

Other mechanisms reinforced even more the concentration of academic power in the United States and the West. Universities play a central role in legitimizing new forms of domination, establishing ties of academic dependence between peripheral and central societies, and creating discourses that justify this new order and serve as the basis for public scientific policies. This dynamic extends beyond the publishing system and encompasses all areas of academia. The paywall system adopted by the majority of "reference" journals imposes structural barriers to access for researchers from the global South (Oliveira et al., 2021). Another example refers to the manner in which international scientific associations are organized. Wiedemann and Meyen (2016) described how the International Communication Association has been controlled for decades by a group of researchers based in the United States, particularly through a network centered around Michigan State University. Field Studies associations also follow a pattern of marginalizing researchers from the global South in the study of their own regions (Albuquerque, 2023). For example, out of the 32 presidents of Latin American Studies Association (LASA) since 1980, only four were affiliated with Latin American institutions at the time they held the position.

Altogether, these mechanisms concentrate the power to produce authoritative interpretations of "reality" in the United States. The effects of this concentration of power in the academic field are tangible. It legitimizes and naturalizes center-periphery relationships, structured around Western countries (Alatas, 2000; Park et al., 2022). Researchers from the Majority World often face challenges in accessing the world of "reference research," having to migrate to research institutions in central countries or integrate into research networks led by Western researchers. Some of the most significant aspects pertain to the price these researchers must pay to gain access to the exclusive world of "reference research." Common strategies include relocating to research institutions situated in central countries or integrating into research networks led by scholars from these countries. Within these networks, the role assigned to these researchers is often to provide specific data from their own country to illustrate conceptual frameworks typically derived from Western premises. At a more micro level, efforts by Majority

World scholars to publish articles in reference journals typically entail strategies such as extensively citing Western authors at the expense of those from their own country and, whenever possible, disguising their origin (Demeter, 2020).

Conversely, for the non-privileged countries, this arrangement resulted in a decline of epistemic sovereignty. This led to a degradation of the status of the Latin American research, as it was not available anymore in the "quality" journals. The analysis of teaching and learning documents in Latin American universities between 2000 and 2015, conducted by Guzmán-Valenzuela (2017), revealed that these curricula tend to prioritize theories produced in the Global North and neglect knowledge from the Global South and its realities. Concrete consequences resulted from this. For instance, the debate on cultural imperialism lost much of its appeal in the global arena. New theoretical concepts, such as "asymmetric dependence" by Straubhaar (1991), and "soft power" by Joseph Nye, Jr. (1990), replaced the classic approach to imperialism in the analysis of communication and culture. These concepts looked more suitable from the perspective of the new academic hegemons.

4 Latin American Resistance to Neoliberal Epistemic Imperialism

Despite this, Latin America has a history of resistance to imperialist practices, which obviously involved significant challenges. The region has created its own research metadata systems and open access scientific publishing channels and adopted state-level diamond models, such as SciELO, Latindex and Redalyc. This was a response to the prevailing English-language model, dominated by authors and editorial board members from Western countries (Albuquerque et al., 2020; Goyanes, 2020; Goyanes & Demeter, 2020), and the commercial model of closed-access science (Alperin, 2015). These pioneering initiatives in Latin America emerged almost a decade before the launch of the Budapest Open Access Initiative (BOAI) in 2002 (Sanchez-Tarragó et al., 2012). Efforts towards open access in Latin America were grounded in the principle that knowledge should be accessible not only through traditional scientific publishing channels but also through alternative spaces that are not as dependent on editorial monopolies, the hegemony of English as a lingua franca, and commercial models (Oliveira et al., 2021). This approach challenged the unequal production and circulation of knowledge in traditional "centers of excellence," which excluded peripheral scientific communities (Beigel, 2016; Vessuri et al., 2014). Latin America established its own transnational and regional channels, creating a publishing system that seeks epistemic sovereignty.

Despite extensive discussions on epistemic violence and academic imperialism in the past (Alatas, 1969; 2003; Castro-Gomes & Martin, 2002), the concept of epistemic sovereignty is relatively recent. One of the earliest discussions on this topic dates back to Karen Litfin's work in 1999, where she explores the impact of rapid advancements in information technologies on knowledge. According to Litfin (1999), knowledge and sovereignty are closely interconnected, as both play a vital role in defining authority and exercising control in the global arena. Science, technology, and the state are emblematic of modernity and greatly influence the configuration of the global landscape. The convergence of these aspects, which Litfin refers to as the "epistemic dimension of sovereignty," is particularly relevant for understanding contemporary global politics. This often overlooked dimension of sovereignty revolves around the control, access, and dissemination of information and knowledge.

Despite the relevance of Latin American initiatives for epistemic sovereignty in scientific publishing, these circuits are still largely invisible due to the dominance of a global model of academic capitalism influenced by university ranking systems that emerged in the 1990s, both in Western and non-Western countries (Mugnaini et al., 2019). This model of university ranking, predominantly produced by institutions based in the United States, perpetuates ethnocentric views and reinforces the notion of a universal standard (Albuquerque & Oliveira, 2020).

The "metric tide," mainly developed by Western and Central countries, has become dominant in discussions on science, technology, and innovation (STI) indicators (Hicks et al., 2015). It emerged alongside a period when Latin America experienced a deep economic crisis and an intense process of de-statization, financed even by the United States through fiscal adjustment policies promoted by the World Bank.

These rankings have been globally criticized for contributing to increased stratification and imperialism among institutions and nation-states (Perna et al., 2021). The use of these rankings has been associated with the Anglicization of scholarly communication, leading to the homogenization and hierarchization of the global field of management, and contributing to academic imperialism (Boussebaa & Tienari, 2019; Suzina, 2020). Furthermore, the impact factor of publication policy has been found to perpetuate academic and economic imperialism, undermining the development of higher education and scholarship in certain regions (Chukwuemeka et al., 2018). The mechanisms of these ranking systems are considered to be entirely Western and neo-colonial, negating the goals of education and scholarship in some contexts.

Despite criticism of this process of scientific measurement, such as the San Francisco Declaration (American Society for Cell Biology, 2012) and the Leiden Manifesto (Hicks et al., 2015), these indicators are still used in the scientific policies of many countries to evaluate the quality of scientific output and make comparisons between countries, industries, and public and private organizations. The use of international ranking systems tends to underestimate the role of scientific production, equating quality and impact with presence in Anglophone and central circuits. Latin America has been excluded from the system because its production was considered invisible by established norms. This does not mean that the region has stopped producing important research or that its quality has decreased. Latin American research has simply been excluded from the circuit that determines global quality, being treated as secondary (Albuquerque & Oliveira, 2020).

5 A Platformization of Science and the Digital Sovereignty Struggle of Universities

In contemporary times, digital transformation has become a priority for many countries in the region, aiming to harness the potential of digital technologies to drive economic and social development. However, such transformation is not without challenges, especially concerning epistemic sovereignty and digital governance in societies grappling with digital exclusion and the dominance of major technological and publishing oligopolies in higher education, for instance.

The increasing presence of tech giants like Google and Microsoft is impacting scientific practices in Latin America and around the world, as these corporations provide their services to universities and research institutions as "free solution" providers. Their business model relies on collecting and analyzing vast amounts of user data and metadata from their educational packages, including email, video conferencing, file exchange, and other services. This results in the platformization and datafication of higher education and universities. The platformization of science entails penetration into all stages of research in higher education systems (van Dijck et al., 2018). According to a study by Amiel et al. (2023), nearly 80% of Latin American universities use the services offered by these tech giants, with eight out of ten institutions relying on their offerings.

In May 2022, Google (which holds a significant 63% share of the market for technological solutions for universities in Latin America, according to Amiel et al. (2022)) modified its contract and imposed policies on academic communities, limiting their service to users. The widespread adoption of these services, accelerated during the Covid-19 pandemic, is largely due to underinvestment in public education in Latin America and poses a threat to the operational autonomy of institutions in research and teaching. Moreover, concerns arise regarding the security of personal data, data control, privacy, and transparency for professors, administrators, and students (Amiel et al., 2022).

Caught between the opportunities to enhance the quality of academic practices and the new risks and threats to security, privacy, freedom, and democracy, these countries face the challenge of developing knowledge, skills, and competencies to effectively leverage these new technologies. The expansion of universities in Latin America, particularly in Brazil, without the concurrent development of their own technological infrastructures, has led public and private actors to accept technological solutions offered by transnational companies in the digital economy without reflecting on the economic, political, institutional, and cultural transformations they may bring about. These public-private partnerships create a strong knowledge dependency and exacerbate the digital divide between Northern and Southern countries.

As discussed by Zuboff (2019), this new economic order asserts human experience as a freely available raw material for covert commercial practices involving extraction, prediction, and sale. They present themselves as providers of "almost free solutions" and use a discourse of democracy, access, work facilitation, security, and transparency to meet the demands for better science. These private companies standardize scientific practices as a way to assign a universal value to the routine of knowledge production. On one hand, the world's major academic publishers, such as Elsevier, Springer, Wiley-Blackwell, Taylor & Francis, and Sage Publishing, have adopted the discourse of "radically collaborative science" to offer services and turn knowledge production into a profitable market segment. On the other hand, the technological oligopoly also self-proclaims to be a provider of digital solutions for problems that have arisen due to a deep de-statization in Latin American countries and a disruption of digital technological systems that were being built until then – even with many criticisms regarding privacy, surveillance, and digital insecurity from researchers (Floridi, 2020; Barbudo, 2020). This phenomenon is known as platformization, defined by Poell et al. (2020) as the infiltration of algorithms and platforms into infrastructures, economic processes, and governmental structures. The platformization of science reinforces the integration of economic processes into scientific practices. It refers to the growing incorporation of digital infrastructures, algorithms, and platforms in scientific practices and processes, impacting the way knowledge is produced, shared, and disseminated.

This notion of the platformization of science is closely linked to challenges that imply how outdated colonial imperialism manifests with the support of the capitalist system (Mejias & Couldry, 2019). Consequently, the agents involved in this sphere face a new challenge in dealing with global structures of information management and their local consequences. To counteract the digital imperialism imposed by the platformization process, it is crucial for a nation to enhance its capability to provide technologies deemed critical for its well-being, competitiveness, and agency – whether by developing them internally or acquiring them from other territories – without unilateral dependence on third parties (IDDLAC, 2022; Gábriš & Hamuľák, 2023). As there is no consensus regarding the concept in the academic literature and efforts are being made to develop indicators of digital sovereignty across countries (Kaloudis, 2022; Dudin et al., 2021), different rhetorics are associated with the notion of digital sovereignty and should be understood and analyzed as geopolitical and political discourses and practices (Pohle & Thiel, 2020; Glasze, 2023; Martin, 2022; Werthner, 2022; Jelinek, 2023). While Russia, China, the United States, and Europe discuss digital sovereignty policies from their respectively distinct perspectives (Gonzalo, 2017; Golubev, 2021; Volodenkov, 2022; Cong & Thumfart, 2022; Wang & Gray, 2022), in a context where Latin American countries are still seeking mechanisms to overcome the digital divide, digital sovereignty is still far from being a reality in the region (López Campa, 2021).

The concept of "digital sovereignty," according to Couture and Toupin (2019), encompasses various forms of independence, control, and autonomy over digital infrastructures, technologies, and data, extending to issues of online informational consumption and proprietary regimes over personal data. Pierre Bellanger (2011) had already pointed out that digital sovereignty implied the ability to control our present and our destiny through the use of digital technologies and networks.

6 Digital and Epistemic Sovereignty in Latin America

Despite being a recent agenda for Latin America, the region has been an example of initiatives for digital sovereignty since the 1990s. For instance, the Lattes Platform, established in 1999, serves as a repository of metadata and information on researchers and their work, and it is widely used in Brazil's academic community for evaluating researchers, programs, and institutions. More recently, on July 19, 2023, the Brazilian Institute of Information in Science and Technology (IBICT), linked to the Federal University of Rio de Janeiro (UFRJ), launched the BrCRIS[1] (acronym for Current Research Information System) project. Its goal is to establish a unique model for organizing scientific information from the entire Brazilian research ecosystem. The initiative involves a network of researchers, projects, infrastructures, laboratories, research institutions, and funders, collecting, processing, and making available on a single platform hundreds of scientific publications, theses, dissertations, scientific datasets, software, and patents in an integrated solution aimed at generating indicators and metrics, analyzing investments, data flows, mapping research, and suggesting ways to optimize investment resources in technologies and scientific production. The proposal, which was allocated R$ 2.2 million through parliamentary amendments and resources managed by the Foundation for the Development of Teaching, Research, and Extension in Brazil, aims to integrate databases that gather various types of information about Brazilian scientific production. Additionally, the Colombia Consortium brings together higher education institutions to facilitate more affordable access to specialized tools, publications, and scientific research. The consortium has established a portal to showcase the strengths and capabilities of its member institutions, promoting collaboration and international networking. Another example is the Latmétricas consortium, which brings together three networks of researchers (LATmetrics,[2] Latin American Symposium on Metric Studies, and ESOCITE), and reflects how academia and researchers' associations seek to consolidate sovereign networks for the creation of evaluation systems based on metrics and indicators that are more consistent with their own realities.

The ability to understand one's own reality through informational metadata and to develop indicators less dependent on international metrics provided by global rankings plays a crucial role in epistemic sovereignty. This means having the control and autonomy necessary to shape the methods of collecting, interpreting, and applying data that reflect the diversity and specificities of a community or nation. Developing indicators less dependent on international metrics, especially those provided by global rankings, is a vital strategy to ensure that assessments accurately reflect the specific contributions and achievements of a community or country.

---

1 https://brcris.ibict.br/pt-BR, accessed on August 4, 2023.

2 https://www.latmetrics.com/, accessed on August 4, 2023.

Access to information is a fundamental condition of epistemic and digital sovereignty. In Brazil, the National Education and Research Network (RNP)[3] in Brazil, a Social Organization (SO) linked to and supported by the Ministry of Science, Technology, and Innovations (MCTI), played a pioneering role in providing internet access to the academic community in Brazil back in May, 1992. Nowadays, it continues to connect with other education and research networks across Latin America, North America, Africa, Europe, Asia, and Oceania through terrestrial and submarine optical fiber cables. Another example is the partnership between Chile and Argentina which, with expressions of interest from other South American countries, has led to the development of the "Digital Gateway to the Pacific." Through its public company Arsat, this project will connect the node in Valparaíso through fiber optics, creating a second international internet access point that directly links to the Asian continent and Australia, bypassing Europe or the United States. Instead, it will pass through Brazil, then connect to Uruguay, Paraguay, Bolivia, and finally to Oceania/Asia. This is a transnational state project, with the participation of academics from the Universidad Tecnológica Metropolitana and the Universidad de Chile.

Infrastructures that consolidate information on the production of science, technology, and innovation while also providing access to digital information less reliant on internet service providers play a crucial role in promoting digital sovereignty. These mechanisms not only contribute to autonomy in terms of information access but also ensure a robust and independent presence in the digital sphere. The creation and reinforcement of such infrastructures are essential strategies for countries and communities aiming to avoid excessive dependence on foreign platforms or multinational companies in the digital realm. By establishing local systems that aggregate data on research, development, and innovation, it becomes possible to preserve and value internally generated knowledge. But digital sovereignty is not solely a matter of technical control; it also relates to a community or country's ability to determine its own narratives, priorities, and policies in the digital environment.

Therefore, broadening discussions on infrastructure and access to digital information not only underscores the importance of digital sovereignty but also highlights the necessity of investing in local resources to ensure active and equitable participation in the digital era. Since 2023, the Brazilian Federal Government has been investing in strengthening university

3 https://www.rnp.br/, Accessed on August 4, 2023.

communication to consolidate an integrated network of communication through educational news agencies, radios and TVs, aiming to offer content production beyond social media digital platforms.[4] Public communication and scientific education are of utmost importance for the country's development, as they stimulate the dissemination of knowledge and raise public awareness of relevant scientific and social issues. Investing in these sectors is essential for advancing society. According to research conducted by the Brazilian Institute of Geography and Statistics (IBGE), television is the most consumed medium in the country, reaching approximately 96% of the national territory. However, despite the approval of concession permits for educational entities' TVs and radios in April 2018, educational channels linked to federal higher education institutions are not yet predominant in the country. The consolidation of an integrated network of educational communication through radios and TVs would significantly contribute to expanding the reach and impact of scientific and educational content production, promoting a more informed and engaged society on academic and social issues.

These examples illustrate how the state, with the support of academic intellectual capital, has been making strides in constructing its own information and communication systems. But digital sovereignty is not confined solely to the state. As discussed by Pohle and Thiel (2020), while the conventional theory of sovereignty, presented in the 16th century by the French political philosopher Jean Bodin, focused on the ruler's authority for final decisions, Jean-Jacques Rousseau reshaped the concept to prioritize popular sovereignty over monarchical sovereignty. Currently, sovereignty is primarily understood as the independence of a state from other states. However, a line of thought has emerged from the perspective of democratic sovereignty, encompassing popular sovereignty and citizens' right to exercise self-determination using their inalienable rights. In the context of informational sovereignty, this pertains to the autonomy of citizens in their roles as employees, consumers, and users of digital technologies and services (Gonzáles et al., 2020). From this perspective, Brazilian academics produced a letter in August 17, 2022 and delivered it to President Luiz Inácio Lula da SIlva, seeking to boost the public debate

4 This is an effort being led by the Communication Managers' Committee of the National Association of Heads of Federal Higher Education Institutions (Cogecom/Andifes), which has been leading the way towards a knowledge sovereignty less dependent on commercial platforms.

agenda for the creation of a federated infrastructure to host data from Brazilian universities and research centers.[5]

7 Final Remarks

The current situation in Latin America and the global scientific system's dependence on oligopolies represent a significant challenge to epistemic sovereignty and access to knowledge. The central question is how to break free from the domination exercised by private technology centers, scientific publishing oligopolies, and industrial conglomerates to consolidate science as a public right. By understanding the dimensions of digital and epistemic sovereignty, it is possible to identify paths to strengthen the role of scientific research in each social sphere, in dialogue with public policies and the circulation of knowledge through the media.

To solidify epistemic sovereignty as a social right, it is crucial to delve into the spaces shaped by digital platforms and enhance the discourse surrounding the shared domain created by media and networks. This will allow the actors involved in knowledge production and dissemination to have a voice and actively participate in shaping collective power, ensuring the legitimacy and recognition of diverse perspectives in the process of constructing information and knowledge.

Despite being a region that produces data, Latin America still suffers from a lack of incentives to develop its own technologies to enter the structure of digital capitalism autonomously. Although it has been a pioneer in various open access initiatives, the creation of digital knowledge infrastructures, and the registration of research metadata, Latin America still faces dependence on oligopolies. Therefore, it is necessary to strengthen epistemic sovereignty. Latin America and the global scientific system have the opportunity to promote a more inclusive, diverse, and democratic science. This will require collaboration between governments, educational institutions, researchers, and civil society, as well as the development of policies and strategies that place knowledge and scientific research in the service of the common good. The struggle for epistemic sovereignty is essential to ensuring that science is accessible to all and contributes to the advancement of society towards a fairer and more sustainable future.

---

5 https://cartasoberaniadigital.lablivre.wiki.br/carta/, Accessed on August 7, 2024.

The study entitled "Digital Technological Sovereignty in Latin America," conducted by Ceballos et al. (2020), highlights the importance of understanding this issue for building sustainable development and social justice in a complex and ever-changing world. The timid approach to implementing public programs to promote and strengthen digital technological capabilities, including software, applications, patents, research, and proper personnel training, and the limited participation of these experts in public decision-making processes, all make it challenging for governments to fully grasp the benefits and underestimate the long-term risks. This often leads to complacency with the status quo and digital technological dependence. As discussed by Ceballos et al. (2020) and the IDDLAC Digital Dossier (2022), digital technological sovereignty is not a permanent and irreversible state. We argue here that, despite Latin America not yet having a Digital Sovereignty Policy, government, associational, and activist initiatives have paved the way for many relevant discussions. However, digital sovereignty has been hijacked in the process of technological imperialism in which much of the Majority World is currently immersed. Reclaiming digital and epistemic sovereignty requires taking back control of production and circulation domains and placing the state in a fundamental role to create contexts that foster digital technological sovereignty through interventions guided by principles of human rights, autonomy, and independence.

The concept of digital sovereignty has been prominent on the agenda and in the speeches of major world powers, such as Russia, the United States, and China, whose President Xi Jinping reiterated this concept on several occasions in recent years while expressing the desire to transform China into the leading superpower in technology and cyberspace. As of 2021, in Latin America, digital sovereignty has not been a priority topic on the agendas of governments in the region.While some electoral programs, such as in Pedro Castillo's election for the Presidency of the Republic of Peru, have mentioned it, it is not currently a prominent subject of debate. Instead, civil society associations have been addressing this issue and taking related action. Since 2022, the topic has gained more presence in government agendas, but it is still mainly limited to the efforts of collectives working towards the implementation of public policies on digital, technological, and communication sovereignty.

Despite Latin America's limited participation in the debate on digital sovereignty, this chapter provides important lessons for understanding the conditions that promote sovereignty in digital policies in the region. In addition to presenting examples of pragmatism in digital policies (Becerra & Waisbord, 2021), the region's experience may resonate in other parts of the world, emphasizing the need to consider how lessons learned in Latin America can be applied in different cultural and historical contexts.

Despite facing a low level of digital literacy and pronounced digital inequality, Latin American countries have tackled digital challenges through solutions arising from collaboration between the state and citizens, with significant involvement by the academic community in shaping digital policies in the region (Antón-Sancho et al., 2023). The Latin American experience also sheds light on the development of the digital ecosystem and its implications for digital and science communication policies, offering significant examples of the accelerated digital development in the region and its potential applicability to other parts of the world facing similar challenges (Katz & Callorda, 2018; Acs et al., 2021). Furthermore, the resistance to the encroachment of the epistemic domain in universities, despite the advancements of neoliberal policies in academia, underscores Latin America's resilience in mitigating the effects of neoliberal digital initiatives on the regional academic landscape. This resilience can provide insights and inform strategies for promoting the strengthening of digital academia in other regions of the Global South.

References

Ács, Z. J., Lafuente, E., & Szerb, L. (2022). A note on the configuration of the digital ecosystem in Latin America. Tec Empresarial, 16(1), 1–19. https://doi.org/10.18845/te.v16i1.5926

Alam, S. (2008). Majority world: Challenging the West's rhetoric of democracy. *Amerasia Journal*, 34(1), 88–98.

Alatas, S. H. (1969). Academic imperialism. Lecture delivered to the History Society, University of Singapore.

Alatas, S. H. (2000). Intellectual imperialism: Definition, traits, and problems. *Southeast Asian Journal of Social Science*, 28(1), 23–45.

Alatas, S. F. (2013). Academic dependency and the global division of labour in the social sciences. *Current Sociology*, 51(6), 599–613.

Albagli, S., & Rocha, L. (2020). Citizen Science on the solid waste issue. Informação & Sociedade: Estudos, 30(4), 1–31.

Albuquerque, A. de. (2021). The institutional basis of anglophone western centrality. Media, Culture & Society, 43(1).

Albuquerque, A. (2023). Towards a multipolar communication international scholarship? World of Media. *Journal of Russian Media and Journalism Studies*, 1(1), 5–18. DOI: 10.30547/worldofmedia.1.2023.1

Albuquerque, A., de Oliveira, T. M., dos Santos Junior, M. A., & de Albuquerque, S. O. F. (2020). Structural limits to the de-westernization of the communication field: The editorial board in Clarivate's JCR system. *Communication, Culture & Critique*, 13(2), 185–203.

Alperin, Juan Pablo. The public impact of Latin America's approach to open access. Stanford University, 2015.

American Society for Cell Biology. (2012). San Francisco declaration on research assessment (DORA).

Amiel, S.A., Potts, L., Goldsmith, K. et al. A parallel randomised controlled trial of the Hypoglycaemia Awareness Restoration Programme for adults with type 1 diabetes and problematic hypoglycaemia despite optimised self-care (HARPdoc). Nat Commun 13, 2229 (2022). https://doi.org/10.1038/s41467-022-29488

Amiel, T., Saraiva, F., Cruz, L. R. D., & Gonsales, P. (2023). Mapping surveillance capitalism in south american higher education. *Latin American Journal of Educational Technology* – RELATEC, 22 (1), 221–239. https://doi.org/10.17398/1695-288X.22.1.221

Antón-Sancho, Á., Vergara, D., Sánchez-Calvo, M., & Fernández-Arias, P. (2023). On the influence of the university tenure on the digital pandemic stress in higher education faculty. *Behavioral Sciences*, 13(4), 335. https://doi.org/10.3390/bs13040335

Arocena, R. and Sutz, J. (2005). Latin American universities: From an original revolution to an uncertain transition. *Higher Education*, 50(4), 573–592. https://doi.org/10.1007/s10734-004-6367-8

Barbudo, Carlos (2020). Hacia una privacidad colectiva: repensar las bases teóricas de la distinción público/privado en la economía de la vigilancia. Teknokultura. Revista de Cultura *Digital y Movimientos Sociales*, 17(1), 69–76. https://doi.org/10.5209/tekn.66844.

Becerra, M. and Waisbord, S. (2021). The curious absence of cybernationalism in Latin America: Lessons for the study of digital sovereignty and governance. *Communication and the Public*, 6(1–4), 67–79. https://doi.org/10.1177/20570473211046730

Beigel, F. (Ed.). (2016). The politics of academic autonomy in Latin America. Routledge.

Bellanger, Pierre. De la souveraineté en général et de la souveraineté numérique en particulier. Les Echos, v. 30, 2011.

Bourdieu, P. & Wacquant, L. 1999. On the cunning of imperialist reason. *Theory, Culture & Society*, 16(1), 41–58.

Boussebaa, M. and Tienari, J. (2019). Englishization and the politics of knowledge production in management studies. *Journal of Management Inquiry*, 30(1), 59–67. https://doi.org/10.1177/1056492619835314

Carlsson, U. (2003). The rise and fall of NWICO. *Nordicom Review* 24(2), 31–67.

Carvalho, G. (2022). The World Bank and healthcare reforms: A cross-national analysis of policy prescriptions in South America. *Social Inclusion*, 10(1), 5–18.

Castro G., Angela Maria. Cidadania e direitos do trabalho. Zahar, 2002.

Ceballos, L. D., Maisonnave, M. A., & Londoño, C. R. B. (2020). Soberanía tecnológica digital en Latinoamérica. Propuestas para el Desarrollo, (IV), 151–167.

Chukwuemeka, E., Chekwubechukwu, O., & Obuteaku, N. (2018). The contributions of impact factor publication policy to academic development in nigerian higher institutions of learning: a critical evaluation. *International Journal of Research -Granthaalayah*, 6(11), 398–408. https://doi.org/10.29121/granthaalayah.v6.i11.2018.1143

Clinio, A. (2019). Ciência aberta na América Latina: duas perspectivas em disputa. Transinformação, 31, e190028.

Cong, W., & Thumfart, J. (2022). A Chinese precursor to the digital sovereignty debate: Digital anti-colonialism and authoritarianism from the Post–Cold War era to the Tunis Agenda. *Global Studies Quarterly*, 2(4). https://doi.org/10.1093/isagsq/ksac059.

Couture, S., & Toupin, S. (2019). What does the notion of "sovereignty" mean when referring to the digital? *New Media & Society*, 21(10), 2305–2322.

Demeter, M. (2020). Academic knowledge production and the Global South. London: Palgrave-Macmillan.

Dezalay, Y., & Garth, B. G. (2002). The internationalization of palace wars: Lawyers, economists, and the contest to transform Latin American states. Chicago and London: University of Chicago Press.

Dorfman, A., & Mattelart, A. (1971). Para leer al Pato Donald [How to Read Donald Duck]. Valparaíso: Ediciones Universitarias de Valparaíso.

Dudin, M. N., Shkodinsky, S. I., & Usmanov, D. I. (2021). Digital Sovereignty of Russia: Barriers and New Development Tracks. Market Economy Problems, 2, 30–49.

Floridi, L. (2020). The fight for digital sovereignty: What it is, and why it matters, especially for the EU. *Philosophy & Technology*, 33(3), 369–378. https://doi.org/10.1007/s13347-020-00423-6.

Freije, V. (2019). The 'emancipation of media': Latin American advocacy for a New International Information Order in the 1970s. *Journal of Global History*, 14(2), 301–320.

Gábriš, T., & Hamuľák, O. (2023). Digital sovereignty or sovereignty with digital elements? In D. R. Troitiño, T. Kerikmäe, & O. Hamuľák (Eds.), *Digital Development of the European Union* (pp. 249–263). Springer International Publishing. https://doi.org/10.1007/978-3-031-27312-4_16.

George Mwangi, C.A., Yao, C.W. (2021). US Higher Education Internationalization Through an Equity-Driven Lens. In: Perna, L.W. (eds) Higher

Education: Handbook of Theory and Research. Higher Education: Handbook of Theory and Research, vol 36. Springer, Cham. https://doi.org/10.1007/978-3-030-44007-7_11.

Golubev, A. (2021). Digitizing archives in Russia: Epistemic sovereignty and its challenges in the digital age. In *The Palgrave Handbook of Digital Russia Studies* (pp. 353–372). Palgrave-Macmillan.

González, J. T. G., Estrada, M. S., & Pargman, T. C. (2022). Construyendo soberanía digital en América Latina: un análisis de las iniciativas de cuatro colectivos sociales. *Chasqui: Revista Latinoamericana de Comunicación*, (149), 227–242.

Gonzalo, T. Á. L. (2017). Despliegue de una Red Inalámbrica Libre Comunitaria: un ejercicio de soberanía digital en el barrio Villa del Río (Bogotá, Colombia). *Hamut'ay*, 4(2), 80–91.

Goyanes, M. (2020). Editorial boards in communication sciences journals: Plurality or standardization? *International Communication Gazette*, 82(4), 342–364.

Goyanes, M., & Demeter, M. (2020). How the geographic diversity of editorial boards affects what is published in JCR-ranked communication journals. *Journalism & Mass Communication Quarterly*, 97(4), 1123–1148.

Guzmán-Valenzuela, C. (2017). Tendencias globales en educación superior y su impacto en América Latina: Desafíos pendientes. Lenguas Modernas, 50(50), 15–32.

Hayler, T. (1971). Aid as imperialism. Middlesex: Penguin Books.

Hicks, D., Wouters, P., Waltman, L., De Rijcke, S., & Rafols, I. (2015). Bibliometrics: the Leiden Manifesto for research metrics. *Nature*, 520(7548), 429–431.

IDDLAC. (2022). Dossier Digital CMDT. Disponível em https://iddlac.org/wp-content/uploads/2022/07/Dossier-Digital-CMDT.pdf

Jelinek, T. (2023). *The digital sovereignty trap: Avoiding the return of silos and a divided world.* SpringerBriefs in International Relations. Springer Nature Singapore. https://doi.org/10.1007/978-981-19-8414-3.

Kaloudis, M. (2021). Sovereignty in the digital age – How can we measure digital sovereignty and support the EU's Action Plan? *New Global Studies*, 16(3), 275–299. https://doi.org/10.1515/ngs-2021-0015.

Katz, R. and Callorda, F. (2018). Accelerating the development of Latin American digital ecosystem and implications for broadband policy. *Telecommunications Policy*, 42(9), 661–681. https://doi.org/10.1016/j.telpol.2017.11.002

Kettell, S. and Sutton, A. (2013). New imperialism: toward a holistic approach. *International Studies Review*, 15(2), 243–258. https://doi.org/10.1111/misr.12033

Lerner, D. (1958). *The passing of traditional society: Modernizing the Middle East.* New York: The Free Press.

Litfin, Karen T. Constructing environmental security and ecological interdependence. Global Governance, v. 5, p. 359, 1999.

López Campa, J. A. (2021). Derechos de la soberanía digital. Universidad Autónoma de Aguascalientes. https://doi.org/10.33064/UAA/978-607-8782-66-6.

Lugo-Ocando, J. (2018). A mouthpiece for truth: Foreign aid for media development and the making of journalism in the global south. *Brazilian Journalism Research*, 14(2), 412–431.

Martin, C. (2022). Geopolitics and digital sovereignty. In H. Werthner, E. Prem, E. A. Lee, & C. Ghezzi (Eds.), *Perspectives on Digital Humanism* (pp. 227–231). Springer International Publishing. https://doi.org/10.1007/978-3-030-86144-5_30.

Meijas, U. A., & Couldry, N. (2019). Datafication. *Internet Policy Review*, 8(4). doi: 10.14763/2019.4.1428

Mugnaini, R., Damaceno, R. J. P., Digiampietri, L. A., & Mena-Chalco, J. P. (2019). Panorama da produção científi ca do Brasil além da indexação: uma análise exploratória da comunicação em periódicos. Transinformação, 31, e190033.

Nye J. R., Joseph S. The changing nature of world power. Political science quarterly, v. 105, n. 2, p. 177–192, 1990. 108, 145–155.

Oliveira, T., Araujo, R. F., Cerqueira, R. C., & Pedri, P. (2021). Politização de controvérsias científicas pela mídia brasileira em tempos de pandemia: a circulação de preprints sobre Covid-19 e seus reflexos. *Revista Brasileira de História da Mídia*, 10(1).

Park, D. W., Pooley, J., & Simonson, P. (2022). Exclusions/Exclusiones: The Role for History in the Field's Reckoning. History of Media Studies, 2.

Parmar, I. (2012). *Foundations of the American Century: The Ford, Carnegie, and Rockefeller Foundations in the rise of American Power*. New York: Columbia University Press.

Poell, T., Nieborg, D., & van Dijck, J. (2020). Plataformização. Revista Fronteiras, 22(1). https://doi.org/10.14763/2019.4.1425.

Pohle, J., & Thiel, T. (2020). Digital sovereignty. *Internet Policy Review: Journal on Internet Regulation*, 9(4), 1–19.

Riveros, P., Meriño, J., Crespo, F., & Baptista, B. (2022). Situated transdisciplinarity in university policy: Lessons for its institutionalization. *Higher Education*, 84(5), 1003–1025. https://doi.org/10.1007/s10734-021-00812-6

Ramiro Beltran, L. (1976). TV etchings in the minds of Latin Americans: Conservatism, materialism, and conformism. *International Communication Gazette* 24(1), 61–85.

Riani, H. (2017). Globalization and american power: (Dis)order through the politics of cultural imperialism. *IOSR Journal of Humanities and Social Science*, 22(2), 52–61. https://doi.org/10.9790/0837-2202035261

Sánchez-Tarragó, N., Fernández-Molina, J. C., & Caballero-Rivero, A. (2012). Reflexiones en torno al acceso abierto a la información en el contexto cubano: el caso del sector Salud. Informação & Sociedade, 22(2), 1–10.

Schiller, D. (1999). *Digital capitalism: Networking the global market system.* MIT Press.

Schwartzman, S. (1991). Latin America: Higher education in a lost decade. *Prospects,* 21(3), 363–373.

Simonson, P., Park, D. W. & Pooley, J. 2022. Exclusions/Exclusiones: The Role for

Steinmetz, G. (2005). Return to empire: The new U.S. imperialism in comparative historical perspective. *Sociological Theory,* 23(4), 339–367. https://doi.org/10.1111/j.0735-2751.2005.00258.x

Straubhaar, J. D. (1991). Beyond media imperialism: Assymetrical interdependence and cultural proximity. *Critical Studies in Media Communication,* 8(1), 39–59.

Suzina, A. C. (2021). English as lingua franca. Or the sterilisation of scientific work. Media, Culture & Society, 43(1), 171–179.

Teichman, J. (2004). The World Bank and policy reform in Mexico and Argentina. *Latin American Politics and Society,* 46(1), 39–74.

van Dijck, J., Poell, T., & De Waal, M. (2018). *The platform society: Public values in a connective world.* Oxford University Press.

Vessuri, H., Guédon, J. C., & Cetto, A. M. (2014). Excellence or quality? Impact of the current competition regime on science and scientific publishing in Latin America and its implications for development. *Current Sociology,* 62(5), 647–665.

Volodenkov, S. V., Fedorchenko, S. N., & Artamonova, Y. D. (2022). Contemporary state in the context of digital technological transformations: Political opportunities, risks, and challenges. *RUDN Journal of Political Science,* 24(3), 351–366.

Wang, Y., & Gray, J. E. (2022). China's evolving Stance against tech monopolies: A moment of international alignment in an era of digital sovereignty. *Media International Australia,* 185(1), 79–92. https://doi.org/10.1177/1329878X221105124.

Werthner, H. (2022). Geopolitics, Digital Sovereignty ... What's in a Word?. Perspectives on digital humanism, 241–248. https://doi.org/10.1007/978-3-030-86144-5_32.

Wiedemann, T., & Meyen, M. (2016). Internationalization through Americanization: The expansion of the International Communication Association's leadership to the world. *International Journal of Communication,* 10, 1489–1509.

Zuboff, S. (2019). Surveillance capitalism and the challenge of collective action. New Labor Forum, 28(1), 10–29.

CHAPTER 3

Experimenting on the Frontier: Imperial Laboratories and Facebook's Political Effects in Myanmar

Stefan Bächtold

1 Introduction

When Myanmar's military embarked on a path of democratic reforms in 2011, it also opened the country's telecommunications market. Within just a few years, mobile internet went from an elusive privilege accessible to the few to a part of everyday life for the majority. The bulk of Myanmar's online activities took place within the algorithm-controlled and feed-based environment of Facebook's[1] platform. Due to its overwhelming popularity with Myanmar users, boosted by the platform's aggressive programs to further expand, Facebook *was* the internet for most people in Myanmar during the 2010s (Thant Sin Oo & Ye Min Thant, 2019). A great deal has been written about Facebook's failure to prevent the spread of hate speech on its platform, and its role in the subsequent genocide against the Rohingya minority.[2] This chapter approaches Facebook's role in Myanmar from a different angle; namely via the notion of the imperial laboratory. Combining the concept of data colonialism (Couldry & Mejias, 2018) with reflections on the governing of populations in (former) colonies and frontiers via the laboratory (Anderson, 1992; Barder, 2015) or experimentation (Aradau, 2020), I approach Facebook's expansion in Myanmar as one that is reminiscent of the experimentality that imperial powers displayed in the colonies.

In this chapter, I take the discourses and knowledge of technology actors about developing countries as an entry point to analyse global tech assemblages, and how power operates within and through the deployment of digital technologies in the Global South. As I will argue with the case of Facebook in Myanmar, the company's discourse and modalities of engagement with developing countries constitute the latter as a digital *terrae*

1 Most references to Facebook in this chapter concern the period before the company changed its name to Meta. For consistency, I use 'Facebook' throughout this chapter.

2 For a comprehensive overview and discussion, see Schissler (2024).

 | DOI:10.1163/9789004711396_005

nullius – as spaces unclaimed by digital technologies that can (or even, *must*) be filled with platform connectivity in the name of progress. My empirical analysis centres on the discourses, practices, and policies (as observable in press statements, policy announcements, and reports) of how Facebook imagined Myanmar – and by extension, its own role in the world. This allows for further conceptual reflection, too: constituting Myanmar as *terra nullius* not only erases Myanmar's pre-existing digital worlds. In doing so, Facebook segments Myanmar from its approaches to Western countries, and makes Myanmar amenable to experimental modalities of governing the digital. In turn, this also constituted Facebook as the sole actor with the means to extract and benefit from the (big) data stemming from the ongoing datafication of Myanmar's population via the Facebook app. Andrejevic (2014) calls this the 'big data divide': a divide that separates Facebook as extracting data – from the millions of people from whom data are extracted from. I will try to make sense of these modalities of experimentation and their effects with the notion of the *imperial laboratory*.

The case of Myanmar is striking in two separate regards: firstly, it highlights the inherent perils of a country where most of the internet is under *de facto* control by a singular, feed-based platform. Secondly, it offers an opportunity to explore the power relations at play when large tech companies expand into developing countries: extractive practices of tech companies are entangled with (bio)political ones. Rather than following a coherent rationale or plan, these processes appear more like the contingent outcomes of large-scale experimentation in places constituted as the frontiers. This was followed by (frantic) improvisation, bricolage, and re-arrangement when things went horribly awry. Particularly, that is, when it became apparent that the imagined big data divide had been breached by the Myanmar military, who was actively using Facebook's platform for its own ends. Akin to historical colonialism, though, these re-arrangements within data-colonialist assemblages do little to change the fundamental relations of power inherent in them, and the big data divide they uphold.

2 Thinking Imperial Laboratories in the 21st-Century Post-colony

When analysing the increasing entanglement of digital technologies and platforms with politics, much of the literature has focused on Western countries where dominating tech giants like Google, Facebook, or Amazon originate. The same applies to the manifold analyses focusing on the sociotechnical imaginaries of Silicon Valley and its proponents, and its specific mix of libertarian ideology with tech-utopian discourse and solutionism

(Haupt, 2021; Morozov, 2013; Rider & Murakami Wood, 2018). Given the size of tech giants and the extent of their grasp on global datafication processes, it is common to begin an analysis of these global, networked assemblages of knowledge, technology, and practices variously known as *platform capitalism* (Srnicek, 2016), *surveillance capitalism* (Zuboff, 2015), or *data colonialism* (Couldry & Mejias, 2019) at their centre.

There are, however, strong arguments in favour of prioritising the analysis of areas that seem peripheral to global digital networks and economies. For one, the diffusionist model underpinning the techno-optimistic Silicon Valley imaginary of its platforms as technology built in Western, developed countries and subsequently diffused into the developing world (Quet & Dahdah, 2020) has long been criticised by scholarship in science, technology and society studies and postcolonial or decolonial literature. As these analyses remind us, spaces considered marginal, peripheral, or liminal are often *central* to the functioning of empires or global political economies; they play a variety of roles in the development and adaptation of new technologies (Anderson, 1992; Anderson & Adams, 2007; Arnold, 2005; Barder, 2015; Khalili, 2010; Lindskov Jacobsen, 2015).

When looking at Myanmar, this becomes particularly evident from the close integration and entanglement of the country (and within the country, the borderlands) in global economies. Although often imagined and portrayed as a place with long histories of autarky, Myanmar and its borderlands are closely integrated into international economies and networks that formed around agricultural investments, land, timber, jade, rare earths, opium, methamphetamines, weapons, or development projects (Bächtold et al., 2020; Buscemi, 2021; Fiskesjö, 2022; Mark, 2023; Meehan & Sadan, 2017; Woods, 2011). In turn, this begs the question of why this image of an isolated Myanmar is continuously reproduced; and, more particularly, how technology actors – and the socio-technical imaginaries (Jasanoff & Kim, 2009) in which their technologies are embedded – are involved in *constituting* countries like Myanmar as isolated and waiting to be connected with the newest technologies. Analysing technologies, knowledges, and practices in liminal spaces can thus reveal much more than their area-specific idiosyncrasies (Anderson & Adams, 2007; Milan & Treré, 2019). Rather, their analysis can illuminate how spaces like Myanmar are intimately tied to what is imagined or constituted as the centre of global tech assemblages.

The current technology behemoths' constant growth and expansion into new areas (Zuboff, 2015), as well as its undertones of older capitalist and colonialist regimes' extractive practices on the so-called 'frontiers' have been noted by several scholars (Couldry & Mejias, 2018; Kwet, 2019;

Madianou, 2019; Ricaurte, 2019). But the parallels with former colonial empires go beyond tech companies' (data) extractivism (cf. Gray, 2023): there are also similarities in the imaginaries they propose – of progress, of technology's role in society, and of the populations in developing countries.

As a range of postcolonial and decolonial scholars have documented, the operation of colonial empires is closely entangled with the production of knowledge that creates a distinction between the metropolis and the periphery that constitutes the 'colonial subject as Other' (Spivak, 1988, p. 280), where the first is associated with rationality and technological progress, and the latter is portrayed as lacking in these traits (Fanon, 1961; Mbembe, 2021; Mignolo, 2007; Said, 1978). Technology, in this discourse, is seen as a 'neutral and inevitably beneficial' force for modernist progress (Escobar, 1995, p. 36; also Scott, 1998). These images are equally observable for the current generation of global technology assemblages (Milan & Treré, 2019), and Facebook in particular (Rider & Murakami Wood, 2018). Silicon Valley's imaginaries of technological futures construct a 'digital universalism' that presents their technologies as apolitical, passive instruments (Lindskov Jacobsen, 2015), as an 'incontrovertible, self-evident, and universally shared necessity' (Chan, 2014, p. 19) that the Global South will have to adopt eventually. However, as scholars have argued both globally (Gillespie, 2010) and for Myanmar in particular (Bächtold, 2022; Sablosky, 2021), platforms' operations are inherently political; and connecting (and disconnecting) are highly political practices (Bächtold, 2023; Lim, 2020).

To highlight the quintessentially political effects of technologies and their entanglements with (colonial) relations of power, scholars have recently started to pay more attention to the laboratory and experimentation in colonial contexts, and to the political effects that their seemingly technical practices and knowledges produce. Anderson (1992), from his analysis of technical reports from colonial archives on laboratory medicine in the Philippines, argues that the apparently technical body of texts on medical technology and knowledge was instrumental for colonial discourse to produce and naturalise 'native bodies' as different from the colonizers, in ways that 'naturalise the power of foreign bodies to appropriate' (Anderson, 1992, p. 507). Even if the technical vocabulary disguises their contribution to the colonial project, technology and experimentation thus appear as parts upholding colonial orders or global relations of power – rather than merely apolitical objects and practices *within* those orders. As Barder (2015) points out, imperial spaces have long histories as sites of experimentation that transformed both periphery and metropole, particularly when it comes to technologies of governing, policing, or counterinsurgency (cf. also Axster et al., 2021; Khalili, 2010; Mbembe, 2003). In turn, the technological and sociopolitical changes in imperial laboratories of governance have significantly impacted how the colonizing states governed the populations in metropolitan states, too.

Arguably, similar dynamics can be observed for today's large tech companies. As Zuboff (2015, p. 84) contends, 'continuous experiments' have been a part of tech companies' repertoires for data extraction for some time – irrespective of geographical place. Facebook's practices in Myanmar, however, shed light on a different kind of experimentation that Aradau (2020, p. 31) calls 'experimentality as a mode of governance': how social and political problems are rendered governable through experiments, bricolage, or improvisation. As I will argue in the following empirical sections, this is visible in how Facebook first ignored how its platform became a salient vector for hate speech in Myanmar, and then started experimenting with automated solutions to the problem – all while maintaining that the company could not have anticipated the effects its technology produced in the country (cf. Deejay et al., 2023). Secondly, it is also apparent in Facebook's shift in constituting its role in Myanmar over the years, and most markedly, with the military coup: from presenting itself as a mere apolitical platform (Gillespie, 2010), the company's discourse notably shifted towards introducing more (experimental) measures to keep its users 'safe' – a shift also observed globally (Haupt, 2021; Rider & Murakami Wood, 2018). Myanmar thus became a 'testing ground' (McNeil & Milko, 2021) for Facebook to introduce and experiment with measures to govern its platform, which could subsequently be deployed to govern the platform in other contexts. I will call it an *imperial laboratory*.

3 Move Fast and Break Things? Facebook's Expansion in Myanmar

In 2011, after decades of military authoritarian rule and waging war against the country's ethnic minorities, Myanmar's generals announced that the country would embark on a path of democratic reforms and peace negotiations (Holliday, 2011; Kyaw Yin Hlaing, 2012). After the military-appointed head of government, Thein Sein, signed bilateral ceasefire agreements, relaxed censorship, and opposition leader Aung San Suu Kyi was released from house arrest (Callahan, 2012), the international community's discourse on Myanmar drastically changed. With the former 'pariah' now suddenly considered a promising frontier market, Myanmar saw a large influx of international development actors and businesses (Bächtold, 2015; Décobert & Wells, 2020).

The surge of digital technologies was among the notable changes in Myanmar's 2010s: the promises of political transition were couched in a socio-technical imaginary that Myanmar as a late adopter would 'leapfrog' other countries by going straight to the newest (digital) technology (Bächtold, 2023). Indeed, after the liberalisation of the telecommunications market in 2013, the country 'leapfrogged'. Whereas only a minuscule percentage of people were able to access the internet before the reforms (Htaike Htaike Aung & Wai Myo Htut, 2019), mobile connectivity via smartphones had become daily life for large parts of the population just a few years later (McCarthy, 2017; Whitten-Woodring et al., 2020; Yan Naung Oak & Brooten, 2019).

In an environment where a large majority uses mobile phones and cellular networks as the primary means of accessing the internet, Facebook's mobile app became the gateway to access any kind of content in Burmese (Leong, 2017). Most businesses or even government services did not have an online presence outside Facebook, with announcements made in posts, contact details provided on profile pages, and menus shared in picture albums. For many in Myanmar, Facebook was synonymous with the internet (Htaike Htaike Aung & Wai Myo Htut, 2019; Whitten-Woodring et al., 2020). As became clear by the mid-2010s, Myanmar had not only leapfrogged cable-based internet lines in favour of 3G and 4G mobile networks. It had also jumped straight to a version of the internet that was algorithm-governed, feed-based, and 'provided' by Facebook.

This is not to suggest that Myanmar's digital environment followed (or leapfrogged) along an inevitable path of digital evolution where large tech companies end up with dominant control over the digital. Quite the contrary: Facebook's dominance in Myanmar was the product of a range of intervening factors like the military's reforms and their timing; the restriction of internet infrastructures before the 2010s; the availability and affordability of smartphones; and Facebook's aggressive moves to capture a dominant share of online activity. Akin to its practices in other developing countries in the mid-2010s, Facebook rolled out its 'free basics' project in Myanmar – a service where internet service providers waive mobile data charges for content that is accessed via the mobile Facebook app (Trautwein, 2016). The same project has faced heavy resistance from activist groups on the grounds of net neutrality in neighbouring India; they argued that 'free basics' constructs a walled garden controlled by Facebook. Indian authorities subsequently banned 'free basics' following these arguments (Prasad, 2017). The service remained active for longer in a range of other countries, including Myanmar. As Nothias (2020) puts it, 'free basics' marked the company's push to become the dominant player in mobile internet in developing countries.

Facebook, on the other hand, portrayed its efforts as humanitarian – as part of its self-declared mission of connecting the world. In a note entitled *Is connectivity a human right?* posted in 2013, Facebook-CEO Mark Zuckerberg stated that 'it may not actually be profitable for [Facebook] to serve the next few billion people for a very long time, if ever' – but that they 'believe that everyone deserves to be connected' (Zuckerberg, 2013, p. 1).

The discursive effect of this framing is twofold: firstly, it obscures Facebook's economic interests (Rider & Murakami Wood, 2018) in continuous expansion – what Gillespie (2020) has called the mentality of 'growth at all costs'; and what has been the main angle of criticism of social media platforms as data colonialism. Secondly, it constitutes populations in the Global South as untouched by digital technology and connectivity, a people who *must* be plugged into global techno-political assemblages on moral grounds. Furthermore, the same note goes on to suggest that a 'knowledge economy is the future'; and that by 'bringing everyone online, we'll not only improve billions of lives, but we'll also improve our own as we benefit from the ideas and productivity they contribute to the world' (Zuckerberg, 2013, p. 3). Myanmar is constituted as a sort of digital *terra nullius* that must be filled with Facebook: an untapped potential that must be tapped in the name of (global) social and economic progress.

Equally noteworthy is the distinct techno-optimism inherent in Zuckerberg's note (cf. Rider & Murakami Wood, 2018), which translates the complex problem of the global digital divide to a simple, techn(olog)ical one that can be solved by diffusing technology to the periphery – a problem that will be solved by Facebook's core operation and expansion. In this framing, the only thing that remains to be sorted is the technical details of making sure that internet service providers have enough financial incentives to expand internet access to the poor. But as the case of Myanmar shows, this framing meant casting aside the complex historical, geopolitical, and socio-technical configurations of factors that Facebook's technology would interact with in the country (cf. Deejay et al., 2023), as well as continuous warnings and cautions from civil society groups about the platform's political effects that started to materialise once public discourse was moving to Facebook (Yan Naung Oak & Brooten, 2019).

4 Will AI 'Solve This'?

Like every country – and particularly, every country in a political transition – Myanmar has a wide range of political actors with conflicting agendas. With Facebook emerging as a digital space for public discourse, these conflicts inevitably became visible on the platform: Buddhist nationalist groups found an expanded stage on Facebook that allowed them to spread pre-existing tropes that constituted Muslims as a violent 'Other' threatening Buddhism (Wade, 2017). They rallied for the protection of Buddhism from the threat of being crowded out, and against Muslim minorities (Cheesman, 2017; McCarthy & Menager, 2017; Schissler, 2016; South & Lall, 2017; Walton & Hayward, 2014; Whitten-Woodring et al., 2020). Facebook was soon overflowing with hate speech, including dehumanisation and calls for violence, against Rohingya and other Muslim minorities (Stecklow, 2018). In 2017, a military campaign of systematic human rights violations and atrocities against the Rohingya population drove more than 700,000 Rohingya over the border into Bangladesh. The UN Office of the High Commissioner for Human Rights (2018, p. 1) stated that the situation 'bears the hallmarks of genocide'. The United Nations, in a report examining the genocidal violence against the Rohingya in Rakhine State, mentions that 'Facebook has been a useful instrument for those seeking to spread hate' (United Nations Human Rights Council, 2018, p. 14). In response, Facebook admitted that its response to hate speech was slow and ineffective (McPherson, 2018).

Although Facebook presents its digital platform as transcending the offline world's borders, fences, and even geographical distances, the experience of using Facebook in an English-speaking country in the Global North compared to using Facebook in a non-English-speaking country in the Global South differs drastically. The algorithm-governed 'global community' Facebook offers is segmented by the socio-technical arrangements of Facebook's business operations. The primary line of segmentation here are languages – and different levels of investments into processing, ordering, and moderating content written in specific languages (cf. Chonka et al., 2022). The case of Myanmar makes this evident: civil society groups in Myanmar had long complained that posts containing hate speech that they flagged via Facebook's reporting features were not taken down (Stecklow, 2018). To blame for this inaction were the scant resources that Facebook invested in removing hate speech on its platform in Myanmar: in 2015, Facebook had only two content moderators who could speak Burmese (McPherson, 2018). Facebook later announced that it would increase that number to 60 (Facebook, 2018b).

Rather than human moderators, Facebook saw the solution to the problem in artificial intelligence that would be able to detect problematic posts pro-actively. Facebook-CEO Mark Zuckerberg thus optimistically proclaimed that AI would fix this, and that the necessary technology to identify even nuanced forms of hate speech would be available over the next five to ten years (Katzenbach, 2021). Relying on the latest technology to address a political problem aligned well with Facebook's hallmark, techno-optimistic discourse – even though substantial doubts on both the feasibility and desirability of fully-automated content moderation to govern platforms had been voiced (Gillespie, 2018; 2020).

Natural language processing requires massive amounts of data to train language models. For languages other than English, there is significantly less data available online, and hence, the doubts that AI can reliably moderate content are exponentially more pronounced. While Facebook promised to invest extensively in automated means of recognising and removing hate speech, its initial attempts in Myanmar seemed haphazard and demonstrated little knowledge of Burmese language. For example, Facebook introduced automated temporary bans for users posting terms in Burmese like 'chair' or 'chickpea' – everyday words, but which in Burmese also contain a racial slur that the automated detection was targeting (Irrawaddy, 2017). In later statements, Facebook claimed that it had improved its proactive detection of hate speech thanks to a combination of developments in artificial intelligence and additional human moderators, and was now able to identify half of the reported hate speech posts proactively (Facebook, 2018b). This, however, says little about the total prevalence of hate speech on the platform. Furthermore, leaked internal Facebook communications investigated by the Wall Street Journal (Seetharaman et al., 2021) suggest that Facebook's own engineers estimate the overall detection rates to be significantly lower.

Both the initial lack of human content moderators, as well as haphazard trial runs with automated content moderation in Myanmar, point to a mindset of experimentality – where Myanmar served as a real-life laboratory for an internet dominated by an algorithm-governed, feed-based platform. In this laboratory, Facebook's discourse and proposed technological solutions constitute the tech company at the helm, and as the sole actor in charge of governing the platform. Andrejevic (2014) calls this the 'big data divide', which separates the relatively small number of people in Facebook who extract and analyse data, and the millions that data are extracted from. This divide is reinforced by the requirements that are imagined to be necessary to make use of big data – expensive processing power, technological infrastructures, or expertise to analyse these data. Facebook's specific problematisation of hate speech as amenable to technological, AI-powered solutions thus excludes local knowledge and context-specific expertise.

This exclusion is also apparent in Facebook's reaction to another incident, when the company claimed that its 'systems' had detected messages aiming to incite violence in Myanmar. As it later turned out, it was not Facebook's systems, but Myanmar civil society organisations who had alarmed Facebook to these messages (Burma Monitor et al., 2018; Hogan, 2018). That Facebook subsumed Myanmar civil society organisations as part of its 'systems', and thus, omitted the agency of actors from Myanmar in governing the platform, is telling of the order that Facebook's techno-optimistic discourse constructed, and the big data divide it upholds.

Upon closer inspection, the discourse formation that ascribes no active roles to Myanmar's people in the roll-out or governance of (Western) digital technology is embedded in the broader construction of Myanmar as digital *terra nullius* – a place void of the digital, but where modern technology is diffused to, unidirectionally. When reading Facebook's statements and reports on Myanmar, the company continuously separates an implied, connected global from 'countries like Myanmar where many people are using the internet for the first time' (Facebook, 2018b). In the latter, documents like Facebook's human right impact assessment for Myanmar (Business for Social Responsibility, 2018, p. 24) identify 'significant shortcomings in the areas of digital literacy, privacy awareness, and critical thinking'. This, in turn, is then used to explain why 'social media can be used to spread hate and fuel tension on the ground' (Facebook, 2018b), a process implied to be less prevalent in other places. Overall, this discourse depicts Myanmar's population as requiring assistance in taking its first digital steps, rather than as actors that could be involved in shaping global tech assemblages (Deejay et al., 2023). In Andrejevic's (2014) terms, Myanmar is firmly placed on the side of the big data divide where data are extracted *from*.

5 Unimaginable Cracks in the Big Data Divide

This (imagined) order, however, is not uncontested. When the civil society organisations mentioned earlier called out Facebook for subsuming them as part of their 'systems', they wrote that '[...] this case exemplifies the very opposite of effective moderation: it reveals [...] a reticence to engage local stakeholders around systemic solutions and a lack of transparency' (Burma Monitor et al., 2018, p. 1). Underpinning this statement is thus a fundamentally different understanding of the role of technology actors: rather than a unidirectional diffusion of technology, it calls for Western tech companies to engage in finding 'systemic' solutions with local actors, and to be transparent about it.

Scholars (Prasse-Freeman, 2021; Schissler, 2016) have also criticised the depiction of Myanmar's population as gullible users directly influenced in their opinions by what they read on Facebook – a trope that obscures the much more complex dynamics at play. What has happened in Myanmar was not the result of people not knowing how to use Facebook or lacking digital literacy when using the internet for the 'first time'. To the contrary, the strategies used to weaponise the platform were low-tech – but far more advanced and effective than Facebook had anticipated.

Facebook's characterisation of its expansion into Myanmar as introducing an inherently neutral and beneficial tool obscures the effects that the platform's dominance has produced in Myanmar's society. As with every other society in which Facebook has been deployed, the company's basic operation as a collector, extractor, and aggregator of (big) data has reconfigured Myanmar's society (cf. Rider & Murakami Wood, 2018) and its population *as data*. It has enabled a process in which large parts of the population are routinely producing data on people's activities, opinions, concerns, or spatial mobilities – in short, datafying their lives and relationships.

This is part of Facebook's basic operation (Couldry & Mejias, 2018; Zuboff, 2015); Andrejevic's (2014) 'big data divide' separates those who extract, analyse, and benefit from these data from the people these data are extracted from. What seemed unimaginable for Facebook is that this divide could be breached without advanced technological infrastructure or extensive processing power – and that this process of datafication could be accessible to other actors to put to their own uses. Myanmar's military, however, has continuously proven that it is very much capable of doing so – without the use of extensive technological infrastructures or admin privileges on the platform.

Throughout Myanmar's most recent history, there have been signs that the prevalence of hate speech on Facebook was linked to orchestrated efforts. In 2018, *The New York Times* (Mozur, 2018) reported that the Myanmar military maintained a facility of soldiers dedicated to astro-turfing – trying to influence public opinion by producing a flurry of posts reproducing the military's narrative, while pretending to be everyday citizens. Facebook confirmed that they had identified a coordinated network of seemingly innocuous accounts that were linked to the military. It comprised hundreds of accounts blending news, lifestyle, celebrity, or beauty topics with hate speech or the military's narrative – reaching several millions of followers (Facebook, 2018a).

It thus becomes clear that in the imperial laboratory of a society dominated by an algorithm-governed, feed-based internet, Facebook was not the only actor experimenting and utilising the extracts of a newly-datafied society. While not necessarily having the technological infrastructure to make full use of the big data Facebook's daily process of datafication produces, nor having full access to it, the military's (and associated Buddhist nationalist groups') strategies made effective use of Facebook's digital infrastructure – its recommendation algorithms privileging the spread of certain topics; its lack of effective content moderation; and its affordances that allowed the military to broadcast its narrative in the guise of inconspicuous citizen posts.

6 Conclusion

Since the *coup d'état* in February 2021, when the Myanmar military stormed government buildings to arrest elected parliamentarians and the heads of government, the military's use of digital technologies has only intensified. While the arrests of elected parliamentarians and the use of brutal violence against protesters followed well-established patterns of Myanmar's military, these struggles quickly went digital with protesters using social media to mobilise for and broadcast resistance to the coup to a global audience as a civil disobedience movement coordinated country-wide strikes (Jordt et al., 2021).

Nonviolent protests against the coup have since given way to a plethora of armed revolutionary groups and (urban) guerrilla warfare against the military regime. But the digital has remained an important part of the military's repertoire in these struggles – in the form of digital surveillance and targeting of individuals online (Bächtold, 2022; Jordt et al., 2021; Ryan & Tran, 2022). Furthermore, the country-wide internet shutdown in February 2021 can even be read as the first announcement of the coup (Bächtold, 2023). Although the military has banned Facebook and other social media platforms, it continues to use data on Facebook for its own purposes. Examples include when the military publishes long lists of Facebook profiles of people expressing their dissent to the coup and marks them for arrest; or when its soldiers engage in 'information combat', where soldiers are ordered to post, like, and share their superiors' content to spread the military's narrative on a variety of platforms (Potkin & Wa Lone, 2021).

Facebook has since revised its ad-hoc approach to what it terms 'countries at risk' with a range of more systematic measures that now prominently feature 'understanding and engaging with local contexts and communities' (Facebook, 2021a) in the context of what Katzenbach (2021, p. 3) has called a 'responsibility turn' among tech companies. In Myanmar, in response to the military's use of Facebook to identify people opposing the coup, it has introduced new safety features that allow for 'locking' one's profile, and it has banned the military and related businesses from the platform (Facebook, 2021b). The coup and its aftermath also reduced Facebook's dominance in Myanmar. While the use of Facebook remains popular despite the military's ban of the app, the coup led more people to take up alternatives like Twitter/X and TikTok, and encrypted messaging apps like Telegram and Signal (Faxon et al., 2023).

The case of Myanmar has once again shown that digital technologies are not the neutral, necessarily beneficial tools of progress that techno-optimistic discourses make them out to be. Nor should developing countries be considered digital *terrae nullius* where technologies can be unilaterally deployed to without considering the political, cultural, economic, or historical configurations and relations of power that these technologies are likely to interact with. The processes of datafication and data extraction that global tech assemblages push forward with their ceaseless expansion have created a plethora of hybridised political effects in Myanmar – just as they have in every other country. Techno-optimistic discourses and their proponents, however, obscure their structural implication in producing these political effects. Instead, they continue to reproduce the tropes of techn(olog)ical solutions to political problems, and pretend that experimentation, bricolage, and tinkering after the catastrophe are all that is needed.

References

Anderson, W. (1992). "Where every prospect pleases and only man is vile": Laboratory medicine as colonial discourse. *Critical Inquiry*, 18(3), 506–529.

Anderson, W., & Adams, V. (2007). Pramoedya's chickens: Postcolonial studies of technoscience. In E. J. Hackett, S. Turner, O. Amsterdamska, W. E. Bijker, M. E. Lynch, J. Wajcman, L. Smith-Doerr, M. Solomon, A. Supnithadnaporn, & C. Thorpe (Eds.), *The Handbook of Science and Technology Studies* (3rd ed., pp. 181–204). MIT Press.

Andrejevic, M. (2014). The big data divide. *International Journal of Communication*, 8, 1673–1689.

Aradau, C. (2020). Experimentality, Surplus data and the politics of debilitation in borderzones. *Geopolitics*, 27(1), 26–46. https://doi.org/10.1080/14650045.2020.1853103

Arnold, D. (2005). Europe, technology, and colonialism in the 20th century. *History and Technology*, 21(1), 85–106. https://doi.org/10.1080/07341510500037537

Axster, S., Danewid, I., Goldstein, A., Mahmoudi, M., Tansel, C. B., & Wilcox, L. (2021). Colonial lives of the carceral archipelago: Rethinking the neoliberal security state. *International Political Sociology*, 15(3), 415–439. https://doi.org/10.1093/ips/olab013

Bächtold, S. (2015). The rise of an anti-politics machinery: peace, civil society and the focus on results in Myanmar. *Third World Quarterly*, 36(10), 1968–1983. https://doi.org/10.1080/01436597.2015.1063406

Bächtold, S. (2022). The smartphone and the coup: How Myanmar's conflicts are entangled with digital technologies, policies and violence. *International Journal of Public Policy*, 16(5/6), 293–310. https://doi.org/10.1504/IJPP.2022.127428

Bächtold, S. (2023). Blackouts, whitelists, and 'Terrorist Others': The role of sociotechnical imaginaries in Myanmar. *Journal of intervention and statebuilding*. https://doi.org/10.1080/17502977.2022.2152940

Bächtold, S., Bastide, J., & Lundsgaard-Hansen, L. (2020). Assembling drones, activists and oil palms: Implications of a multi-stakeholder land Platform for state formation in Myanmar. *The European Journal of Development Research*, 32(2), 359–378. https://doi.org/10.1057/s41287-020-00267-y

Barder, A. D. (2015). *Empire within: International hierarchy and its imperial laboratories of governance*. Routledge.

Burma Monitor, Center for Social Integrity, Equality Myanmar, Myanmar Human Rights Educator Network, MIDO, & Phandeeyar (2018). *Myanmar Civil Society Organization's Open Letter to Mark Zuckerberg – Progressive Voice Myanmar*. https://progressivevoicemyanmar.org/2018/04/05/myanmar-civil-society-organizations-open-letter-to-mark-zuckerberg

Buscemi, F. (2021). The art of arms (not) being governed: Means of violence and shifting territories in the borderworlds of Myanmar. *Geopolitics*, 28(1), 282–309. https://doi.org/10.1080/14650045.2021.1901083

Business for Social Responsibility (2018). *Human Rights Impact Assessment – Facebook in Myanmar*. https://about.fb.com/wp-content/uploads/2018/11/bsr-facebook-myanmar-hria_final.pdf

Callahan, M. P. (2012). The generals loosen their grip. *Journal of Democracy*, 23(4), 120–131. https://doi.org/10.1353/jod.2012.0072

Chan, A. S. (2014). *Networking peripheries: Technological futures and the myth of digital universalism*. MIT Press.

Cheesman, N. (2017). Introduction: Interpreting communal violence in Myanmar. *Journal of Contemporary Asia*, 47(3), 335–352. https://doi.org/10.1080/00472336.2017.1305121

Chonka, P., Diepeveen, S., & Haile, Y. (2022). Algorithmic power and African indigenous languages: search engine autocomplete and the global multilingual

Internet. *Media, Culture & Society*, 45(2), 246–265. https://doi.org/10.1177/01634437221104705

Couldry, N., & Mejias, U. A. (2018). Data colonialism: Rethinking big data's relation to the contemporary subject. *Television & New Media*, 20(4), 336–349. https://doi.org/10.1177/1527476418796632

Couldry, N., & Mejias, U. A. (2019). *The costs of connection: How data is colonizing human life and appropriating it for capitalism.* Stanford University Press.

Décobert, A., & Wells, T. (2020). Interpretive Complexity and Crisis: the History of International Aid to Myanmar. *The European Journal of Development Research*, 32(2), 294–315. https://doi.org/10.1057/s41287-019-00238-y

Deejay, A., Wells, T., Henne, K., & Bächtold, S. (2023). Bad adopters or bad proponents of technology? Facebook and the violence against Muslims in Myanmar. *Third World Quarterly*, 45(8), 1309–1324. https://doi.org/10.1080/01436597.2023.2285808

Escobar, A. (1995). *Encountering development: The making and unmaking of the Third World.* Princeton University Press.

Facebook (2018a). *Removing Myanmar Military Officials From Facebook.* https://about.fb.com/news/2018/08/removing-myanmar-officials

Facebook (2018b). *Update on Myanmar.* https://about.fb.com/news/2018/08/update-on-myanmar

Facebook (2021a). *Our Approach to Maintaining a Safe Online Environment in Countries at Risk.* https://about.fb.com/news/2021/10/approach-to-countries-at-risk

Facebook (2021b). *An Update on the Situation in Myanmar.* https://about.fb.com/news/2021/02/an-update-on-myanmar

Fanon, F. (1961). *Les damnés de la terre.* Librairie François Maspero.

Faxon, H. O., Kintzi, K., Tran, V., Wine, K. Z., & Htut, S. Y. (2023). Organic online politics: Farmers, Facebook, and Myanmar's military coup. *Big Data & Society*, *10* (1). https://doi.org/10.1177/20539517231168101

Fiskesjö, M. (2022). The Wa of the Burma-China borderlands: Identities and polities in the maelstrom of world-system cycles. In J. J. P. Wouters & M. T. Heneise (Eds.), *Routledge Handbook of Highland Asia* (pp. 339–350). London: Routledge. https://doi.org/10.4324/9780429345746-31

Gillespie, T. (2010). The politics of 'platforms'. *New Media & Society*, 12(3), 347–364. https://doi.org/10.1177/1461444809342738

Gillespie, T. (2018). *Custodians of the internet: Platforms, content moderation, and the hidden decisions that shape social media.* Yale University Press.

Gillespie, T. (2020). Content moderation, AI, and the question of scale. *Big Data & Society*, 7(2). https://doi.org/10.1177/2053951720943234

Gray, C. (2023). More than Extraction: Rethinking data's colonial political economy. *International Political Sociology*, 17(2). https://doi.org/10.1093/ips/olad007

Haupt, J. (2021). Facebook futures: Mark Zuckerberg's discursive construction of a better world. *New Media & Society*, 23(2), 237–257. https://doi.org/10.1177/1461444820929315

Hogan, L. (2018, April 6). Myanmar groups criticise Zuckerberg's response to hate speech on Facebook. *The Guardian*. https://www.theguardian.com/technology/2018/apr/06/myanmar-facebook-criticise-mark-zuckerberg-response-hate-speech-spread

Holliday, I. (2011). *Burma redux: Global justice and the quest for political reform in Myanmar*. Columbia University Press.

Htaike Htaike Aung, & Wai Myo Htut (2019). From blogging to digital rights: Telecommunication reform in Myanmar. In L. Brooten, J. McElhone, & G. Venkiteswaran (Eds.), *Myanmar media in transition: Legacies, challenges, and change* (pp. 366–376). ISEAS Publishing.

Irrawaddy (2017, May 31.). Facebook ban of racial slur sparks debate in Burma. *Irrawaddy*. https://www.irrawaddy.com/news/burma/facebook-ban-of-racial-slur-sparks-debate-in-burma.html

Jasanoff, S., & Kim, S.-H. (2009). Containing the atom: Sociotechnical imaginaries and nuclear power in the United States and South Korea. *Minerva*, 47(2), 119–146. https://doi.org/10.1007/s11024-009-9124-4

Jordt, I., Than, T., & Sue Ye Lin (2021). *How generation Z galvanized a revolutionary movement against Myanmar's 2021 military coup*. ISEAS Publishing. https://www.iseas.edu.sg/articles-commentaries/trends-in-southeast-asia/how-generation-z-galvanized-a-revolutionary-movement-against-myanmars-2021-military-coup-by-ingrid-jordt-tharaphi-than-and-sue-ye-lin/

Katzenbach, C. (2021). "AI will fix this" – The technical, discursive, and political turn to AI in governing communication. *Big Data & Society*, 8(2). https://doi.org/10.1177/20539517211046182

Khalili, L. (2010). The location of Palestine in global counterinsurgencies. *International Journal of Middle East Studies*, 42(3), 413–433. https://doi.org/10.1017/s0020743810000425

Kwet, M. (2019). Digital colonialism: US empire and the new imperialism in the Global South. *Race & Class*, 60(4), 3–26. https://doi.org/10.1177/0306396818823172

Kyaw Yin Hlaing. (2012). Understanding recent political changes in Myanmar. *Contemporary Southeast Asia: A Journal of International and Strategic Affairs*, 34(2), 197–216. https://doi.org/10.1353/csa.2012.0015

Leong, L. (2017). Mobile Myanmar: The development of a mobile app culture in Yangon. *Mobile Media & Communication*, 5(2), 139–160. https://doi.org/10.1177/2050157917689950

Lim, M. (2020). The politics and perils of dis/connection in the Global South. *Media, Culture & Society*, 42(4), 618–625. https://doi.org/10.1177/0163443720914032

Lindskov Jacobsen, K. (2015). *The politics of humanitarian technology: Good intentions, unintended consequences and insecurity*. Routledge.

Madianou, M. (2019). Technocolonialism: Digital innovation and data practices in the humanitarian response to refugee crises. *Social Media + Society*, 5(3). https://doi.org/10.1177/2056305119863146

Mark, S. (2023). *Forging the nation: Land struggles in Myanmar's transition period.* University of Hawai'i Press.

Mbembe, A. (2003). Necropolitics. *Public Culture*, 15(1), 11–40.

Mbembe, A. (2021). *Out of the dark night: Essays on decolonization.* Columbia University Press. http://www.jstor.org/stable/10.7312/mbem16028

McCarthy, G. (2017). Cyber-Spaces. In A. Simpson, N. Farelly, & I. Holliday (Eds.), *Routledge Handbook of Contemporary Myanmar* (pp. 92–105). Routledge.

McCarthy, G., & Menager, J. (2017). Gendered rumours and the Muslim scapegoat in Myanmar's transition. *Journal of Contemporary Asia*, 47(3), 396–412. https://doi.org/10.1080/00472336.2017.1304563

McNeil, S., & Milko, V. (2021, November 18). Hate speech in Myanmar continues to thrive on Facebook. *AP News*. https://apnews.com/article/technology-business-middle-east-religion-europe-a38da3ccd40ffae7e4caa450c374f796

McPherson, P. (2018, August 16). Facebook says it was 'too slow' to fight hate speech in Myanmar. *Reuters*. https://www.reuters.com/article/us-myanmar-facebook-rohingya/facebook-says-it-was-too-slow-to-fight-hate-speech-in-myanmar-idUSKBN1L1066

Meehan, P., & Sadan, M. (2017). Borderlands. In A. Simpson, N. Farelly, & I. Holliday (Eds.), *Routledge Handbook of Contemporary Myanmar* (pp. 83–91). Routledge.

Mignolo, W. D. (2007). Delinking: The rhetoric of modernity, the logic of coloniality and the grammar of de-coloniality. *Cultural Studies*, 21(2–3), 449–514. https://doi.org/10.1080/09502380601162647

Milan, S., & Treré, E. (2019). Big data from the South(s): Beyond data universalism. *Television & New Media*, 20(4), 319–335. https://doi.org/10.1177/1527476419837739

Morozov, E. (2013). *To save everything, click here: The folly of technological solutionism*. Public Affairs.

Mozur, P. (2018, October 15). A genocide incited on Facebook, with posts From Myanmar's military. *New York Times*. https://www.nytimes.com/2018/10/15/technology/myanmar-facebook-genocide.html

Nothias, T. (2020). Access granted: Facebook's free basics in Africa. *Media, Culture & Society*, 42(3), 329–348. https://doi.org/10.1177/0163443719890530

Office of the High Commissioner of Human Rights. (2018). *Myanmar: UN expert calls for accountability over violence in Rakhine State*. https://www.ohchr.org/en/NewsEvents/Pages/DisplayNews.aspx?NewsID=22793&LangID=E

Potkin, F., & Wa Lone (2021, November 1). 'Information combat': Inside the fight for Myanmar's soul. *Reuters*. https://www.reuters.com/world/asia-pacific/information-combat-inside-fight-myanmars-soul-2021-11-01/

Prasad, R. (2017). Ascendant India, digital India: How net neutrality advocates defeated Facebook's Free Basics. *Media, Culture & Society*, 40(3), 415–431. https://doi.org/10.1177/0163443717736117

Prasse-Freeman, E. (2021). Hate Bait, Micro-publics, and National(ist) Conversations on Burmese Facebook. *Independent Journal of Burmese Scholarship*, 1(December), 144–204.

Quet, M., & Dahdah, M. A. (2020). Technologies without borders? The digitization of society in a postcolonial world. *Science, Technology and Society*, 25(3), 363–367. https://doi.org/10.1177/0971721820912894

Ricaurte, P. (2019). Data epistemologies, the coloniality of power, and resistance. *Television & New Media*, 20(4), 350–365. https://doi.org/10.1177/1527476419831640

Rider, K., & Murakami Wood, D. (2018). Condemned to connection? Network communitarianism in Mark Zuckerberg's "Facebook Manifesto". *New Media & Society*, 21(3), 639–654. https://doi.org/10.1177/1461444818804772

Ryan, M., & Tran, M. V. (2022). Democratic backsliding disrupted: The role of digitalized resistance in Myanmar. *Asian Journal of Comparative Politics*, 9(1), 133–158. https://doi.org/10.1177/20578911221125511

Sablosky, J. (2021). Dangerous organizations: Facebook's content moderation decisions and ethnic visibility in Myanmar. *Media, Culture & Society*, 43(6), 1017–1042. https://doi.org/10.1177/0163443720987751

Said, E. W. (1978). *Orientalism*. Routledge.

Schissler, M. (2016). On Islamophobes and Holocaust deniers: Making sense of violence, in Myanmar and elsewhere. In N. Cheesman & N. Farelly (Eds.), *Conflict in Myanmar: War, Politics, Religion* (pp. 283–309). ISEAS Publishing.

Schissler, M. (2024). Beyond Hate Speech and Misinformation: Facebook and the Rohingya Genocide in Myanmar. *Journal of Genocide Research*, 1–26. https://doi.org/10.1080/14623528.2024.2375122

Scott, J. C. (1998). *Seeing like a state: How certain schemes to improve the human condition have failed*. Yale University Press.

Seetharaman, D., Horwitz, J., & Scheck, J. (2021, 17.10.). Facebook says AI will clean up the platform: Its own engineers have doubts. *Wall Street Journal*. https://www.wsj.com/articles/facebook-ai-enforce-rules-engineers-doubtful-artificial-intelligence-11634338184?mod=article_inline

South, A., & Lall, M. (Eds.). (2017). *Citizenship in Myanmar: Ways of being in and from Burma*. ISEAS – Yusof Ishak Institute.

Spivak, G. C. (1988). Can the subaltern speak? In C. Nelson & L. Grossberg (Eds.), *Marxism and the interpretation of culture* (pp. 271–313). University of Illinois Press.

Srnicek, N. (2016). *Platform capitalism*. Polity.

Stecklow, S. (2018, August 15). Why Facebook is losing the war on hate speech in Myanmar. *Reuters*. https://www.reuters.com/investigates/special-report/myanmar-facebook-hate

Thant Sin Oo, & Ye Min Thant (2019). *Exploring digital & mobile cultures in Myanmar 2019*. Document no longer available online.

Trautwein, C. (2016, June 6). Facebook Free Basics lands in Myanmar. *Myanmar Times*. https://www.mmtimes.com/business/technology/20685-facebook-free-basics-lands-in-myanmar.html

United Nations Human Rights Council (2018). *Report of Independent International Fact-Finding Mission on Myanmar*. https://www.ohchr.org/EN/HRBodies/HRC/MyanmarFFM/Pages/ReportoftheMyanmarFFM.aspx

Wade, F. (2017). *Myanmar's enemy within: Buddhist violence and the making of a Muslim 'other'*. Zed Books.

Walton, M. J., & Hayward, S. (2014). *Contesting Buddhist narratives: Democratization, nationalism, and communal violence in Myanmar*. East-West Center.

Whitten-Woodring, J., Kleinberg, M. S., Thawnghmung, A. M., & Thitsar, M. T. (2020). Poison if you don't know how to use it: Facebook, democracy, and human rights in Myanmar. *The International Journal of Press/Politics*, 25(3), 407–425. https://doi.org/10.1177/1940161220919666

Woods, K. (2011). Ceasefire capitalism: military–private partnerships, resource concessions and military–state building in the Burma–China borderlands. *Journal of Peasant Studies*, 38(4), 747–770. https://doi.org/10.1080/03066150.2011.607699

Yan Naung Oak, & Brooten, L. (2019). The tea shop meets the 8 o'clock news: Facebook, convergence and online public spaces. In L. Brooten, J. McElhone, & G. Venkiteswaran (Eds.), *Myanmar Media in Transition: Legacies, Challenges, and Change* (pp. 325–365). ISEAS Publishing. http://portal.igpublish.com/iglibrary/search/ISEASB0000754.html

Zuboff, S. (2015). Big other: surveillance capitalism and the prospects of an information civilization. *Journal of Information Technology*, 30(1), 75–89. https://doi.org/10.1057/jit.2015.5

Zuckerberg, M. (2013). *Is connectivity a human right?* Facebook. https://www.facebook.com/isconnectivityahumanright

CHAPTER 4

From Datafication to Interpellation: Becoming a Data Subject in Contemporary Surveillance Cultures

Bjorn Beijnon

1 Introduction

In this paper I provide a conceptual framework to understand how people in contemporary surveillance are transformed into what I call data subjects. A *data subject* is a subject that actively datafies itself into abstract quantities by responding to triggering actions of big tech platforms – ranging from smart speakers, smart watches, and social media platforms, to streaming platforms, internet browsers, and supermarkets – while considering itself in control over its behaviours, attitudes, and future choices. Through unprecedented datafication practices, i.e. quantification processes of people and their lives into digital data (Mejias & Couldry, 2019), people in contemporary surveillance cultures are culturally expected to each be a subject that undergoes, self-initiates, and advocates for such practices. Big tech companies normalize their datafication practices in contemporary surveillance cultures by tying this data subject to their identity so that it subliminally becomes a part of people's self-knowledge and consciousness (Foucault, 1997). When people in these cultures become this subject, it becomes easier for big tech companies to legitimize their intrusive data collection on people, normalize the black boxing of their algorithmic operations, and control people's perceptions of socio-political issues (Nguyen & Beijnon, 2024). By conceptually unpacking how exactly people become a data subject today, I construct a conceptual framework that can help researchers to systematically delineate key procedures and steps through which data subjects come into being today.

As a starting point for this framework, I use the theories and concepts of the French postmodern philosophers Michel Foucault (1988), Giles Deleuze & Félix Guattari (1972), and Louis Althusser (1968). Although none of these philosophers had the opportunity to write about data, algorithms, and platforms in the 21st century, more Critical Data and Platform Studies scholars

 | DOI:10.1163/9789004711396_006

have started to use their works on the study of power relations in their research on platforms (Törnberg & Uitermark, 2020), apps (Lupton, 2019), biometric data (Sanders, 2017), and interfaces (Stanfill, 2015). In the last two decades, research on the harvesting of user data for profit by big tech companies has become more prominent in these fields of study (cf. Morozov, 2014; Zuboff, 2019; Wark, 2019). This research not only focuses on the surveillance practices that these companies enforce on their users (Lyon, 2018), but also on the agency of users and their resistant practices (cf. Milan, 2015; Lovink, 2019; Dencik et al., 2019). By combining postmodern theories and concepts on power with the contemporary research on datafication practices, I systematically unpack how people in surveillance cultures are today involved in these practices and positioned in asymetrical power relations.

Although many sorts of datafication practices can be found across the world, I limit the examples of such practices in this paper to those conducted by the five biggest North American tech companies: Alphabet, Amazon, Apple, Meta, and Microsoft. Furthermore, when examining these companies, I confine my conceptual framework to a global northern context. Even though the framework that I sketch is also applicable outside this context, it is important to keep in mind that in different circumstances other big tech companies often dominate the market, such as Tencent in China or Mohalla in India. Research on the platforms of such companies in the global south often focusses on the data politics that emerge in the countries in which these platforms are dominant, such as WeChat in China (Chen et al., 2018) or VK in Russia (Enikolopov et al., 2020). This chapter might not draw from examples and experiences from a global southern context, but the framework I lay out does provide a general outline which can be used as a starting point for analysis of similar datafication practices in other regions, while one should remain sensitive to the many cultural differences per region.

2 Modulated Media Ecologies

First, it is important to point out that people who live in surveillance cultures today can be considered as users of platforms. José van Dijck, Thomas Poell, and Martijn de Waal (2018) argue that a platform is a programmable architecture designed to organize interactions between users. To summarize, they explain that 'a platform is fuelled by *data*, automated and organized through *algorithms* and *interfaces*, formalized through *ownership* relations driven by *business models*, and governed through *user agreements*' (van Dijck, Poell & de Waal, 2018, p. 9; emphases in original). Platforms are thus data circulation systems that are owned and ruled by big tech companies; they invite users to datafy themselves and actuate this data within an interconnected system of other platforms that a company owns. Securing the process of user data collection through surveillance practices (Lyon, 2014) is therefore essential for the existence and profitability of big tech companies; users are monitored via the data that these companies have of them so that they can continually pull users back to their platforms (van Dijck, 2014). Data is therefore not just fuel for platforms (like gasoline for a car), but an indispensable part that platforms need to be operational (like the tire and wheel for a car); platforms would not just be non-functioning without data, they would be incomplete.

Each big tech company seeks to lock users into their own platform ecosystem to sustain their harvesting of user data. They strategically enclose users in an immersive platform ecosystem through the foundational infrastructure that they have been able to lay out over the last two decades (Bolter, 2019). Take for example the platform ecosystem of Alphabet: users can mail via Gmail, write documents in Google Docs, navigate using Google Maps, etc. To harvest as much data as possible, Alphabet would like to lock the user into its ecosystem. However, users do not confine themselves to the services of one platform ecosystem but move between them (Clauss et al., 2019): they use Microsoft Outlook for email, but Google Calendar for keeping track of their agenda. Users thus tend to move across company platforms and, consequentially, find themselves in a bricolage of platforms (Gerlitz & Helmond, 2013) that datafy, commodify, and circulate data on them (Nieborg & Poell, 2018).

Despite the fact that users move across different platforms they are still able to find consistency and logic within the bricolage of platform collisions. This consistency is provided by what I will refer to as users' *modulated media ecology*: a complex and chaotic – yet clear – arrangement of a personal and personalized datafied milieu in which users' platform interactions are assembled. The term is derived from three concepts: modulation, ecology, and media. *Modulation* refers to a numerical mode of control, while simultaneously pointing to the fluidity of such controlling practices (Deleuze, 1992, p. 4). An *ecology* is a register that can be captured by capitalistic power formations although it can simultaneously question these formations through the degree of creative autonomy an individual has within it (Guattari, 1989, p. 31). The term *media* stands for the representational logic behind the capturing practices of media technologies (Hepp, 2020). A combination of the three concepts in *modulated media ecology* therefore refers to platform users' emergent registers that can capture them through fluid power practices of control that are defined by the representational logic of media technologies.

I argue that in today's modulated media ecologies, users become a data subject through repeatedly recognizing themselves in *data profiles* that they confront. Data profiles are constructed identities that are derived from data that are circulated across the many platforms that the user has interacted with (see Figure 4.1).[1] Within the modulated media ecology, users are confronted with such data profiles. For example, the data of the conversation that a user had with a friend over Meta's Messenger about their desperate need for a holiday rearranges their feed on Facebook and Instagram to display holiday pictures of tropical islands from friends, family, and influencers. Within their modulated media ecology, this type of data profile – that is, bored of their life at work, feels stuck in their day-to-day life, emotionally a bit unstable – is represented to the user in the form of recommendations and other related topics. Temporarily, the user is part of an emergent ecology in which they are invited to recognize themselves in the data profile that they are confronted with.

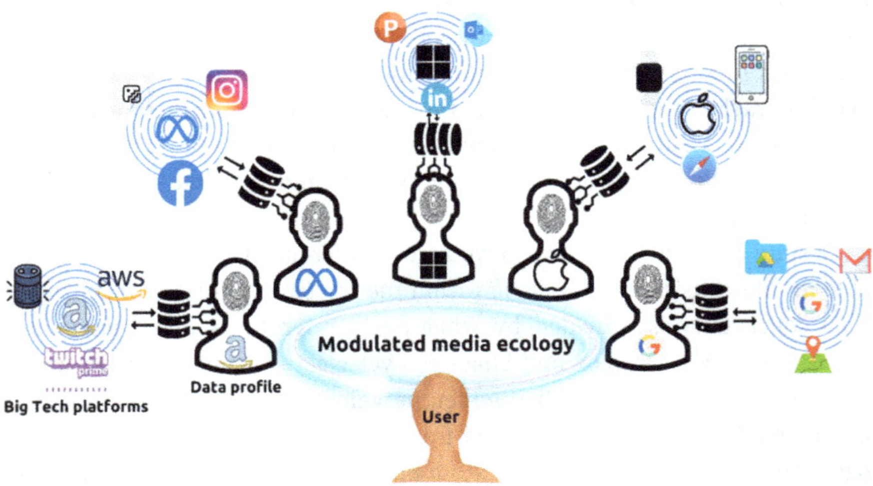

FIGURE 4.1 Schematic illustration of the modulated media ecology

1 When this ecology emerges, the user is confronted with data profiles of each big tech company that are based on circulated user data within the platform ecosystem of the same big tech company. Figure designed by Bjorn Beijnon (2024).

Modulated media ecologies are not only confronting users with one data profile at a time. Because of the plurality of platforms that the user interacts with, and thus the various data circulation systems, users are constantly invited to recognize themselves in numerous data profiles simultaneously. The data profile that is offered by Meta might therefore be different from the one that is offered by other big tech platforms. Users' modulated media ecologies therefore use an implied form of address that eventually shapes them into a particular subject. Considering data profiles as such an address, I argue later in this chapter that recognizing oneself in data profiles is a prerequisite to becoming a data subject. It is for this reason that I move to the practices that these companies use in monitoring and controlling the constant emergence of users' modulated media ecology.

3 Algorithmic Governmentality

As modulated media ecologies emerge for each user individually, users are themselves regulating the emergence of their ecology: they shape and guide their own conduct when it comes to constructing their modulated media ecology. Consequently, big tech companies seek to establish organized practices through which they can monitor and control this self-governance. This conduct of conduct, or governance from a distance, is what Michel Foucault called *governmentality*: the process through which a form of government with specific goals, with a particular type of knowledge, and with the technical means, achieves these goals (Mazel, 2019, p. 51).

Before examining governmentality in today's surveillance cultures, one first has to recognize that the data on which data profiles are based are not verisimilitudinous to the lived life of the user. This other reality is what Jonathan Gray calls a *data world*: a constructed horizon of intelligibility that can be used as a practical utility in which AI-driven algorithms attempt to produce a data profile by reindexing the categorical meaning of human-centred contexts into measurable datafied elements (Gray, 2018, p. 6). Data profiles are therefore 'empirical assessments of prejudice and/or patterns that work, according to a particular confidence percentage, at a particular moment and by a particular set of rules' (Cheney-Lippold, 2017, p. 89). For example, Meta's data profile of a user that is in desperate need for a holiday does not necessarily reflect users' actual desire for a holiday in their lives, as the data profile is based on circulated data in Meta's platform ecosystem, not on users' actual perceived and experienced desires. Data profiles can only reflect such desires, but do require indexical accuracy in the harvested user data and algorithmic precision in the construction of the data profile. Users are thus governed by the statistical principles by which platform algorithms operate.

Governmentality can only occur when specific techniques are employed on individuals that support them in understanding themselves as self-governing subjects. Governmentality arises in the contact between technologies of power and technologies of the self. *Technologies of power* 'determine the conduct of individuals and submit them to certain ends or domination, an objectivizing of the subject' (Foucault, 1988, p. 18). Such technologies lay out the infrastructures that by which people must abide, which transforms them into subjects. *Technologies of the self*, however,

> permit individuals to effect by their own means or with the help of others a certain number of operations on their own bodies and souls, thoughts, conduct, and way of being, so as to transform themselves in order to attain a certain state of happiness, purity, wisdom, perfection, or immortality.
>
> FOUCAULT, 1988, p. 18

In these kinds of technologies, individuals have the possibility to enact power on themselves to change themselves into a type of subject that they want themselves to be. When these two technologies coincide, governmentality comes into being: the immersion of individuals' bodies of knowledge, belief, and opinion in a controlled system that they willingly inhabit and contribute to (Dean, 2010, p. 24). Through the connection between technologies of power and technologies of the self people become conditioned to a type of governance that imposes the possibilities and limits of what an individual can do or be, such as the constant construction of a modulated media ecology.

As platforms are the mechanisms that bring about data profiles, they can be considered as both technologies of power and technologies of the self. First, platforms are technologies of power as their infrastructure is enforced on users; they are unable to deviate from paths and spaces that have already been laid out for them within and between platforms owned by the same tech company. For example, users are not able to reroute the results of their verbal search request to Amazon Alexa's smart speaker system via a personal home data storage cloud that is not supported by Amazon. Second, platforms are technologies of the self as they offer people a space in which they get to transform themselves into 'interpretations of data that stand in as digital containers of categorical meaning' (Cheney-Lippold, 2017, p. 19); platforms datafy users into algorithmic caricatures of themselves (i.e.

data profiles) that within the data world of platform infrastructures start to have their own histories, logics, and rationales. For example, the way that Meta's platform algorithms construct a data profile for a user based on circulated data within the company's platform ecosystem, such as likes of other users' holiday pictures, steers users to recognize and transform themselves as a person that needs to go on a holiday. As both technologies of power and of the self, platforms are thus both a space of domination and self-transformation.

Considering that platforms are both these technologies, platforms contribute to a kind of governance over their users from a distance in which users govern themselves in generating their modulated media ecology. This new conduct of conduct is what Antoinette Rouvroy calls *algorithmic governmentality*: 'a mode of government motivated almost exclusively by the goal of fluidification (or liquidation) of existences, requiring the suppression of all that would oppose the indistinctiveness of the world and a numerical, calculable reality' (Rouvroy, 2013, p. 162). Algorithmic governmentality therefore seeks to invisibly activate and motivate people to subordinate themselves to datafied power relations. This subordination would mean that users acknowledge that the data that platforms have on them not only reflects a correct indexation of their lived experiences, but that this data *is* what they perceive, experience, or are.

For example, on Meta's social media platforms Facebook and Instagram, users are invited to interact with other users' content on these platforms: they can like, comment, repost, or just scroll over it. In the case of 'Stories', users can interact with a vertical image or video of another user that takes up the full screen of their smartphone. Users can quickly click through this content, pause it, like it, or comment on it via a private message. The algorithm behind these platforms organizes these stories on the likelihood that a user would be interested in the content. Stories from someone's best friend would, for example, appear first in the stories feed. Between stories, advertisements have been embedded for users to see, like an advertisement with possible holiday destinations sponsored by a travel agency. For Meta, it is important that users recognize themselves in the data profile that the company constructs of its users, which can only be done in a modulated media ecology. Thus, Meta steers users to construct their modulated media ecology any time that they start to interact with a feed of stories – which has been made quite addictive by design (Bayer et al., 2022). Whenever users engage with stories on Meta's platforms, the company implicitly demands that users construct their modulated media ecology, which conditions them to comply to an algorithmic governmentality.

In short, algorithmic governmentality is about surpassing individuals' consciousness, rationality, and reflexivity, and having them unconsciously generate their modulated media ecology. Algorithmic governmentality 'anticipatively 'adapt[s]' the informational and physical environment of persons according to what these persons are susceptible to do or wish to do' (Rouvroy, 2013, p. 157). So, although it is important to recognize that data profiles are produced in a data world (Cheney-Lippold, 2017), users live in a reality that runs on self-enforcing, statistically established, and algorithm-driven norms that are also fluid, real time adaptive, immediate, and dynamic. Thus, in this new type of governmentality, users are not becoming subjects that are derived from knowledge *about* who they are in their lived lives, but from data *from* a data world. Algorithmic governmentality therefore shows how the modulated media ecology has not become a reflection of the lived lives of users, but a product of a data world that acts as if it is such a reflection using statistical analyses by platform algorithms.

4 Stratified Data Assemblages

So far, I have laid out how big tech companies use algorithmic governmentality to make users govern themselves by generating their modulated media ecology in which they are confronted with data profiles. But how exactly are such data profiles constructed and how do they relate to users? To unpack how data profiles are assembled, I move to Gilles Deleuze's and Felix Guattari's assemblage theory and particularly their conceptualization of *desire*. For Deleuze and Guattari (1972), desire is not an inert feeling to want something, such as love or sex. Instead, they propose that desire is productive, like a free-flowing stream of intensities subject to processes of capture and coagulation (Buchanan, 2020). Desires always flow within an ensemble of various factors and intensities (Pamart, 1988). For example, when I say that 'I desire to watch a film,' I am not desiring the object of the film itself, but I am desiring the ensemble of which the film is a part: I am desiring a specific genre, a sort of atmosphere, the feeling of endless scrolling on the video streaming platform Amazon Prime Video, the company of my friends, etc. Desire thus flows within an ensemble of various factors and intensities of which the actual film that I am going to watch is just one factor. This complex arrangement through which desires flow is called an *assemblage*: a composition of desire that as a unity produces new desires (Deleuze & Guattari 1980, p. 31). Saying that you desire something is therefore indicating that you are in the process of constructing an assemblage.

I argue that in modulated media ecologies, desires are, on the one hand, suppressed in formatted possibilities of assemblages in the form of data profiles, while, on the other hand, desires are primed and nudged by platform algorithms through their transformation into impulses of consumption. The platform infrastructures of big tech companies are therefore exploiting users' desires through data profiles for the sake of their own financial profit, and users become – ultimately as data subjects – surveillance capitalist commodities.

The assembling of desires on platforms is carefully controlled and owned by big tech companies. Platform algorithms assemble users' desires through their formatting principles of *stratification*: 'the process which consists in "giving form to matters" by 'imprisoning intensities or locking singularities into systems of resonance and redundancy' (Buchanan, 2020, p. 38). In the case of platform algorithms, stratification is a practice that puts different desires – materialized in user data – together into *stratified data assemblages*. For example, when users use Alphabet's Google search engine to look for their next, still unknown holiday destination, the algorithm uses any data that it already has within its platform ecosystem to compile the search results that will appear for the user. This involves users' search and browser history, locations that they have visited, books that they have read, series that they have watched, etc. The analysis of all this information is then weighted by the Alphabet's algorithm to predict which possible search results would, according to the judgement of the algorithm, best fit the user. Users might thus think that their choice for a holiday destination is based on their own experienced desires that connect with an ensemble of other desires that together form an assemblage of what the holiday destination might be. However, by considering user data as materializations of their desires, Google's search engine algorithm created a stratified data assemblage through which Alphabet controls the data that is served back to the user.

Although stratified data assemblages might not seem like a bad thing – no one has time to go through all possible holiday destination anyway – this process does become worrisome when it leads to manipulation of users. Platform algorithms of big tech companies can decide, based on the data that they have on their users, what data they would like to stratify together in an assemblage so the users can be presented with a data profile in their modulated media ecology. For example, a big tech company like Amazon can stratify users' data into assemblages in which new desires emerge while also being able to control what users' future desires might be. Let's imagine that Amazon picked up the following data of the user via within its platform ecosystem: the purchase of avocados at Whole Foods, watching of a

Mexican film on Amazon Prime Video, and the search input 'Learn Spanish for Adults' in the webstore (see Figure 4.2).[2] With this data, Amazon can construct its own stratified data assemblage into a data profile that would like to consume products and services linked to Spanish culture. Subsequentially, Amazon can now control this stratified data assemblage by using its platforms to accommodate this platform-prompted user desire: it rearranges the webstore's interface to Spanish books and series, gives discounts for more ingredients for Spanish dishes at Whole Foods, and recommends more Spanish movies and series on Amazon Prime Video. The user is therefore through the data profile unconsciously guided into preprogrammed desires that have become materialized in data that they might not have been interested in if it had not been provided by Amazon's platforms.

Stratified data assemblages thus not only format how users' desires as materialized data should flow in the form of a data profile, but also produce new desires that are not coming from the users themselves but from the stratified data assemblage that is constructed by big tech platform algorithms. I therefore argue that the ways in which platforms manipulate desires that users connect with amounts to a *priming of desire*: the social shaping of conditions in which users have to account for the consequences of their desires, although these were carefully crafted and placed in the platforms by the companies behind these platform algorithms. Within the discourse of psychology, *priming* involves the unconscious guiding of behaviour through exposing people to stimuli that share similar semantics,

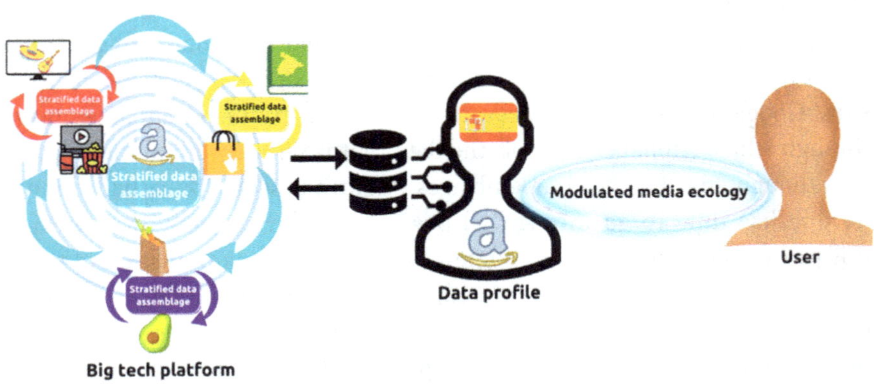

FIGURE 4.2 Schematic illustration of user data in stratified data assemblages

2 Stratified data assemblages construct a data profile that users confront in their modulated media ecology. Figure designed by Bjorn Beijnon (2024).

perceptions, or affects (Kahneman et al., 1992). Using stratified data assemblages, platform algorithms prime their constructed desires as if they are the users' own. What big tech companies therefore do is 'deliberately organizing wants and needs amid an abundance of production [...] and making the object dependent upon a real production that is supposedly exterior to desire' (Deleuze & Guattari, 1972, p. 28). A company like Amazon thus stratifies data assemblages for a user in a data profile and regenerates them into new desires as if users have produced these desires themselves. Such primed desires might seem harmless when they involve the recommendation of a film genre but become problematic when they, for example, manipulate people's political preferences with individual-targeted content (Kruikemeier et al., 2016). In the datafied logic of platforms, desires thus lose their assembling freedom, but are presented as if they are free.

To summarize, considering desires as productive, provides a way to examine the assembling logic behind data profiles. User data can be considered as stratified data assemblages in the form of data profiles that platform algorithms construct. Big tech companies use these data profiles to create and prime new desires that benefit the business behind the platform; primed desires might therefore draw users into a loophole in which they are confronted with data profiles that the company behind the platform algorithms wants the user to be.

5 Algorithmic Interpolation

Up to this point, I have unpacked how people self-govern the constant construction of their modulated media ecology and how the data profiles in this ecology are stratified data assemblages that steer users' desires. But how much control do users have over recognizing themselves in such data profiles and how do they become a data subject? To examine the power of data profiles in the process of becoming a data subject, I move to Louis Althusser's work on how people recognize and reproduce themselves into a subject. Althusser (1968) called this process *interpellation*: the act of recognizing oneself as a subject that comes to think of itself in terms of the social categories in which it is addressed. For example, interpellation occurs in the moment when a user is hailed by their Apple Watch: 'Hey, you there! Look how many steps you took today.' When the user looks at their watch and recognizes themselves as the 'you' that is addressed as the person that walked a lot that day, they have interpellated themselves into the type of subject that is expected from them by the watch: a subject that recognizes themselves as a bundle of biometric data that was tracked, analysed, and commodified by Apple. Recurring hails like these over time become rituals that make users recognize themselves in the subject that they are addressed as. Interpellation is then the becoming of the imposed subject that comes to think of themselves in terms of the social categories in which they are addressed in such rituals.

Coming back to modulated media ecologies in which users are confronted with data profiles, I argue that these ecologies provide a moment in which a data subject can be called out. Modulated media ecologies prompt users to imagine themselves as a data subject through repeatedly recognizing themselves in algorithmically generated data profiles. Modulated media ecologies therefore gestate users not to relate themselves to their actual lived conditions, but to algorithmically generated datafied representations of these conditions in the form of data profiles.

To better understand interpellation in the modulated media ecology, it is important to examine more closely the hail to the users in this ecology. Ronald Day argues that the indexical signs that are operational for algorithms of which data profiles are constructed, can only be acknowledged by users if they take place within a documentary logic, 'which structures the self as a *responsible* subject that *responds* not simply to an other, but first of all to the symbolic order itself and its logic of identity and expression in terms of the presenting of evidence' (Day, 2014, p. 81; emphasis in original). The documentary logic within the modulated media ecology thus requires users to believe that the data profile is based on reliable information that the platform algorithm has on the user.

As algorithms will always have to construct data profiles based on indexical evidence – and therefore on data that it has on each user within big tech company's platform ecosystem – the process of recognizing oneself in a data profile is not one of interpellation, as a data profile is not the same as the culturally expected data subject. Instead, algorithms make users recognize themselves in interpolated data profiles. Known in mathematics and data sciences as a method for constructing new data points based on a discrete set of known data points (Korstanje, 2022), *interpolation* by platform algorithms suggests following the logic of statistical relatability and connecting the dots between data points. This means that in the modulated media ecology users become the data profile they are confronted with due to users' acknowledgment of the accuracy of the predicted preferences. As put by Day,

> we are first of all hailed by the sociotechnical information system itself (as representative and mediator of a political economy) as a fulfiller of an information need. We come to hail ourselves as certain representations of persons by our very searching and communicating online.
>
> DAY, 2014, p. 82

When users thus generate their modulated media ecology, they are automatically called out as data profiles that have been interpolated for the user (see Figure 4.3 Left). As users are repeatedly recognizing themselves into data profiles in their modulated media ecology, this constant interpolation becomes a ritual. Therefore, I argue that the ritual of interpolating data profiles in modulated media ecologies is constituting the process of users' interpellation into a data subject; *algorithmic interpolation of data profiles is a condition for the interpellation of a data subject.*

To study how interpolation is a condition for interpellation, it is important to examine the rituals of the interpellation, which means zooming in on materialized traces of the actualizations of the data profiles interpolations. Personalized attentional triggers, such as notifications, pop-ups, or discounts, are a good example of such materializations. Take, for example, the personalized attentional triggers that are generated by Microsoft's social media platform LinkedIn. These triggers can only be effective if they embody the data profile that has been interpolated for the user. A pop-up notification by LinkedIn on career advice only makes sense if users have already recognized themselves as someone who seeks such advice; interpolation of the data profile must therefore have occurred before the trigger was pushed by the platform with the intention to keep users in a continuous cycle in which they recognize themselves in this data profile (see Figure 4.3 Middle). Personalized attentional triggers via LinkedIn are therefore not only materializations of the interpolation of data profiles, but they are also a way to hail the user as the data profile LinkedIn wants the user to be. As personalized attentional triggers are materializations of users' recognition of the interpolated data profile, these triggers have through their repetitive occurrence become rituals for how users interpellate themselves in a data subject (see Figure 4.3 Right).[3]

3 Left: user recognizes the interpolated data profile by answering to its hail through a job recommendation on LinkedIn. Middle: user recognizes the interpolated data profile by answering to the hail of a personalized attentional trigger in the form of a notification. Right: user has become a data subject through the ritual of repetitively answering to

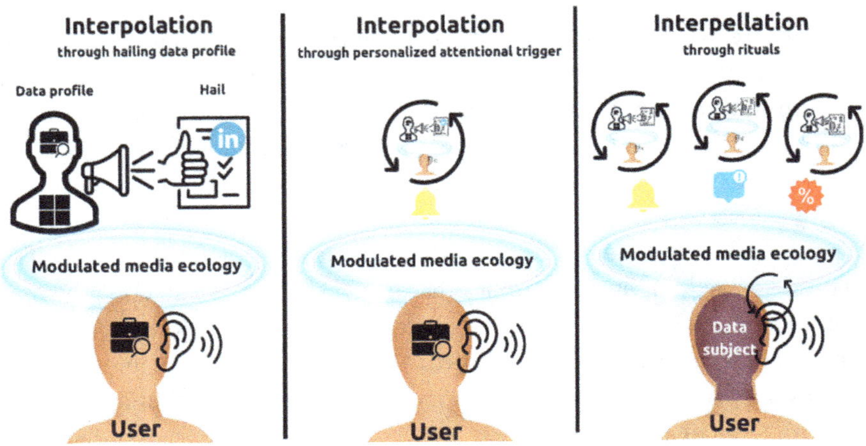

FIGURE 4.3 Schematic illustration of the interpolation/interpellation cycle

Expanding interpellation to the notion of algorithmic interpolation demonstrates how users come to recognize themselves in data profiles in their modulated media ecology. Users are steered to follow the guiding mechanisms of platform algorithms under the strict conditions of platforms. Through the continuous flow of data that these algorithms gather on their users, they can enclose users in a never-ending cycle of interpolation/interpellation. Users will keep on recognizing themselves in interpolated data profiles, while personalized attentional triggers that are based on this interpolation become rituals that make users interpellate themselves into a data subject: a subject that actively datafies itself into abstract quantities by responding to triggering actions of big tech platforms while considering itself in control over its behaviours, attitudes, and future choices.

6 Conclusion

In this chapter I have provided a conceptual framework to understand how people in contemporary surveillance cultures become a data subject. By introducing the concept of modulated media ecologies, I demonstrated how the becoming of a data subject can be examined on three conceptual levels. First, the level of algorithmic governmentality: big tech companies have

personalized attentional triggers, like notifications, pop-ups, and personal discounts. Figure designed by Bjorn Beijnon (2024).

been able to normalize a power practice in which users conduct their own conduct in constructing their modulated media ecology and considering the data world of data profiles as an accurate indexical reflection of their lived lives. Second is the level of stratified data assemblages: user data is formatted by the platform algorithms of big tech companies into stratified data assemblages that use data profiles to prime algorithmically generated desires to users, in ways that are profitable for the company behind the algorithm and platform ecosystem. Third, the level of algorithmic interpolation: users repetitively recognize themselves in interpolated data profiles when they interact with platforms by listening to the hail of personalized attentional triggers which ultimately makes them interpellate themselves into a data subject.

Datafication in current surveillance cultures has made platform users govern their own interpellation by acknowledging that data profiles indexically reflect their past, present, and future. As elucidated in the introduction, scholars in Critical Data and Platform Studies underscore the agency retained by users within platform ecosystems, emphasizing instances of resistance in which users refuse to recognize themselves in the data profiles that platforms algorithms construct for them. Despite these user-driven nuances, big tech companies persistently advance the precision, detection, and prediction capabilities of their algorithms, which should make the data profiles of their platform users indexically more accurate and appealing. This continuous enhancement serves the overarching goal of retaining users within platform ecosystems, prompting these companies to invest in robust methods of user data collection and analysis to mitigate any dissonance or dissatisfaction in users' recognition of their data profiles. The conceptual framework presented in this chapter establishes a robust foundation for critically examining these practices and envisioning alternative ways of living with data and platforms.

References

Althusser, L. (2004) [1968]. Ideology and ideological state apparatuses. In J. Rivkin & M. Ryan (Eds.), *Literary Theory: An Anthology* (pp. 768–777). Blackwell Publishing.

Bayer, J. B., Anderson, I. A., & Tokunaga, R. S. (2022). Building and breaking social media habits. *Current Opinion in Psychology*, 45, 279–288. https://doi.org/10.1016/j.copsyc.2022.101303

Bolter, J.D. (2019). *The digital plenitude: The decline of elite culture and the rise of new media.* MIT Press.

Buchanan, I. (2020). *Assemblage theory and method.* Bloomsbury Publishing.

Chen, Y., Mao, Z., & Qiu, J. L. (2018). Super-Sticky design and everyday cultures. In *Super-Sticky Wechat and Chinese Society* (pp. 47–76). Emerald Publishing Limited. https://doi.org/10.1108/978-1-78743-091-420181004

Cheney-Lippold, J. (2017). *We are data.* New York University Press.

Clauss, T., Harengel, P., & Hock, M. (2019). The perception of value of platform-based business models in the sharing economy: Determining the drivers of user loyalty. *Review of Managerial Science*, 13(3), 605–34. https://doi.org/10.1007/s11846-018-0313-0

Day, R.E. (2014). *Indexing it all: The subject in the age of documentation, information, and data.* MIT Press.

Dean, M. (2010). *Governmentality: Power and rule in modern society.* SAGE.

Deleuze, G. (1992). Postscript on the societies of control. *October* 59, 3–7.

Deleuze, G., & Guattari, F. (2004) [1980]. *A thousand plateaus.* A&C Black.

Deleuze, G., & Guattari, F. (2009) [1972]. *Anti-Oedipus: Capitalism and schizophrenia.* Penguin.

Dencik, L., Hintz, A., Redden, J., & Treré, E. (2019). Exploring data justice: Conceptions, applications and directions. *Information, Communication & Society*, 22(7), 873–881. https://doi.org/10.1080/1369118X.2019.1606268

Dijck, J. van. (2014). Datafication, dataism and dataveillance: Big Data between scientific paradigm and ideology. *Surveillance & Society*, 12(2), 197–208. https://doi.org/10.24908/ss.v12i2.4776

Dijck, van J., Poell, T., & de Waal, M. (2018). *The platform society: Public values in a connective world.* Oxford University Press.

Enikolopov, R., Makarin, A., & Petrova, M. (2020). Social media and protest participation: Evidence from Russia. *Econometrica*, 88(4), 1479–1514. https://doi.org/10.3982/ECTA14281

Foucault, M. (1988). *Technologies of the self: A seminar with Michel Foucault.* The University of Massachusetts Press.

Foucault, M. (2003) [1997]. *Society must be defended: Lectures at the College de France, 1975–76.* Penguin.

Gerlitz, C., & Helmond, A. (2013). The like economy: Social buttons and the data-intensive web. *New Media & Society*, 15(8), 1348–65. https://doi.org/10.1177/1461444812472322

Gray, J. (2018). Three aspects of data worlds. *Krisis*, 38(1), 4–17.

Guattari, F. (2005) [1989]. *The three ecologies.* Bloomsbury Publishing.

Hepp, A. (2020). *Deep mediatization.* Routledge.

Kahneman, D., Treisman, A., & Gibbs, B. J. (1992). The reviewing of object files: Object-Specific integration of information. *Cognitive Psychology*, 24(2), 175–219. https://doi.org/10.1016/0010-0285(92)90007-O

Korstanje, J. (2022). *Machine learning on geographical data using python*. Apress.

Kruikemeier, S., Sezgin, M., & Boerman, S. C. (2016). Political microtargeting: Relationship between personalized advertising on Facebook and voters' responses. *Cyberpsychology, Behavior, and Social Networking*, 19(6), 367–372. https://doi.org/10.1089/cyber.2015.0652

Lovink, G. (2019). *Sad by design: On platform nihilism*. Pluto Press.

Lupton, D. (2019). The thing-power of the human-app health assemblage: Thinking with vital materialism. *Social Theory & Health*, 17(2), 125–139. https://doi.org/10.1057/s41285-019-00096-y

Lyon, D. (2014). Surveillance, Snowden, and big data: Capacities, consequences, critique. *Big Data & Society*, 1(2), 2053951714541861. https://doi.org/10.1177/2053951714541861

Lyon, D. (2018). *The culture of surveillance: Watching as a way of life*. John Wiley & Sons.

Mazel, A.G. (2019). *Governing food: Media, politics and pleasure*. [PhD dissertation, University of Amsterdam]. University of Amsterdam Digital Academic Repository.

Mejias, U.A., & Couldry, N. (2019). Datafication. *Internet Policy Review*, 8(4), 1–10. https://doi.org/10.14763/2019.4.1428

Milan, S. (2015). When algorithms shape collective action: Social media and the dynamics of cloud protesting. *Social Media + Society*, 1(2), 1–10. https://doi.org/10.1177/2056305115622481

Morozov, E. (2014). *To save everything, click here*. PublicAffairs.

Nguyen, D., & Beijnon, B. (2024) The data subject and the myth of the 'black box' data communication and critical data literacy as a resistant practice to platform exploitation. *Information, Communication & Society*. 27(2), 333–349. https://doi.org/10.1080/1369118X.2023.2205504

Nieborg, D.B., & Poell, T. (2018). The platformization of cultural production: Theorizing the contingent cultural commodity. *New Media & Society*, 20(11), 4275–92. https://doi.org/10.1177/1461444818769694

Pamart, M. (Director). (1988). D comme Désir [TV series episode]. In Boutang. P. (Executive Producer), *L'Abécédaire de Gilles Deleuze*. La Fémis; Sodaperaga Productions.

Rouvroy, A. (2013). The end(s) of critique: Data behaviourism versus due process. In M. Hildebrandt & K. de Vries (Eds.), *Privacy, Due Process and the Computational Turn* (pp. 143–168). Routledge.

Sanders, R. (2017). Self-tracking in the Digital Era: Biopower, Patriarchy, and the New Biometric Body Projects. *Body & Society*, 23(1), 36–63. https://doi.org/10.1177/1357034X16660366

Stanfill, M. (2015). The interface as discourse: The production of norms through web design. *New Media & Society*, 17(7), 1059–1074. https://doi.org/10.1177/1461444814520873

Törnberg, P., & Uitermark, J. (2020). Complex control and the governmentality of digital platforms. *Frontiers in Sustainable Cities*, 2(6), 1–16. https://www.frontiersin.org/articles/10.3389/frsc.2020.00006

Wark, M. (2019). *Capital is dead: Is this something worse?* Verso.

Zuboff, S. (2019). Surveillance capitalism and the challenge of collective action. *New Labor Forum*, 28(1), 10–29. https://doi.org/10.1177/1095796018819461

PART 2

Digital Cultures and Digital Politics

CHAPTER 5

'BM Girl' Influencers on Xiaohongshu: Tracing Beauty Discourse, Social Media Challenges, and Consumption Practices in Chinese Society

Shen Sijun and Crystal Abidin

1 Introduction: The BM Trend on Xiaohongshu

Xiaohongshu (小红书), also known as Red (红; *hóng*), is an integrated social platform that combines social media with e-commerce, using community-generated content to promote commercial products. One of the fashion trends popularised by Xiaohongshu since 2020 is BM Style (BM 风; BM *fēng*). BM, short for Brandy Melville, is an Italian fashion brand specialising in slim-sized clothes with the slogan 'one size fits most'. In China, the trend began when celebrities posted street snaps while dressed in BM clothes. The impact of such popular culture has continued onto Xiaohongshu, where content creators followed suit to post images of themselves clad in BM clothes. This ignited the new social media trend of the BM girl, which quickly came to collectively represent the gold standard of body, face, and beauty ideals as encapsulated by influencer role models on Xiaohongshu.

In response to the growing impact of Xiaohongshu influencers like BM girls on the national discourse about beauty trends, bodily standards, and consumption practices, in this chapter, we interrogate how Xiaohongshu has facilitated influencer promotional cultures, or the colonisation of (digital) culture by commercial advertising (Wernick, 1991). Specifically, we survey the features and affordances of Xiaohongshu that have directly shaped the rising growth of influencer cultures and their entanglements with product and brand placements that have facilitated fast consumerism among users. Methodologically, we draw on textual analysis conducted among BM girls and the wider ecology of BM accounts on Xiaohongshu, whom we deem to be gatekeepers and originators of the new discourse on beauty standards. Our corpus focuses on a purposive sampling of key search terms that span the BM ecology on Xiaohongshu, including BM girl influencers, BM employee accounts ('BrandyMelville 店员'), and general conversations

 | DOI:10.1163/9789004711396_007

about the BM beauty discourse and ideology. In so doing, we trace how discourses about standards, values, and beliefs are shaped on Xiaohongshu and travel to the wider digital media ecology in Chinese society.

In this chapter, we deploy the term 'influencer' over '*wǎnghóng*' (literal trans. 'internet red'; hereafter *wanghong*) to highlight the e-commerce feature on Xiaohongshu, i.e. advertisements and product sales. *Wanghong*, which refers to the technology-driven celebrification of the ordinary on the internet (J. Xu & Zhao, 2019), is approximate to a broader combination of 'internet celebrity,' 'influencer,' and 'microcelebrity' (Abidin et al., 2021). In comparison, influencers refer specifically to online celebrities who 'accumulate a following on blogs and social media through the textual and visual narration of their personal, everyday lives, upon which paid advertorials ... for products and services are premised' (Abidin, 2016, p. 26).

2 Xiaohongshu's Role in China's Digital Cultures and Fashion Trends

Xiaohongshu (小红书), China's most well-known social media e-commerce platform, markets itself as empowering Chinese consumers to establish unique personal styles (Ju, 2022; Ye & Rodriguez, 2023). This platform is of particular interest to this socio-cultural inquiry into beauty trends, bodily standards, and consumption practices in Chinese digital cultures for four reasons. First, its user base is comprised of disproportionately younger women who are interested in beauty/fashion which makes them more susceptible to being influenced by related commercial messages embedded in social media posts. Second, Xiaohongshu has established one of the most successful business models in monetising online posts, through its *zhòngcǎo* (literal translation: 'planting grass or seedings', hereafter *zhongcao*) culture, which will be discussed below. Third, in comparison to other digital platforms in China, Xiaohongshu is disproportionately formed by influencers, i.e. users with clear commercial agendas. Finally, the platform's strategic employment of traditional celebrities has given it significant social power in steering opinions, attitudes and trends related to beauty and fashion in Chinese society.

Moreover, in comparison to other Chinese or global social media platforms, Xiaohongshu is often known as a representation of the so-called 'she-economy' (Gong, 2022). More so than having a higher proportion of women as influencers, which is similar to other digital platforms and cultures in China and worldwide, Xiaohongshu is a platform that focuses particularly on women consumers with the claim of empowering women, providing them with a voice and means to express themselves (Gong, 2022). Xiaohongshu's users are younger women interested in fashion brands or products. In 2022, Xiaohongshu had 200 million monthly active users, 70% of whom were women; the gender ratio for the fan base of active influencers is 88.8% (women) and 11.2% (men) (qiangua, 2022). The majority of Xiaohongshu users are young: 72% were born after 1990, and from top-tier cities in China (qiangua, 2022). Xiaohongshu labels its user base with six labels including 'Gen-Z', 'urban fashion guru', 'white collar elites', 'rich singles', 'elegant mothers' and 'devotees to pleasure', and the top five topics that they care about are makeup, clothes, skin-care, food, and fitness/weight management (qiangua, 2022).

In addition, Xiaohongshu is considered to have the highest advertising value among digital media platforms in the Chinese market (Lian et al., 2021). Begun as a travel recommendation application in China (Fan & Zhang, 2022; Zheng, 2022), the platform has successfully transformed and extended into China's digital (fashion) market and gained its footing as a uniquely 'one-stop' platform where the nurturing of consumer needs and final purchasing can be completed within the same platform (Xiaohongshu, n.d.). This successful platform transformation can be largely attributed to its collection of user data and targeted ads, exposing users to carefully calculated and selected media content to optimise sales conversion (He, 2022; Hu et al., 2022). More significantly, contributing to this success is its use of influencer promotional culture. Influencer promotional culture, in this context, refers to everyday, ordinary Internet users who monetise their accumulated, relatively large following through personalised, opinion-laden commercial advertising for 'products/services that Influencers personally experience and endorse for a fee' (Abidin, 2015, p. 1). Nonetheless, influencers often obscure their promotional intention while exploiting the commercial potential in 'persuading' platform users and followers (Arnesson, 2023).

In the vernacular, the influencer promotional model Xiaohongshu has created is known as *zhongcao,* which can take on various forms, such as product evaluation, shopping guides, and experience sharing, and can be uploaded onto Xiaohongshu as 'journal notes' – a popular format on the platform comprising a combination of features from blogs, Pinterest, and Instagram (C. Chen et al., 2022; Sun & Ly, 2022; Y. Wang et al., 2022). The monetisation of these journal notes can be done through three primary means. The first is 'customer divergence', where Xiaohongshu users guide their followers to their business accounts on other platforms. The second uses the platform's 'in-built functions', such as livestreaming sales, shopping mall sales, and product endorsement. The third is to 'sell established accounts' for commercial use.

In addition, in comparison to other digital platforms, Xiaohongshu users are dispositionally influencers (Abidin, 2015, 2016), in the sense that they are not merely everyday internet users but entrepreneurs with clear business agendas to upload the narration of their personal experiences to accumulate large followings to profit from direct product sales, endorsements, or to themselves be sold as narrative products. As of 2022, 80% of the platform's revenue is from advertisements (Sang, 2022) and according to QianGua (n.d.), its revenue includes commercial fees for promoting brands and products, promoting influencers, and facilitating and managing influencer-business collaborations. Secondly, Xiaohongshu is particularly known for utilising traditional celebrities as digital influencers to promote fashion trends to commercially advertise certain brands and products.

Last but not least, unlike the influencer cultures that are formed by ordinary users (Abidin, 2016), Xiaohongshu has launched a new strategy to invest in traditional celebrities to promote their platform. Since 2018, celebrities handpicked by the platform from the mainstream entertainment industry are paid for uploading journal notes on Xiaohongshu, presumably enticing their fanbase to join the platform to access these contents. This initiative has continued alongside Xiaohongshu's existing programmes for its influencers and ordinary users (R. Lin & Feng, 2021; J. Wang & Gao, 2022). According to Tencent (2019), one of the biggest tech companies in China, these celebrities were paid more than RMB 1 million (approx. AUD216,000) for uploading a single post on Xiaohongshu. As more traditional celebrities joined the platform, the number of registered and monthly active users increased fivefold within a year (Yi, 2018). The trendsetters that played significant roles in starting and promoting BM style in China, including prominent film stars Yang Mi (杨幂), Ouyang Nana (欧阳娜娜), and Song Zu'er (宋祖儿), are all well-known Xiaohongshu fashion leaders. In fact, among the earlier posts and discussions about BM, instead of BM style, it is often referred to as *mìshìchuāndā* (meaning Yang Mi style fashion; hereafter *mishichuanda*).

3 Discourses about Beauty in Chinese Digital Cultures

Research findings confirm that there has been a long-lasting tendency in East Asian countries, especially in China, to pursue slim body figures and lower body weight; this occurs more so among women than men (Kennedy et al., 2004; K. Wang et al., 2020; Xie et al., 2006). In contemporary China, having a 'thin' body is popularly equated with perceptions of being healthy (Jackson et al., 2021; Jung, 2018), successful and unique (Hua, 2013), empowered (Luo, 2012), and having self-control (Luo, 2013). However, historically, thinness was not always considered to be desirable in China (Jung, 2018). During the Warring States Period, being thin and frail was indeed considered by the ruling class as a standard of beauty for women (Chō, 2012; Leung et al., 2001), whilst obesity was regarded as a sign of prosperity and beauty in the Tang Dynasty (Kawamura, 2002).

The observation that social media have significant impacts on beauty standards and users' body image is not a new thesis, although it is growing in complexity and nuance. On the one hand, research worldwide presents and cautions against social media's adverse effects on body image-related mental issues (Frederick et al., 2017; Widdows, 2018). Holland and Tiggemann (2016) observe that viewing and uploading photos and status updates contribute significantly to the normalisation of unrealistic beauty standards, which are increasingly based on particular physical characteristics such as a slim figure (Chua & Chang, 2016). Research shows that by overwhelmingly portraying extremely slim figures, social media creates strong pressure on users to pursue such ideals (Ando et al., 2021). The concept of 'digitised dysmorphia' (Coy-Dibley, 2016) underscores the gravity of collective practices, where filtered self-representations are created via digitally modified images to correct and manipulate one's appearance to fit into a normative ideal. Similarly, Ramphul et al. (2018) focus on 'Snapchat dysmorphia', where Snapchat filters are used as plastic surgery templates. The pressure to compare oneself with others is a main driving force in internalising unrealistic beauty norms.

On the other hand, social media is also believed to hold subversive potential against the hegemonic beauty ideal (Rodgers et al., 2020). For example, the body positivity (or BoPo) movement, which promotes appreciation and acceptance of all body shapes and sizes, challenges and rejects conformity to narrow perceptions of beauty norms (Cohen et al., 2019; Sastre, 2014). Despite the framing of the scholarship cited in our brief review thus far, concern over social media's impacts on beauty ideals is not limited to Western contexts. However, comparatively less attention is paid to non-Western contexts.

In China, the beauty standard is significantly impacted by the homogenised Western standard of beauty generated chiefly by the American media discourse (Lotti, 2018; T. W. Wang, 2018; G. Xu & Feiner, 2007). However, Westernization should not be considered the only, or the primary, factor as to why

Chinese women have developed such beauty norms (T. Chen et al., 2020). First, researchers argue that Asian, especially Chinese, rather than Western media, are more likely to lead Chinese women to internalise beauty standards because of their preoccupation with appearances (Jackson et al., 2016). Chinese social media and celebrity cultures contribute to the homogenisation and normalisation of beauty standards and the internalisation of unrealistic and unhealthy beauty standards in China (Y. Lin et al., 2023; A. Y. Peng, 2021; Zhang, 2012). In addition, as Leung et al. (2001) argue, the expectation of being thin is a cultural phenomenon that existed in China before and after its contact with the Western media.

In addition, (social) media, albeit having a significant impact on beauty standards, should not be considered as the only cause. In the Chinese context, the contemporary beauty standard fixated on being thin is regarded to be a result of economic development and social hierarchies. China's climb in economic power since the end of the Cultural Revolution is considered by many researchers as one of the driving forces behind contemporary Chinese women's pursuit of beauty (Hua, 2013; Johansson, 1998; Lotti, 2018). Studies confirm the tangible connections between beauty or being able to fit the beauty norm, and one's enhanced ability to secure job opportunities, higher wages, social networks, and even romantic partners in contemporary Chinese society (Gu & Ji, 2019; Hamermesh et al., 2010; Ma, 2022). Research shows that the 'beauty premium', i.e. the more attractive an employee is, the higher their income is, is even bigger for women in roles that require higher skills and higher education levels (L. Peng et al., 2020). As Qi and Cui (2018) argue regarding the impact of the media on beauty discourse, it is the socio-economic status associated with a slim figure, rather than having a slim figure itself, that impacts women's internalisation of unrealistic body images.

4 Methodology: Purposive Sampling and Content Analysis

This study is conducted based on a qualitative content analysis of a corpus of Xiaohongshu posts and comments. The data is collected manually using a newly registered Xiaohongshu account through purposive sampling of key search terms. The search terms include 'BM 女孩' (BM girl), 'BM 店员' (BM employee accounts) and 'BM'/'BrandyMelville', to offer insights into the general conversations about the BM beauty discourse and ideology. Specifically, we collected the top 50 most popular posts from the search results of each key search term on Xiaohongshu between 2020 and 2022. The years 2020 and 2022 are chosen as the collection cut-off dates because the former marks the time when the brand began to be promoted on Xiaohongshu and the latter was when it became prevalent beyond the platform and the brand. The ranking of popularity is based on the platform's built-in filter, 'sort by popularity'. Because of the inclusion of the same term BM (Brandy-Melville), in all searches, the results yielded much repetition. After a thorough process of deleting repeated content, we assembled an original corpus of 122 posts and their corresponding comments. One unit of coding as shown in Table 5.1 below is constituted by either a single post or each comment. This dataset allows us to understand how Xiaohongshu influencers facilitated promotional consumer cultures and informs us about the broader BM ecology regarding Xiaohongshu users' responses.

We use purposive sampling to identify three groups of Xiaohongshu influencers: 'BM 女孩' (BM girl), 'BM 店员' (BM employee accounts) and 'BM'/'BrandyMelville' influencers. These keywords are chosen based on influencers' knowledge and/or experience which demonstrates their close connection to research inquiries (Robinson, 2014). The purposive sampling process involves several rounds of observations, selections, and reflections (Schutt, 2018). BM employee accounts are run by people who work in BM retail shops and these BM promotional posts are part of their '(in)visibility labour', i.e. their labour to make their posts seen is hidden behind the brand name, BM (Duffy et al., 2021; Duffy & Sawey, 2022). These Xiaohongshu users are deidentified and the names we use in this paper will not be traced back to the original accounts. The analysis where the user names are the examples is an exception.

Data in this research is collected using a new Xiaohongshu account explicitly registered for this purpose; we ensure the search results are not tainted by previous viewing habits. However, it is essential to note that Xiaohongshu's functionalities make it difficult to avoid algorithm recommendations altogether because, in the registration process, the platform collects data on the user's age, gender, and 'preferences' – in particular, up to three 'preferences' must be selected from a predetermined list of 32 genres. For this research, we selected 'fashion', 'beauty', and 'fitness' as our topical preferences, as these relate most closely to the discussions around BM and the thin body ideal. As per the popular conventions on Xiaohongshu, all data collected was in Chinese. Coding was carried out in Chinese, and select data and quotes were translated into English for this chapter. Both authors are fluently bilingual in English and Chinese.

Data analysis was conducted using qualitative inductive content analysis (Schreier, 2019), where researchers analysed the corpus without predefined codes and derived different categories from the texts. Inductive content analysis is done through several rounds of coding to merge categories until reaching the point of saturation, i.e. the categories can no longer be merged (Schreier, 2019). For example, many comments following the 122 selected posts focus on the BM girl/influencer's body shape: some of the general categories of the comments include long legs, skinny waist, tall, look like a model, white/pale skin, etc. After coding each of the comments individually, we have merged these comments into one single category, 'commenting on looks' which cannot be merged with other categories under body image comments. Following this approach, we compiled a Google spreadsheet that contains all texts. The metadata collected includes user name, text content, date, and number of likes. Based on the reviewed literature in the studies of beauty discourse in relation to social media, the analysis focuses specifically on the role of Xiaohongshu influencers in facilitating BM trends, and how the influencer promotional culture on the platform shapes and (re)produces the discourse around beauty and body images. A summary of our coding schema is as follows.

TABLE 5.1 Summary of results from content analysis

	n	%
Text: Fashion style		
purchasing tips	745	22.3%
dress-up tips	398	12%
BM	357	10.7%
sweet and/or spicy style	274	8.2%
Text: Body image		
commenting on looks	437	13.3%
idolising celebrities	262	8%
admiring BM employees	211	6.3%
informing the post author for taking screenshots of their post images	159	4.7%
Image: Celebrity style		
copying the celebrities	132	30.4%
#ootd	124	28.5%
introducing celebrity fashion	89	20.5%

The analysis of BM posts and comments on Xiaohongshu was accompanied by a further examination of recent trendy beauty challenges. We also incorporated the 2021 trend of adult Chinese women posting Xiaohongshu images of themselves wearing children-sized clothes from Asian brands like Uniqlo and 'Sweet and Spicy Girl' trend (#甜心辣妹风; *tiánxīn làmèi fēng*) on Taobao. This further examination of discourse related to beauty and BM allows us researchers to unpack the dominant social values and perceptions of beauty in Chinese society and to trace how discourses about standards, values, and beliefs are shaped on Xiaohongshu, traveling to the wider digital media ecology in Chinese society.

5 BM and Xiaohongshu's Multi-layered Influencer Promotion Tactic

Table 5.1 illustrates the analysis of texts and imagery content. In terms of texts, the analysis yields two categories of comments and posts: fashion style and body image. This is the point of saturation where the two categories are distinctive and significant for further discussion and thus cannot be combined. Under each category, there are subcategories of topics that represent distinctive content of discussions. 'Purchasing tips', which refers to comments and posts that include specific shopping links, price guides and instructions on online searches for a particular product(s), account for the highest percentage of the posts. In terms of images, 'copying celebrities' accounts for nearly one-third of imagery content. Overall, the findings demonstrate how Xiaohongshu's recent strategic use of traditional celebrities interplays with influencer cultures and its economic cultural implications.

First of all, as shown in the Table, while the text analysis reveals two overarching categories of content, the images accompanying both categories focus on celebrity style, over three-quarters of the analysed images. In this analysis, the celebrities refer to traditional celebrities who are well-known film stars, singers, and/or models. This result highlights and is coherent with Xiaohongshu's commercialisation strategy – focusing on content produced by traditional celebrities alongside those by influencers and ordinary users (Zhu, 2022). The BM trend on Xiaohongshu began when fashion influencer accounts posted pictures of various traditional celebrities' BM outfits, including those from Gu Ailing (古爱凌), Liu Yifei (刘亦菲), Ouyang Nana (欧阳娜娜), Yang Ying (杨颖) and Yang Mi (杨幂). Among these traditional celebrities, Yang Mi (杨幂), especially, is one of the most prominent film stars and trendsetters in China. The equivalence between BM style and *mishichuanda* contributed significantly to the early popularisation of this brand, as many Xiaohongshu users were enthusiastic imitators of Yang Mi's fashion style. Sentiments expressing loyalty and faith in copying *mishichuanda* are prevalent throughout BM circles on Xiaohongshu:

> You can forever believe in Yang Mi's style.
>
> Xiaohongshu user 'Erin'

The 'celebrity effect' of *mishichuanda* in China entails traditional celebrities promoting a brand by wearing their clothes, and young Chinese girls subsequently aspiring to copy the fashion styles of the celebrities to feel attractive (Zhang, 2012). Such promotional strategies have been studied elsewhere among personal style bloggers in Asia, who often inspire imitations of their style among social media users (Pham, 2015).

Secondly, as shown in Table 5.1, nearly one-third of the images include someone copying traditional celebrities' outfits or fashion styles. These images were posted by BM girls and BM employees through a process of 'proximate microcelebrification' (Abidin, 2015), where these ordinary users inherited some celebrity and fame through their access to traditional celebrities like Yang Mi, and by their repeated exposure to Xiaohongshu as mediators of Yang Mi's and BM's fashion styles. For example, BM employee accounts draw popularity to themselves by imitating the fashion styles of film star Yang Mi. Some of these BM employee accounts would simply post a picture of a slim girl (implied as Yang Mi) wearing a cap and mask with the caption *#mishichuanda* associating the photographed person with the BM brand:

> Sister Mi in person is the prettiest in the world
>
> Xiaohongshu user 'huyunen'

> Mimi took the clothes I gave her.
>
> Xiaohongshu user 'huyunen'

These posts often send ambiguous messages regarding whether the person in the photo is indeed Yang Mi. They work well as a commercial tactic, as in response to these posts the comment section is often filled with envy and the prospect of meeting Yang Mi should fans drop by the BM stores.

Evidently, the success of BM's branding strategy via its deployment of sponsored traditional celebrity journal entries inevitably led to the rise of BM employees becoming among the brand's most sought-after influencers on Xiaohongshu. In the long run, the promotional labours undertaken by

BM employee-influencers were more cost-effective for the brand, as they were paid for their work as retail employees but not remunerated for their digital labour on Xiaohongshu. These employee-influencers are 'aspirational labourers' who are motivated to obtain social and economic capital through activities they feel passionate about, although the majority of them fail to gain visibility for the brand (Duffy, 2016). What this means is that while some BM employee influencers went on to acquire a sizable following on Xiaohongshu, the majority of the BM employees remain in the state between visibility and invisibility doing 'free labour' for BM while they are working at BM retail stores (Duffy, 2022). Besides the impacts on BM employee influencers, Xiaohongshu's multi-layered promotion tactic has long-term effects on beauty discourse and body images on social media.

Combined, this finding demonstrates that Xiaohongshu's strategic use of traditional celebrities creates a multi-layered influencer promotion effect. What this means is that the traditional celebrities perform the opinion leader role whereas the Xiaohongshu influencers act as mediators and imitators. The latter rise into celebrity positions because of their proximity to traditional celebrities – in other words, their ability to follow and copy celebrities. BM girls and employees' roles lie in the subsequent layer of influencer promotion, as semi-public figures who are comparatively more accessible. In this sense, the BM employees become 'employee influencers' who are at once formal retail employees at the BM shops and informal influencers promoting BM culture online and their social media labour is hidden behind BM (Duffy & Sawey, 2022).

In addition, as discussed earlier, the two highest subcategories for text and imagery, respectively, are 'purchasing tips' and 'copying celebrities'. The two top subcategories demonstrate content that relates precisely to using celebrity images to attract a large base of followers and then provide specific purchasing tips with details like direct links to guide the followers to purchase the particular product(s). This particular finding suggests Xiaohongshu's reliance on traditional celebrities, but also the streamlined conversion of traditional celebrities' social power into monetary value through proximate micro-celebrities.

Xiaohongshu's strategic use of traditional celebrities should not be considered as a trivial difference from that on other platforms, nor, in some way, a move back/away from social media influencer cultures. Instead, as this research demonstrates, such a tactic significantly elevates commercial brands'/businesses' power in steering and even controlling social media discourses, providing them with a greater capacity to steer public opinion for commercial gain, even if it is at the cost of the public. Unlike ordinary-user-turned-influencers, who need to spend time and effort to accumulate an online following, traditional celebrities, especially the ones who are handpicked by the Xiaohongshu platform, already have a substantial social influence and have already created an existing discourse that suits certain brands' commercial agenda. On Xiaohongshu's business site, Pugongying (n.d.), this is referred to as a promotional package. This is perhaps why Xiaohongshu has long been deemed as a social media platform that particularly generates homogeneity around beauty standards in China (Qiu, 2023).

6 Conclusion

While the BM trend is not chiefly responsible for the origins of the desire for slim body figures in China, it is yet another fashion trend that reproduces existing beauty standards established through several beauty challenges, which are also popularised through circuits of traditional celebrities and influencers on Xiaohongshu. Our findings confirm that celebrity cultures and the digital cultures of influencers intertwine to contribute significantly to the homogenisation of beauty standards in China (A. Y. Peng, 2021; Zhang, 2012).

Xiaohongshu may brand itself as a platform that empowers women to achieve both self-expression and income (Sun & Ly, 2022). Nonetheless, as our analysis shows, the platform's branding is in fact already a reproduction of the myth that pursuing beauty and physical alteration is to achieve self-empowerment, falsely promoted by cosmetics and medical institutions over the years (Luo, 2012). In the case of BM style, by following similar claims of empowerment, BM girls are viewed as role models who teach and provide guidance to ordinary Xiaohongshu users who need help to obtain beauty. In the BM ecology, BM employees are facilitators who help women who idolise *mishichuanda* to achieve their dream of wearing the 'exact same' clothes as Yang Mi. Within this context, Xiaohongshu influencers publish content that is already popular and accepted under the norm to draw followers (W. Wang, 2021).

This promotion of imitation under the guise of empowerment is at the centre of Xiaohongshu's most well-known business model: the *zhongcao* economy. As noted earlier, the *zhongcao* economy is a business model particularly associated with Xiaohongshu, where influencers share their 'real-life' experiences as a means to nurture or generate consumption, and thus monetise their following (Jiang, 2019; Pemarathna, 2019). As Xiaohongshu's business service site, *Pugongying* (n.d.) indicates on its 'about' page, Xiaohongshu is a trading platform for influencers to attract potential customers for businesses. One of Xiaohongshu's selling points is thus to provide platform-specific reports on influencers' trends and trendy posts based on big data calculations, to assist businesses in selecting the 'right candidate' to promote their products on their site (*Pugongying,* n.d.).

Collectively, BM girls on Xiaohongshu – whether they be traditional celebrities paid for their sponsored journal entries, BM employee influencers, BM influencers at large, or ordinary users who are fans of BM – constitute an ecology of women who leverage fashion and consumption trends as a way of policing their own and each other's body images. They drive up the barometer for the ideal 'thin body' in Chinese society. Despite some awareness from reflexive Xiaohongshu users that the BM trend fixates on one's social capital and its associated capacity to pursue thinness – rather than a grassroots interest in fashion trends per se – Xiaohongshu ultimately platforms the discourse on thinness as a positive movement, framed in the language of empowerment. As evidenced through the algorithmically recommended users and posts, and the platform's business collaboration directory offered to brands, Xiaohongshu also rewards BM girls with more in-platform visibility, and this increase in viewer traffic and follower favour can tangibly translate to monetisation opportunities with brands. As such, it is difficult to resist social norms and peer pressure, when the prospect of accumulating social capital online extends to tangible financial outcomes – so long as one 'fits in' with the ideal.

References

Abidin, C. (2015). Micro-microcelebrity: Branding Babies on the Internet. *M/C Journal,* 18(5), Article 5. https://doi.org/10.5204/mcj.1022

Abidin, C. (2016). Visibility labour: Engaging with Influencers' fashion brands and #OOTD advertorial campaigns on Instagram. *Media International Australia,* 161(1), 86–100. https://doi.org/10.1177/1329878X16665177

Abidin, C., Lee, J., Barbetta, T., & Miao, W. S. (2021). Influencers and COVID-19: Reviewing key issues in press coverage across Australia, China, Japan, and South Korea. *Media International Australia,* 178(1), 114–135. https://doi.org/10.1177/1329878X20959838

Ando, K., Giorgianni, F. E., Danthinne, E. S., & Rodgers, R. F. (2021). Beauty ideals, social media, and body positivity: A qualitative investigation of influences on body image among young women in Japan. *Body Image,* 38, 358–369. https://doi.org/10.1016/j.bodyim.2021.05.001

Arnesson, J. (2023). Influencers as ideological intermediaries: Promotional politics and authenticity labour in influencer collaborations. *Media, Culture & Society,* 45(3), 528–544. https://doi.org/10.1177/01634437221117505

Chen, C., Ouyang, H., Tan, J., Wu, F., & Zhan, Y. (2022). Based on COVID-19, analysis of the strategic model of e-commerce enterprises – A case study of Xiaohongshu (RED). *Journal of Psychology Research*, 12(8): 628–636. https://doi.org/10.17265/2159-5542/2022.08.008

Chen, T., Lian, K., Lorenzana, D., Shahzad, N., & Reinesse Wong. (2020). *Occidentalisation of beauty standards: Eurocentrism in Asia* 1(2): 1–11. https://doi.org/10.5281/ZENODO.4325856

Chō, K. (2012). *The search for the beautiful woman: A cultural history of Japanese and Chinese beauty*. Rowman & Littlefield.

Chua, T. H. H., & Chang, L. (2016). Follow me and like my beautiful selfies: Singapore teenage girls' engagement in self-presentation and peer comparison on social media. *Computers in Human Behavior*, 55(A), 190–197. https://doi.org/10.1016/j.chb.2015.09.011

Cohen, R., Fardouly, J., Newton-John, T., & Slater, A. (2019). #BoPo on Instagram: An experimental investigation of the effects of viewing body positive content on young women's mood and body image. *New Media & Society*, 21(7), 1546–1564. https://doi.org/10.1177/1461444819826530

Coy-Dibley, I. (2016). "Digitized Dysmorphia" of the female body: The re/disfigurement of the image. *Palgrave Communications*, 2(1), Article 1, 1–9. https://doi.org/10.1057/palcomms.2016.40

Duffy, B. E. (2016). The romance of work: Gender and aspirational labour in the digital culture industries. *International Journal of Cultural Studies*, 19(4), 441–457. https://doi.org/10.1177/1367877915572186

Duffy, B. E., Pinch, A., Sannon, S., & Sawey, M. (2021). The Nested Precarities of Creative Labor on Social Media. *Social Media + Society*, 7(2), 1–12. https://doi.org/10.1177/20563051211021368

Duffy, B. E., & Sawey, M. (2022). In/Visibility in social media work: The hidden labor behind the brands. *Media and Communication*, 10(1), 77–87. https://doi.org/10.17645/mac.v10i1.4460

Fan, L., & Zhang, D. (2022). Study on the hotspots of urban tourism spaces based on Instagram-Worthy locations data: Taking Beijing as an example. *Environment and Planning B: Urban Analytics and City Science* 50(7), 1822–1837. https://doi.org/10.1177/23998083221146542

Frederick, D. A., Daniels, E. A., Bates, M. E., & Tylka, T. L. (2017). Exposure to thin-ideal media affect most, but not all, women: Results from the Perceived Effects of Media Exposure Scale and open-ended responses. *Body Image*, 23, 188–205. https://doi.org/10.1016/j.bodyim.2017.10.006

Gong, S. (2022). Research on the digital business model of Xiaohongshu under the Background of the SHE Economy. Proceedings of the 2022 7th International Conference on Social Sciences and Economic Development (ICSSED 2022), Advances in Economics, Business and Management Research, 215, 1241–1244. https://doi.org/10.2991/aebmr.k.220405.205

Hamermesh, D. S., Meng, X., & Zhang, J. (2010). Discrimination and development: The case of beauty in China. In *Investing in Human Capital for Economic Development in China* (pp. 377–390). WORLD SCIENTIFIC. https://doi.org/10.1142/9789812814425_0021

He, Z. (2022). How to get out of the commercialization maze of a professional "grass-planting" content community: The commercialization of Xiaohongshu's "Filter." *Frontiers in Business, Economics and Management*, 4(2), Article 2. https://doi.org/10.54097/fbem.v4i2.788

Holland, G., & Tiggemann, M. (2016). A systematic review of the impact of the use of social networking sites on body image and disordered eating outcomes. *Body Image*, 17, 100–110. https://doi.org/10.1016/j.bodyim.2016.02.008

Hu, X., Jin, T., Lu, Y., & Zhong, S. (2022). *The interest-based communities on Xiaohongshu recreate the era of "tribalization."* Proceedings of the 2022 6th International Seminar on Education, Management and Social Sciences (ISEMSS 2022), Advances in Social Science, Education and Humanities Research 687, 711–720. https://doi.org/10.2991/978-2-494069-31-2_86

Hua, W. (2013). *Buying beauty: Cosmetic surgery in China*. Hong Kong University Press.

Jackson, T., Jiang, C., & Chen, H. (2016). Associations between Chinese/Asian versus Western mass media influences and body image disturbances of young Chinese women. *Body Image*, 17, 175–183. https://doi.org/10.1016/j.bodyim.2016.03.007

Jackson, T., Ye, X., Hall, B. J., & Chen, H. (2021). "Have you taken the A4 Challenge?" Correlates and impact of a thin ideal expression from Chinese social media. *Frontiers in Psychology*, 12, 1–13. https://www.frontiersin.org/articles/10.3389/fpsyg.2021.669014

Jiang, Y. (2019). "小红书" 的市场营销策略研究_江洋.pdf. *Yingxiaotanmi*, 8, 252.

Johansson, P. (1998). White skin, large breasts: Chinese beauty product advertising as cultural discourse. *China Information*, 13(2–3), 59–84. https://doi.org/10.1177/0920203X9801300204

Ju, R. (2022). Producing entrepreneurial citizens: Governmentality over and through Hong Kong influencers on Xiaohongshu (Red). *Policy & Internet*, 14(3), 618–632. https://doi.org/10.1002/poi3.324

Jung, J. (2018). Young women's perceptions of traditional and contemporary female beauty ideals in China. *Family and Consumer Sciences Research Journal*, 47(1), 56–72. https://doi.org/10.1111/fcsr.12273

Kawamura, K. Y. (2002). Asian American body image. In T. Cash & T. Pruzinsky (Eds.), *Body image: A handbook of theory, research, and clinical practice* (pp. 243–249). Guilford Press.

Kennedy, M. A., Templeton, L., Gandhi, A., & Gorzalka, B. B. (2004). Asian body image satisfaction: Ethnic and gender differences across Chinese, Indo-Asian, and European-Descent Students. *Eating Disorders*, 12(4), 321–336. https://doi.org/10.1080/10640260490521415

Leung, F., Lam, S., & Sze, S. (2001). Cultural expectations of thinness in Chinese women. *Eating Disorders*, 9(4), 339–350. https://doi.org/10.1080/106402601753454903

Lian, K., Chen, Z., & Zhang, H. (2021). *From the perspective of feminism: Market positioning of Xiaohongshu*. 190–194. https://doi.org/10.2991/assehr.k.210806.037

Lin, R., & Feng, Z. (2021). Research on the application strategy of online opinion leaders in virtual community brand word-of-mouth marketing. *Tourism Management and Technology Economy*, 4(1), 7–12. https://doi.org/10.23977/tmte.2021.040102

Lin, Y., Lu, C., Huang, Z., Barnhart, W. R., Cui, T., & He, J. (2023). Exploring the links between celebrity worship, body dissatisfaction, and disordered eating among young adult celebrity worshippers in China. *Body Image*, 45, 210–218. https://doi.org/10.1016/j.bodyim.2023.03.006

Lotti, V. (2018). The image of the beautiful woman: Beauty ideals in modern urban China. *ASIEN: The German Journal on Contemporary Asia*, 147, Article 147, 92–104. https://doi.org/10.11588/asien.2018.147.14492

Luo, W. (2012). Selling cosmetic surgery and beauty ideals: The female body in the web sites of chinese hospitals. *Women's Studies in Communication*, 35(1), 68–95. https://doi.org/10.1080/07491409.2012.667055

Luo, W. (2013). Aching for the altered body: Beauty economy and Chinese women's consumption of cosmetic surgery. *Women's Studies International Forum*, 38, 1–10. https://doi.org/10.1016/j.wsif.2013.01.013

Ma, H. (2022). Inner and outer beauty: Exploring female beauty in contemporary China. *Journal of Gender Studies*, 0(0), 1–13. https://doi.org/10.1080/09589236.2022.2070463

Pemarathna, R. (2019). *Impact of Xiaohongshu on its user based and society: A review. Iconic Research And Engineering Journals* 2(11), 285–294.

Peng, A. Y. (2021). A techno-feminist analysis of beauty app development in China's high-tech industry. *Journal of Gender Studies*, 30(5), 596–608. https://doi.org/10.1080/09589236.2021.1929091

Peng, L., Wang, X., & Ying, S. (2020). The heterogeneity of beauty premium in China: Evidence from CFPS. *Economic Modelling*, 90, 386–396. https://doi.org/10.1016/j.econmod.2019.12.014

Pham, M.-H. T. (2015). *Asians wear clothes on the internet: Race, gender, and the work of personal style blogging*. Duke University Press. https://doi.org/10.2307/j.ctv11g96t3

Qi, W., & Cui, L. (2018). Being successful and being thin: The effects of thin-ideal social media images with high socioeconomic status on women's body image and eating behaviour. *Journal of Pacific Rim Psychology*, 12, e8. https://doi.org/10.1017/prp.2017.16

Qian-Gua (2022). *2022* 年活跃用户画像趋势报告(小红书平台). [2022 Active User Portrait Trend Report (Xiaohongshu Platform)]. CBN Data, 23 March. https://www.cbndata.com/report/2891/detail?isReading=report&page=14

Ramphul, K., Mejias, S. G., Ramphul, K., & Mejias, S. G. (2018). Is "Snapchat Dysmorphia" a Real Issue? *Cureus*, 10(3). https://doi.org/10.7759/cureus.2263

Rodgers, R. F., Meyer, C., & McCaig, D. (2020). Characterizing a body positive online forum: Resistance and pursuit of appearance-ideals. *Body Image*, 33, 199–206. https://doi.org/10.1016/j.bodyim.2020.03.005

Sang, M. (2022). 我为什么越来越看不懂小红书了 | 人人都是产品经理 [Woshipm]. [Why I can't understand Xiaohongshu anymore]. Woshipm, 12 July. https://www.woshipm.com/it/5522369.html

Sastre, A. (2014). Towards a radical body positive: Reading the online "body positive movement." *Feminist Media Studies*, 14(6), 929–943. https://doi.org/10.1080/14680777.2014.883420

Schreier, M. (2019). Content analysis, Qualitative. In P. Atkinson, S. Delamont, A. Cernat, J. W. Sakshaug, & R. A. Williams (Eds.), *SAGE Research Methods Foundations*. SAGE Publications Ltd. https://doi.org/10.4135/9781526421036753373

Sun, Y., & Ly, T. P. (2022). The influence of word-of-web on customers' purchasing process: The case of Xiaohongshu. *Journal of China Tourism Research*, 0(0), 1–24. https://doi.org/10.1080/19388160.2022.2057378

Tencent (2019). 明星小红书带货、*KOL* 小红书带货哪个更适合你的产品 – 腾讯云开发者社区-腾讯云 [Which one is more suitable for your product: celebrity promotion or KOL promotion?] Tencent.com, 20 August. https://cloud.tencent.com/developer/news/421352

Wang, J., & Gao, M. (2022). *Innovative Analysis of Internet Enterprise Management Mode Under the Background of Big Data – Taking Xiaohongshu as an Example*. 232–237. https://doi.org/10.2991/aebmr.k.220603.040

Wang, K., Liang, R., Yu, X., Shum, D. H. K., Roalf, D., & Chan, R. C. K. (2020). The thinner the better: Evidence on the internalization of the slimness ideal in Chinese college students. *PsyCh Journal*, 9(4), 544–552. https://doi.org/10.1002/pchj.346

Wang, T. W. (2018). Children's Westernized beauty ideals in China: Notions of feminine beauty. In A. M. Kraehe, R. Gaztambide-Fernández, & B. S. Carpenter ii (Eds.), *The Palgrave Handbook of Race and the Arts in Education* (pp. 423–439). Springer International Publishing. https://doi.org/10.1007/978-3-319-65256-6_24

Wang, W. (2021). The factors affecting the willingness of e-commerce consumer information sharing in the new media era – Take "Xiao Honshu" for example. *2021 3rd International Conference on Economic Development and Management Science (EDMS 2021)*. International Conference on Economic Development, Dalian, China.

Wang, Y., Fardouly, J., Vartanian, L. R., Wang, X., & Lei, L. (2022). Body talk on social networking sites and cosmetic surgery consideration among Chinese young adults: A serial mediation model based on objectification theory. *Psychology of Women Quarterly*, 46(1), 99–110. https://doi.org/10.1177/03616843211026273

Wernick, A. (1991). *Promotional culture: Advertising, ideology and symbolic expression*. SAGE.

Widdows, H. (2018). *Perfect me: Beauty as an ethical ideal*. Princeton University Press. https://doi.org/10.2307/j.ctvc77j2q

xiaohongshu (n.d.). *Content Marketing Platform | Influencer Marketing | Xiaohongshu*. Retrieved May 29, 2023, from https://www.xiaohongshu.com/en/business

Xie, B., Chou, C.-P., Spruijt-Metz, D., Reynolds, K., Clark, F., Palmer, P. H., Gallaher, P., Sun, P., Guo, Q., & Johnson, C. A. (2006). Weight perception and weight-related sociocultural and behavioral factors in Chinese adolescents. *Preventive Medicine*, 42(3), 229–234. https://doi.org/10.1016/j.ypmed.2005.12.013

Xu, G., & Feiner, S. (2007). Meinü Jingji/China's beauty economy: Buying looks, shifting value, and changing place. *Feminist Economics*, 13(3–4), 307–323. https://doi.org/10.1080/13545700701439499

Xu, J., & Zhao, X. (2019). Changing platformativity of China's female Wanghong: From Anni Baobei to Zhang Dayi. In S. Cai (Ed.), *Female Celebrities in Contemporary Chinese Society*. Springer Singapore. https://doi.org/10.1007/978-981-13-5980-4

Ye, P., & Rodriguez, C. Q. (2023). *Personal Style Promotional Strategies of Chinese Fashion Micro-influencers*.

Yi, S. (2018). 明星批量入驻小红书,到底有什么隐情? | 人人都是产品经理. [What's the secret behind the large number of celebrities joining Xiaohongshu?] *Woshipm.com*, 19 October. https://www.woshipm.com/it/1531253.html

Zhang, M. (2012). A Chinese beauty story: How college women in China negotiate beauty, body image, and mass media. *Chinese Journal of Communication*, 5(4), 437–454. https://doi.org/10.1080/17544750.2012.723387

Zheng, Y. (2022). Research on the branding development of rural tourism destinations from the perspective of product level. *Frontiers in Business, Economics and Management*, 5(1), Article 1. https://doi.org/10.54097/fbem.v5i1.1576

Zhu, Y. (2022). Brief analysis on the development of grass-knitting industrialization. *Asian Journal of Social Science Studies*, 7(4), Article 4, 94–96. https://doi.org/10.20849/ajsss.v7i4.1098

CHAPTER 6

Privacy Expectations and Norms: Perceptions of Individual Digital Activists in Turkey's Xsphere (Twittersphere)

Yusuf Yüksekdağ and Sarper Durmuş

1 Introduction

As we discuss informational privacy in the digital age within higher education, we often encounter a mix of apathetic acceptance and discontent among students regarding the constant monitoring and surveillance of their lives. A similar sentiment toward this dialectical digital culture was raised by a digital activist who once lamented the demise of their private life on X (formerly Twitter), attributing it to the socio-political pressures in Turkey. This remark was made in response to criticism for not being in the streets, highlighting what they give up even for their digital endeavours. This sentiment reflects a debate on the trade-off between the value of the activist pursuit and caring less for one's informational privacy. However, it begs the question: is this a shared balancing and norm attribution for digital activists in Turkey using popular digital platforms? There have been several studies on how different groups of digital users (e.g. young groups, high school students, Facebook users, non-Western users) perceive informational privacy (Gogus & Saygın, 2019), its social value (Steeves & Regan, 2014), the effects of cultural values on privacy settings and perceptions (Liang et al., 2017; Abokhodair & Vieweg, 2016) and the contextual concerns surrounding it (Steeves & Regan, 2014). Additionally, studies explore the proactive measures users undertake to safeguard their privacy or circumvent data capture (Paine et al., 2007; Zhang et al., 2020; Wulf et al., 2022).

In this study, we take another perspective focusing on how individual users perceive the *value* of their informational privacy. The perception of informational privacy among digital activists, who face heightened risks to their privacy and security, requires further exploration to underscore its urgency (Sanfilippo & Strandburg, 2021). Arguably, the platforms the activists use might have implications on how this transpires. This is the core concern of the technological mediation approach that takes technology not only as objects but as affordances through which we mediate our

 | DOI:10.1163/9789004711396_008

understandings of the world (Verbeek, 2016). This constitutive role of technology has hitherto been questioned in the works of Heidegger, Ihde and Latour (Verbeek, 2009). Mainly focusing on norms and values, Verbeek (2016) also suggests how technologies mediate our moral values and norms in relevant spaces. For example, pervasive data capture can simultaneously diminish and reinvigorate the value placed on privacy.[1] Specifically, platforms like X might prompt reconsideration of what information is private or public (Steeves & Regan, 2014). The latest perspectives on technology's mediating role and the empirical shift in the philosophy of technology emphasise the unique influence of each technological artefact, moving away from broad generalisations like *social media* (Verkerk et al., 2016). Perceptions toward privacy might then be shaped depending on the activity one pursues as much as where one pursues it, be it the socio-political setting or the particular digital platform.

This chapter aims to understand how individual digital activists using X in Turkey perceive the value of informational privacy and how they balance privacy norms with other norms in pursuing activism. While concerns of activist groups over their privacy as a collective or privacy risk of marginalised groups have been surveyed (Lerner et al., 2020; Jia & Baumer, 2022), individual perceptions toward, in particular, the value of privacy are still somewhat uncharted territory, especially in non-Western contexts. Similar to Steeves and Regan's (2014) framework on the social value of privacy for young digital users, we rely on existing (normative) accounts (e.g. control, ownership, ontological and contextual) on the value of privacy. Leaning on these accounts, particularly on Nissenbaum's (2010) contextual integrity, we discuss how activists' perceptions resonate with the accounts.

We should note that privacy, or any other norm, is considered to be culturally shaped and understood (Liang et al., 2017; Yao-Huai, 2005). Accounts emphasising control and ownership are based upon universal and liberal takes on the value of privacy (Masur, 2020). On the other hand, Nissenbaum's (2010) contextual integrity approach focuses on context-relative privacy norms. This approach considers the relevance of the specific activity of sharing, the platform where the information flow transpires, and the existing socio-cultural meanings attributed to it by the subjects of the flow. As Steeves and Regan (2014) put forward, social meanings attributed to various

---

1 This example is taken from John Danaher's (2021, February 25) blog entry. See https://philosophicaldisquisitions.blogspot.com/2021/02/the-technological-mediation-of-morality.html

digital interactions and the information flow also tend to shape privacy norms.

Our focus on the digital activist in Turkey attempts to contribute to understanding non-Western settings, not simply as a matter of comparing them with Western counterparts. We are merely interested in what these particular interactions – taking on a particular activity (e.g. digital activism) and in a particular context (e.g. being a digital activist using X in Turkey) – bring about regarding how users perceive privacy norms. Our focus on X and Turkey is not only based on taking on yet another case but also on questioning the particular norm-balancing X might bring about for the concerned activists in Turkey. This aligns with the call for decolonising privacy studies, suggesting an epistemological break away from essentialist takes on the value of privacy (Arora, 2019). We attempt to avoid demographic generalisations or comparisons with naturalised Western-based, middle-class and ethnocentric perceptions by focusing on the particular perceptions of eight queer, green, feminist or migrant activists in Turkey.

In section one, we delve into revisiting the concept of privacy, its different conceptions and values while emphasising the significance of individual perceptions. In section two, we take on the digital activist-privacy nexus while contextualising such concerns for X and non-Western settings like Turkey. In sections three and four, we set out our methodology and analysis, emphasising the contextual privacy norms and factors shaping identity formation, activist pursuits, and the informational flow of digital activists. Our interviews reveal an interplay between theoretical constructs of the value of privacy and digital activists' lived experiences. While the narratives somewhat align with established theories, such as control and restricted space, they mostly echo Floridi's (2005) ontological perspective and Nissenbaum's (2010) contextual integrity approach, highlighting context-based considerations over the value of privacy embedded in the activists' roles and the socio-political landscape of Turkey.

2 Revisiting (Informational) Privacy

Digital activists' perceptions of privacy and its value are not merely based on personal preferences but are intricately intertwined with their activist pursuits, making their insights particularly valuable. What, then, exactly are the different reasons for valuing privacy? Revisiting the accounts on the value of informational privacy provides the thematic guidance for this study.

2.1 *Informational Privacy and Its Value*

While there are physical, decisional and psychological conceptions of privacy concerned with being free of any form of intrusion regarding those dimensions, since the 1960s, the concept of informational privacy has received critical attention given the emergence of Information Communication Technologies (ICTs) (Tavani, 2007; Mai, 2016). The concept has been mainly discussed regarding (i) its conceptual features (Introna & Pouloudi, 1999), (II) its value from a normative standpoint (Solove, 2008; Tavani, 2007; Nissenbaum, 2010; Marmor, 2015), and (III) the understandings it puts forth about loss or protections of privacy considering the constant flow of information in digital and data-driven communication platforms (Floridi, 2006; Mai, 2016).

Regarding its conceptualisation, informational privacy generally represents an individual's ability and capacity to control information about themselves. The definition opens up specific conceptual challenges regarding what counts as information, what is considered personal, and how to understand interspatiality between public and private (Mai, 2016). Personal data, for instance, refers to anything individuals deem confidential or sensitive, or it might be subject to the question of whether the access to that information is public or not (Nissenbaum, 2004).

Privacy is a normative buzzword. It is assumed to have deep normative significance, fuelling continuous efforts to refer to how it will be respected and augmented. What exactly are the values attributed to informational privacy? Many accounts refer to the notion and value of privacy as messy, multi-faceted and complex (Floridi, 2006; Nissenbaum, 2010). There are many different values and rights associated with the notion of privacy, sometimes to the point of making privacy devoid of any particular and tangible concern or mixing up the protection of privacy with the protection of certain forms of conduct. For the latter, as Marmor (2015) argues, the right to privacy should be targeted to protect particular interests resting on informational flow that does not concern arguably private conduct such as having children, using contraceptives, or pursuing an intimate relationship. Such issues of conduct are better understood as matters of personal autonomy. If not, privacy interests might boil down to not relying on particular interests (Marmor, 2015, p. 25). It is best at this point to delineate some of the norm-based accounts of privacy in the pursuit of such particular interests.

The perspectives based on control and restricted access posit that informational privacy is fundamentally about an individual's ability to control and limit access to their personal information or personal space (Solove, 2008; Tavani, 2007; Mai, 2016). It emphasises the autonomy and agency of the individual in determining who gets access to their data and under what conditions (Becker, 2019). Whether digital activists emphasise control over their personal data or question the lack of autonomy in sharing, it would align with this norm-based perspective on privacy. This viewpoint also intertwines with demands for enhanced digital and data literacy to enhance users' privacy control abilities and capacities (Carmi & Yates, 2020). On the other hand, the ownership perspective views personal data as a property belonging to the individual (Floridi, 2006). From this viewpoint, individuals should have the same rights over their data as any personal possession. Unlike the control and restricted access theory, this approach frames data explicitly as property. The extent to which digital activists view privacy as an extension and protection of the ownership of their online data would fall under this framework. One potential setback of these approaches is that they rely on a universal understanding of the value of privacy. This is a matter of universality or contextuality in ethics and normative theorising that goes even to the point of being subjected to the criticism of being Western-based and hegemonic (Donnelly, 2007).

Notably, some perspectives take privacy beyond the scope of solely how data is handled. Floridi's (2005) ontological approach suggests that privacy is essential for personal identity and self-construction. It is not just about data but the self and one's relationship with society. This conception moves beyond data as a tangible entity, focusing on the existential implications of privacy given that one's personal information is not considered external to the self but a constitutive part of it (Floridi, 2006). How would privacy concerns or lack of privacy impact an activist's sense of self or identity? We are therefore also interested in whether and how their online presence contributes to their identity and self-construction. No doubt, being a digital activist is not merely an external pursuit but propagated as part of one's identity. The issues of anonymity, local norms and challenges, and the private-public dichotomy might all play a role in how privacy norms are understood and balanced, with respect to the particular socio-political agenda an activist embraces. How the identity is constructed and performed, and the extent to which privacy concerns take shape, might also depend on the purpose of digital activism and the socio-political background in which the activism is performed. To that end, we also emphasise the need to explore taking privacy as contextual integrity, an approach developed by Nissenbaum (2010).

This theory posits that privacy expectations and norms are context-dependent (or pursuit-dependent), and it is essential to consider the context of information flow when evaluating the appropriateness of a privacy norm (Nissenbaum, 2010). According to Nissenbaum (2010), context refers to a structured social setting that changes over time and is influenced by various contingencies such as function, space, culture, and historical events (Haltaufderheide et al., 2023). What is considered a privacy violation in one context might be acceptable in another, given that norm appropriateness would depend on the activity. So, the norms relevant to the context of the information flow (e.g. hospital, confession) would feed into the privacy norms. This approach does not view privacy in absolute terms but within specific situations and contexts. It should be noted that Nissenbaum's context does not directly refer to the wider socio-political context but rather the activity/pursuit in question. At the same time, existing socio-cultural shared meanings play a role as part of that very context. Nissenbaum's perspective posits the risk of disregarding allegedly universal values attributed to privacy by focusing on some other pursuits and their values. This can reinforce existing social norms without sufficient scrutiny (Rule, 2019). However, this presents an opportunity to understand how digital activists balance privacy values in specific interactions on a particular platform.

Both the thicker normative frameworks, such as control/restricted access, and more flexible ones, such as ontological and context-dependent approaches, form the basis of our study for surveying the reasons digital activists value informational privacy. Nevertheless, we especially value the latter approaches as they are more grounded in user-oriented and interaction-based perspectives through which individuals' perceptions matter more. While generally conceptualised as an ability and capacity, informational privacy can also be considered a process through which individuals make sense of the value of privacy within the context of a particular activity. Individuals pursue digital activism via a respective technological affordance such as X; the governing nature of their activities and information flow gains importance and is relevant to individual interactions within such technologies.

3 Digital Activism and Privacy Concerns

Digital activism refers to the numerous ways of engaging in political participation, protesting or creating social movements, and creating subcultures that revolve around novel ideas achieved through digital technologies. The phenomenon has been criticised in terms of how it can create bubbles with low reach (Karpf, 2010) and how it is still a challenge for critical voices to be heard unless they are being 'reposted' by mainstream media journalists (Poell, 2014). It nevertheless posits opportunities for self-representation, especially for underprivileged groups (Cammaerts, 2015). Most definitions of contemporary activism underline digital platforms and touch upon how digital activism is contentious, particularly relative to geographies and political settings (Kaun & Uldam, 2018).

Digital activism might be conceptualised as digitally saturated collective actions and social movements, as in the case of #EndSARS in Nigeria, which culminate in organised street protests (Dambo et al., 2022). Another example is WhatsApp's increasing use, especially in the Global South, for political conversations and coordinating resistance movements (Milan & Barbosa, 2020). There is also the phenomenon of individual curators and personas in X. Such individualised endeavours are subject to criticisms of 'slacktivism', mass amateurisation, or lack of collective action (Shirky, 2008; Madison & Klang, 2020). As Madison and Klang (2020) suggest, accusations of slacktivism might downplay the reality of the everyday interplay in political action between online and offline spheres. It might also understate the limited spaces where activists might be subjugated in non-Western contexts (Milan & Barbosa, 2020). While not entirely disregarding such criticisms, our core concern is the individual perceptions of informational privacy and its value, usually understood as an individual's capacity and freedom. That is why we are primarily interested in individual digital activists who mediate their personas and personal endeavours in X.

Digital activism and privacy perceptions are essential research areas, given the role of online platforms in activism and the risks activists face in online spaces. Surveillance can lead to self-censorship and decreased online participation in contentious issues (Penney, 2017). The use and effectiveness of privacy tools like VPNs, Signal, and Tor are other points of contention (Lokot, 2020). However, the socio-political configuration and the platform in use do matter concerning how platform policies function (Stern & Kumar, 2014), how digital activists utilise particular platforms in a specific context like Turkey (Tufekci, 2017), and the regional challenges faced by the activists (Howard & Hussain, 2013). To that end, the question of digital activism in non-Western contexts regarding its potential connections to differing understandings of privacy warrants a little more scrutiny.

3.1 *Privacy Concerns (for Digital Activists) in Non-Western Contexts*

One particular study on the perceptions of privacy of digital activists in the U.S. is by Sanfilippo and Strandburg (2021), who use Facebook as a case in point. They discuss privacy governing knowledge in public Facebook groups for political activism and rely on Solove's (2008) taxonomy to measure the values behind privacy. However, the crux of the study is how privacy perceptions shape digital activism rather than vice versa. Regarding non-

Western contexts, these studies point out the language barriers, public-private dichotomy, and trade-offs between privacy and outreach as primary privacy-related issues for digital activists.

In non-Western contexts, navigating the privacy settings of social media platforms, particularly when translated into languages other than English, seems frustrating (West, 2017). Considering how the protection of digital privacy is usually considered an individual responsibility, cultural determinants of data literacy also gain importance (Kazansky, 2015). To that end, privacy and security awareness tools have also been tested and discussed in polarised settings (Tadic et al., 2018). In more authoritarian contexts, privacy seems to be associated with breaking the wall of fear in the pursuit of activism (West, 2017). Milan and Barbosa (2020) point out how what is considered public and private for individuals who use WhatsApp for social mobilisation is also a relevant affordance effect to explore. While their study focuses on underprivileged young individuals in the Global South, Bernadas and Soriano (2018) take on the significance of access to diverse forms of connectivity and social determinants of privacy behaviour.

While these studies focusing on privacy behaviours are important, we would like to take a step back and survey the perceptions of the reasons digital activists in Turkey value informational privacy as they construct it themselves. This would be a significant addition to digital users' perspectives on privacy in non-Western contexts.

4 *X*, Data Practices and Privacy Concerns

What about the potential implications of X regarding how the very design of their affordances impacts how digital activists perceive their informational privacy? The platform's expansive reach allows activists to amplify their voices, galvanise support, and engage with a global audience. However, the platform's situatedness in data capitalism and data-sharing practices merits a brief critical take.

X, after all, is a private enterprise prioritising profit; it is in complete control of its opaque algorithms. This has implications for its role in shaping digital public spheres and the reach of digital activism. It is not a neutral platform that happens to collect data about various speech acts, meanings and behaviour. Its entire logic and reason for existence is turning data into profit in one way or the other. Zuboff (2019) extensively documents the business model of interest targeting, asserting that surveillance capitalism appropriates human experiences as raw material for generating behavioural data. In their work, Couldry and Mejias (2019), adopting the Marxist notion of relations of production, examine how the conversion of everyday life into data hints at data relations that justify and perpetuate exploitative, profit-driven socio-economic dynamics, such as commodification.

According to Zuboff (2019), the platforms established by tech companies do not function as social, inclusive, or democratic spaces; instead, surveillance capitalism has successfully converted various forms of digital interactions into avenues for commercial profit by the tech companies. This perspective aligns with other studies that view data collection as an inherent aspect of the business logic of Silicon Valley companies (boyd & Crawford, 2012; Kitchin, 2014). Accordingly, criticism has been directed at the wide-ranging and omnipresent algorithms of these corporations, often referred to as black boxes (Pasquale, 2015), essentially constituting data industries that continuously gather and analyse individual-level data on people's actions, behaviours, and expressions (Couldry & Powell, 2014).

Furthermore, these platforms might simultaneously dictate what forms of communication and interactions are possible through the very design of their affordances. Users may find ways to utilise affordances unexpectedly, but ultimately, what is possible is determined by the owners of a given platform. Notably, Verbeek (2009), following his technological mediation approach based on Ihde's works, accommodates this concern that technologies are not neutral instruments and carry out certain intentionalities. At the same time, being sceptical toward full-blown technological determinism, Verbeek (2009, p. 229) argues that the intentionalities of artefacts are not fixed but are shaped by humans' relationships with them.

It is challenging to pinpoint at this stage how the structure of this medium has an effect on user privacy norm balancing from the way private messages are received to character count, and from hashtag searches to correlate one's ideological disposition to media sharing practices (Cristofaro et al., 2012). This is one of the aims of this study, after all. While X provides features like protected tweets and controlled visibility, the default public nature of the platform can pose challenges for activists who may sometimes require anonymity or discretion due to the sensitive nature of their work. Additionally, X's data – from content interactions to geolocation – can be particularly concerning for activists at risk of surveillance or targeting by state actors or opposing groups.

Data Protection Regulation (GDPR) in the EU and the California Consumer Privacy Act (CCPA) have limited data collection by tech companies in the Western world. As GDPR became a benchmark, it raised issues regarding how it can flatten privacy values across cultures and contexts

(Arora, 2019). However, the Turkish Law on Protection of Personal Data (KVKK) enacted in 2016 lacks clarity on what happens after a user consents to a privacy agreement. Users are unaware of the data collected and traded with third parties. KVKK grants the right to inquire, request redaction, and demand removal, but these are often disregarded. Government officials have also used the law to omit their names from the web or news articles to avoid public scrutiny.

4.1 *A Brief Take on Turkey*

A brief discussion on the socio-political setting in Turkey over the last decade is also needed to put things into perspective. The erosion of Turkish democracy has been rapidly worsening for over a decade (Özbudun, 2015). Critics attribute this trend to the global rise of authoritarianism following the 2008 financial crisis, which has undermined liberal reform agendas (Hellmeier et al., 2021). However, internal factors such as power struggles among the political elite, culminating in the failed coup attempt of 2016, have significantly weakened the rule of law and restricted freedom of expression during the Justice and Development Party's (AKP) second decade in power. Many observers have labelled this as deepening authoritarianism (Yılmaz & Turner, 2019), using explanatory frameworks like competitive authoritarianism and authoritarian neoliberalism (Altınörs & Akçay, 2022). The state's heavy-handed response to political activism and civil society activities that do not align with the government's ideology has also severely impacted civil society and all forms of protest (Yabanci, 2019; Zihnioğlu, 2019).

In Turkey, freedom of expression, media independence, and journalistic integrity have always come with limitations. The state and its power holders frequently depict routine scrutiny of institutions or policies as actions that threaten the state's unity, a narrative that spans the entirety of the republican period (Alemdar, 2014). While Turkey's record in this regard was never a source of international pride, developments in the 2000s made it arguably worse. In 2002, when the AKP assumed power, Turkey held the 100th position in the World Press Freedom Index (Reporters Without Borders, 2016). However, by 2021, Reporters Without Borders had ranked Turkey 153rd out of 180 countries, accompanied by a cautionary note regarding 'the risk of imprisonment and the fear of being subjected to judicial control or stripped of one's passport is ever-present' (Reporters Without Borders, 2021).

While media organisations faced escalating pressures during the AKP's rule, the internet has provided a relative haven for various political voices and activists (Akser & Baybars, 2023). This latter group is of particular significance because the crackdown on civil society and most non-governmental organisations (NGOs) has reduced civil society's traditional role as an outlet for dissent to becoming a conveyor of the government's official narrative. This led to the emergence of what can be described as a tamed civil society (Yabanci, 2019). On the other hand, similar to how government critics and journalists operated on social media, various forms of dissident activism also gained momentum on digital platforms. Diverse voices from marginalised groups, such as feminists, LGBTI+ individuals, and other minority community members, could express their grievances about the state on social media.

5 Methodology

We have conducted semi-structured interviews in Turkish to explore the viewpoints of activists from Turkey. In selecting our interviewees, we employed a purposive sampling method along with a criterion sampling approach to identify activists who met the criteria of being part of the civil opposition and having at least more than 7,000 followers on X. This group encompasses eight individuals engaged in human rights advocacy, feminism, women's empowerment, LGBTI+ and queer activism, environmental and climate activism, and migrant solidarity activism. Given the sensitive endeavours the participants deal with – which also accounts for the low participation number – we choose not to disclose more demographic information. While one might contend that eight interviews could be perceived as a limited number, we observed recurring themes and patterns, suggesting that data saturation was partially reached (Guest et al., 2006). We should also note that since the study population is generally relatively small and difficult to recruit for this sensitive research, this resulted in a low participation rate.

The interviews focused on several key themes related to informational privacy in the digital age. Firstly, we aim to gain insights into their general views on privacy. We are interested in how they define and value it and its role in online interactions. Next, we explore how these activists respond to privacy regulations on X. How do they interpret these rules, and how do they affect their online behaviours? Third, we consider how these activists balance their privacy concerns with the values and demands of their activism within the Xsphere. It is essential to understand how they prioritise and navigate these potentially conflicting interests mediated via interactions in X. How do routine online engagements shape their views and values about privacy? We also look into activists' challenges as they move between the digital and physical worlds. This dual existence, including safety, arises from many issues that warrant attention. Overall, we also evaluate how these activists perceive the broader digital environment. We want to understand how they use and view these digital spaces in the context of their activist goals.

After the initial theory-driven code list, an inductive coding process was implemented during the analysis, creating additional codes. We conducted calibration sessions where coders independently coded the same subset of data and discussed discrepancies to align their understanding of the coding categories. The inductive coding revealed both explicit and implicit themes in the data. Semantic codes captured direct expressions of privacy management, such as 'deliberate personal information control', while latent codes unveiled deeper insights into identity construction amidst social pressures, symbolised by 'constructing digital personas.' After a discussion, both authors verified the accuracy of the codes, compiling them into a comprehensive list. This bottom-up approach was necessary to make sense of the unique dataset specific to the Turkish context, where the activists meet various expectations such as public visibility, state pressure, and data collection. Several subsequent rounds of analysis were conducted via a qualitative data analysis software to examine the data thoroughly.

6 Analysis

Our examination of digital activists' privacy perceptions in Turkey is somewhat informed by theoretical insights, notably those concerning the ontological approach and contextual integrity. These frameworks ground our analysis and highlight how activists pursue their goals in X.

6.1 *Theme 1: Theoretical Constructs and Privacy Norms in Practice*

Participants articulated an understanding of privacy, echoing theoretical constructs of control and restricted space as discussed in Tavani's (2007) and Solove's (2008) works, without explicitly discussing ownership. Participants' deliberate management of personal information reflects a meaningful application of these theories, underscoring an acute awareness of the stakes involved in informational control. For instance, *Participant B* highlighted the selective sharing of personal information as a boundary-setting mechanism, stating, "Privacy is a bit like this. There are some private matters ... I do not share them, but I know it is no longer private once I share something there." This also reflects an awareness of privacy as dynamic and negotiable rather than static or absolute. It echoes Mai's (2016) concerns about the interspatiality between public and private realms, with our findings further exemplifying the lived realities of these conceptual challenges.

Similarly, *Participant G's* proactive approach to blocking over 1,000 individuals, including family members, underscores a deliberate strategy to maintain control over personal space and aligns with Floridi's (2005) ontological perspective on privacy as integral to one's identity. Participant F also highlights this association:

> I do not think of privacy solely regarding the state. I consider it on psychological and psychosocial grounds. The real issue is, who am I? What details of my life will my digital presence encompass? [...] Ultimately, I have to create a persona there, which should not contain certain details that might hinder or become an obstacle to my fight for rights.

6.2 *Theme 2: Identity Construction and Digital Personas*

As informed by Floridi's (2005) ontological approach, our findings reveal a relationship between privacy and identity among digital activists. As both Participant G and Participant F demonstrated, the strategic construction of online personas illustrates the ontological dimension of privacy, where personal data and identity are inseparable.

Participant G stated, for example:

> I do not want them to see just a single identity of mine. It is about everything I do and even more. That is why I want people to see those things, too. The connections I make with other struggles, with this life, and from which perspective I look at issues.

Participant F elaborated:

> For the sake of the consistency of that public persona, I try to live as holistically as possible. That also prevents quite a few things. In other words, it prevents me from living as my true self [...] Of course, being online prevents me from speaking as my true self, living as my true self, and sharing the details of my life [...] Society wants me in a certain way.

Somewhat contrary to Goffman's (1959) theory of identity as a performance tailored to societal expectations, participants exhibited a keen awareness of the limitations and opportunities presented by digital platforms that have been emphasised in some theories, especially for non-Western contexts (Zhao et al., 2008). In particular, most participants discussed a similar holistic approach to public persona management, noting its constraints on authentic self-expression. This reflects a critical understanding of digital platforms not as liberating spaces but as environments where identity construction is influenced by medium-specific norms and affordances, particularly within the Turkish context of heightened state surveillance and socio-political pressures.

This might also resonate with Nissenbaum's (2010) notion of contextual integrity, suggesting that the activists' digital self-presentation strategies – to the extent they are embedded in Turkey's socio-political and cultural fabric – would challenge the universal applicability of Western-centric privacy theories.

6.3 *Theme 3: Self-Disclosure as a Threat or a Tool?*

Participants' strategic use of self-disclosure highlights the dual role of personal information as both a potential threat and a powerful tool for activism. *Participant B's* decision to include their real name on their profile as a means of deterrence and *Participant D's* experience of sharing personal anecdotes for support illustrate the careful calibration of privacy risks against the benefits of authenticity and engagement.

Participant D said, for instance:

> Truly, digitally, sometimes I want to share my happiness. When I am having fun, sad, and so on, I shared once [tellings about an intimate anecdote]. From there, I received a lot of supportive messages.

The activists' adept use of self-disclosure, as highlighted by *Participant D's* experiences, reflects a keen understanding of the dual nature of personal information as both a tool and a threat. This duality aligns with Nissenbaum's (2010) theory of privacy as contextual integrity, where the appropriateness of information flow is contingent upon the specific context. All participants share concerns over security threats to their family members to the extent that they share personal details. However, contextual considerations over the purpose and function of activism sometimes seem to alleviate such concerns. Furthermore, data literacy is highlighted as a need or a lack of information for most participants, making the existing and emerging tool-based studies in non-Western contexts more relevant and vital than ever (Tadic et al., 2018; Wulf et al., 2022).

6.4 *Theme 4: Evolving Norms Based on the Platform*

The differentiation in platform use among participants underscores Nissenbaum's (2010) theory of privacy as contextual integrity. The variation in personal sharing across platforms like X and Instagram reflects an adaptive response to the affordances and risks of each medium.

Participant B explained: '*Because of this privacy issue, I hesitated to post on Instagram for a while.*'

Participant D said: '*I share very few personal things on X. My shares/posts tend to be more private on Instagram.*'

This adaptability is crucial in Turkey, where digital surveillance and political-cultural dynamics significantly influence digital practices and privacy considerations (Oz & Yanik, 2024).

7 Discussion: Other Theoretical Implications and Future Research

As Participant F put it:

> When I am texting with my friends or arguing with my girlfriend, in the end, I know this: one of these conversations can surely be in the possession of these people. We live a life where our insides and outsides are one with the state. [LAUGHS]. Moreover, I am not bothered by this.

We observed in all the interviews that there is a datafied indifference in that participants are accustomed to or indifferent towards the constant data capture of personal data. This resonates with the data relations approach of Couldry and Mejias (2019), as if the personal data is merely there to be captured, which implies that data capture is a natural process. An alternative explanation would be that the activists prioritise having a space for expression in the face of eroding public space in Turkey as they frequently refer to. Interestingly, this is also the case regarding perspectives toward Turkey and its public officials. As more saturated data, or a particular study and bigger sample size, would be needed to account for this question fully, the fact that the participants do not consider the state as a relevant figure here either implies that they think the state's interventions relate to more important things than privacy, or because of the deeply internalised surveillance culture in Turkey. Alternatively, they may be primarily interested in how they pose themselves to the public via X. While Oz and Yanik (2024) argue that perceived surveillance leads to reduced opinion sharing among X users in Turkey, it seems difficult to isolate concerns about state surveillance from individual experiences of aggression for the digital activists, as our interviewees more frequently indicated the latter.

The activists' privacy-related decisions seem rooted in their pursuits and are often influenced by external factors such as cyber-harassment. The rampant rise of online harassment seems to significantly shape activists' online behaviour, making them more selective about what they share in the digital realm. Moreover, cultural and ideological codes in Turkey might play a pivotal role in shaping their digital footprint. Activists may have to steer through complex cultural landscapes depending on the region, community, or specific cause they relate to. Anonymity – as opposed to sharing one's name and profession – is sometimes described as a reason for attracting more cyber-harassment, as norms attributed to certain professions in Turkey tend to cultivate a certain respect from the citizens. These codes might additionally dictate not only what is appropriate to share but also the manner and tone of the discourse – something cited by many of the respondents.

However, amidst these external factors, digital activists seem to maintain an internal compass guided mainly by contextual privacy concerns. Recognising that the digital realm can be both a tool for empowerment and a potential threat, they adopt privacy norms for their safety and as a shield for the vulnerable communities and individuals they advocate for. They understand that any breach in their privacy could have ripple effects, potentially jeopardising the very groups they aim to protect. This is closely related to emerging social conceptualisations of privacy that diverge from mainstream liberal understandings and are increasingly prevalent, especially in non-Western contexts (Jia & Baumer, 2022). For instance, Western-style values embedded in understanding privacy as an individual right are open to questioning in non-Western settings such as China, where collective values and social benefits of privacy are mostly considered (Yao-Huai, 2005).

Overall, how the digital activist perceives the value of informational privacy seems to be closer to ontological and contextual integrity accounts than control/restricted access or ownership accounts (Floridi, 2006; Nissenbaum, 2010; Solove, 2008; Tavani, 2007; Mai, 2016). This chapter echoes the idea that privacy considerations seem contextual, depending on the flow of information and the sender's pursuit, as well as ontological implications of how the information flow is constitutive of one's identity (Floridi, 2006; Nissenbaum, 2010). These considerations seem to impact how digital activists make trade-offs between different values in consideration or against the value of privacy in their pursuits. The concept of contextual integrity might be adapted to incorporate the dynamics of digital activism, where norms around privacy and information sharing are frequently negotiated against a backdrop of visibility, social pressures, authenticity and self-representation. Contextual integrity in Turkey's setting would also have to account for the rapidly changing norms around, for instance, privacy and information sharing, influenced by the very politicised and polarised public arena (Dogu & Mat, 2019).

Two significant events during our research period – the takeover and transformation of Twitter into X and the 2023 presidential election in Turkey – add another layer of context. The change in platform policies and the continuation of the AKP rule could have both immediate and long-term implications on digital activism and perceptions of privacy. Additionally, future work could focus on collective forms of activism and how they intersect with individual privacy concerns and norms.

8 Concluding Remarks

This chapter highlights perceptions towards the value of privacy and its practice that can also inform the critical data literacy literature regarding digital activists in Turkey. Our study offers insights into the relations between privacy, identity, and activism in non-Western contexts. However, more substantial takes would need to include a bigger sample size and, if possible, an ethnographic approach that also considers the interactions of digital activists in real-time with the specific technological affordances of the platforms. The way platforms mediate these interactions would be more accurately understood through research emphasising daily interaction, thereby enriching our understanding of digital activists' perceptions. Future research should also explore the implications of these findings in other settings, examining how different political regimes, cultural contexts, and digital platforms shape privacy norms and activism strategies.

References

Abokhodair, N., & Vieweg, S. (2016). Privacy & social media in the context of the Arab Gulf. In *Proceedings of the 2016 ACM Conference on Designing Interactive Systems* (DIS '16) (pp. 672–683). ACM.

Akser, M. & Baybars, B. (2023). Repressed media and illiberal politics in Turkey: The persistence of fear. *Southeast European and Black Sea Studies*, 23(1), 159–177.

Alemdar, Z. (2014). 'Modelling' for democracy? Turkey's historical issues with freedom of speech. *Middle Eastern Studies*, 50(4), 568–88.

Altınörs, G. & Akçay, Ü. (2022). Authoritarian neoliberalism, crisis, and consolidation: The political economy of regime change in Turkey. *Globalizations*, 19(7), 1029–53.

Arora, P. (2019). Decolonizing privacy studies. *Television & New Media*, 20(4), 366–378.

Becker, M. (2019). Privacy in the digital age: comparing and contrasting individual versus social approaches towards privacy. *Ethics and Information Technology*, 21 (4), 307–317.

Bernadas, J. M. A. C., & Soriano, C. R. (2018). Online privacy behavior among youth in the Global South: A closer look at diversity of connectivity and information literacy. *Journal of Information, Communication and Ethics in Society*, 17(1), 17–30.

boyd, d., & Crawford, K. (2012). Critical questions for big data. *Information, Communication & Society*, 15(5), 662–679.

Cammaerts, B. (2015). Social media and activism. In P.H. Ang & R. Mansell (Eds.), *The International Encyclopedia of Digital Communication and Society* (pp. 1–8). John Wiley & Sons.

Carmi, E. & Yates, S. J. (2020). What do digital inclusion and data literacy mean today? *Internet Policy Review*, 9(2), 1–14.

Couldry, N., & Mejias, U. A. (2019). *The costs of connection: How data is colonizing human life and appropriating it for capitalism*. Stanford University Press.

Couldry, N., & Powell, A. (2014). Big data from the bottom up. *Big Data & Society*, 1(2), 1–5.

Cristofaro, E. D., Soriente, C., Tsudik, G., & Williams, A. (2012). Hummingbird: Privacy at the time of Twitter. In *2012 IEEE Symposium on Security and Privacy* (pp. 285–299). IEEE.

Dambo, T. H., Ersoy, M., Auwal, A. M., Olorunsola, V. O., Olonode, A., Arikewuyo, A. O., & Joseph, A. (2022). Nigeria's #EndSARS movement and its implication on online protests in Africa's most populous country. *Journal of Public Affairs*, 22(3), 1–11.

Dogu, B., & Mat, H. O. (2019). Who sets the agenda? Polarization and issue ownership in Turkey's political Twittersphere. *International Journal of Communication*, 13, 229–250.

Donnelly, J. (2007). The relative universality of human rights. *Human Rights Quarterly*, 29(2), 281–306.

Floridi, L. (2005). The ontological interpretation of informational privacy. *Ethics and Information Technology*, 7(4), 185–200.

Floridi, L. (2006). Four challenges for a theory of informational privacy. *Ethics and Information Technology*, 8(3), 109–119.

Goffman, E. (1959). *The presentation of self in everyday life*. New York: Doubleday.

Gogus, A., & Saygın, Y. (2019). Privacy perception and information technology utilization of high school students. *Heliyon*, 5(5), 1–9.

Guest, G., Bunce, A., & Johnson, L. (2006). How many interviews are enough?: An experiment with data saturation and variability. *Field Methods*, 18(1), 59–82.

Haltaufderheide, J., Viero, D. & Krämer, D. (2023). Cultural implications regarding privacy in digital contact tracing algorithms: Method development and empirical ethics analysis of a German and a Japanese approach to contact tracing. *Journal of Medical Internet Research*, 25, e45112.

Hellmeier, S., Cole, R., Grahn, S., Kolvani, P., Lachapelle, J., Lührmann, A., Maerz, S. F., Pillai, S. & Lindberg, S. I. (2021). State of the world 2020: Autocratization turns viral. *Democratization*, 28(6), 1053–74.

Howard, P. N. & Hussain, M. M. (2013). *Democracy's fourth wave?: Digital media and the Arab Spring*. Oxford University Press.

Introna, L. D. & Pouloudi, A. (1999). Privacy in the information age: Stakeholders, interests and values. *Journal of Business Ethics*, 22(1), 27–38.

Jia, H., & Baumer, E. P. S. (2022). Birds of a feather: Collective privacy of online social activist groups. *Computers & Security*, 115, 1–13.

Karpf, D. (2010). Online political mobilization from the advocacy group's perspective: Looking Beyond Clicktivism. *Policy & Internet*, 2(4), 7–41.

Kaun, A., & Uldam, J. (2018). Digital activism: After the hype. *New Media & Society*, 20(6), 2099–2106.

Kazansky, B. (2015). FCJ-195 Privacy, responsibility, and human rights activism. *Fibreculture*, 26, 189–207.

Kitchin, R. (2014). *The data revolution: Big data, open data, data infrastructures and their consequences* (1st ed.). SAGE Publications Ltd.

Lerner, A., He, H. Y., Kawakami, A., Zeamer, S. C., & Hoyle, R. (2020). Privacy and activism in the transgender community. In *Proceedings of the 2020 CHI Conference on Human Factors in Computing Systems* (CHI '20) (pp. 1–13). Association for Computing Machinery.

Liang, H., Shen, F., & Fu, K. (2017). Privacy protection and self-disclosure across societies: A study of global Twitter users. *New Media & Society*, 19(9), 1476–1497.

Lokot, T. (2020). Data subjects vs. people's data: Competing discourses of privacy and power in modern Russia. *Media and Communication*, 8(2), 314–322.

Madison, N., and M. Klang. 2020. The case for digital activism: Refuting the fallacies of slacktivism. *Journal of Digital Social Research*, 2(2), 28–47.

Mai, J. (2016). Big data privacy: The datafication of personal information. *The Information Society*, 32(3), 192–199.

Marmor, A. (2015). What is the right to privacy? *Philosophy & Public Affairs*, 43(1), 3–26.

Masur, P. (2020). How online privacy literacy supports self-data protection and self-determination in the age of information. *Media and Communication*, 8(2), 258–269.

Milan, S., & Barbosa, S. (2020). Enter the WhatsApper: Reinventing digital activism at the time of chat apps. *First Monday*, 25(1).

Nissenbaum, H. (2004). Privacy as contextual integrity. *Washington Law Review*, 79(1), 119–158.

Nissenbaum, H. (2010). *Privacy in context: Technology, policy, and the integrity of social life*. Stanford: Stanford University Press.

Oz, M. & Yanik, A. (2024). Fear of surveillance: Examining Turkish social media users' perception of surveillance and willingness to express opinions on social media. *Mediterranean Politics*, 29(1), 1–25.

Özbudun, E. (2015). Turkey's judiciary and the drift toward competitive authoritarianism. *The International Spectator*, 50(2), 42–55.

Paine, C., Reips, U., Stieger, S., Joinson, A., & Buchanan, T. (2007). Internet users' perceptions of 'privacy concerns' and 'privacy actions.' *International Journal of Human-Computer Studies*, 65(6), 526–536.

Pasquale, F. (2015). *The black box society: The secret algorithms that control money and information* (1st ed.). Harvard University Press.

Penney, J. W. (2017). Internet surveillance, regulation, and chilling effects online: A comparative case study. *Internet Policy Review*, 6(2), 1–39.

Poell, T. (2014). Social media and the transformation of activist communication: exploring the social media ecology of the 2010 Toronto G20 protests. *Information, Communication & Society*, 17(6), 716–731.

Reporters Without Borders (2021). Turkey 2021. Retrieved from https://rsf.org/en/analyse_regionale/531

Reporters Without Borders (2016, April 19). Reporters Without Borders publishes the first worldwide press freedom index (October 2002). Retrieved from https://rsf.org/en/reporters-without-borders-publishes-first-worldwide-press-freedom-index-october-2002

Rule, J. B. (2019). Contextual integrity and its discontents: A critique of Helen Nissenbaum's normative arguments. *Policy & Internet*, 11(3), 260–279.

Sanfilippo, M. R., and Strandburg, K. J. (2021). Privacy governing knowledge in public Facebook groups for political activism. *Information, Communication & Society*, 24(7), 960–977.

Shirky, C. (2008). *Here comes everybody: The power of organizing without organization*. New York: Penguin Publishing.

Solove, D. (2008). *Understanding privacy*. Harvard University Press.

Steeves, V., & Regan, P. (2014). Young people online and the social value of privacy. *Journal of Information, Communication and Ethics in Society*, 12(4), 298–313.

Stern, T., & Kumar, N. (2014). Improving privacy settings control in online social networks with a wheel interface. *Journal of the Association for Information Science and Technology*, 65(3), 524–538.

Tadić, B., Rohde, M., & Wulf, V. (2018). CyberActivist: Tool for raising awareness on privacy and security of social media use for activists. In G. Meiselwitz (Ed.), *Social Computing and Social Media. User Experience and Behavior* (pp. 498–510). Springer International Publishing.

Tavani, H. T. (2007). Philosophical theories of privacy: Implications for an adequate online privacy policy. *Metaphilosophy*, 38(1), 1–22.

Tufekci, Z. (2017). *Twitter and tear gas: The power and fragility of networked protest.* New Haven & London: Yale University Press.

Verbeek, P. P. (2009). Moralizing technology: On the morality of technological artifacts and their design. In D. Kaplan (Ed.), *Readings in the philosophy of technology* (2nd ed., pp. 226–243). Lanham, MD: Rowman & Littlefield.

Verbeek, P. P. (2016). Toward a theory of technological mediation: A program for postphenomenological research. In J. K. Berg, O. Friis, & R. C. Crease (Eds.), *Technoscience and Postphenomenology: The Manhattan Papers* (pp. 189–204). Lexington Books.

Verkerk, M. J., Hoogland, J., Van Der Stoep, J., & De Vries, M. J. (2016). *Philosophy of technology: An introduction for technology and business students.* Abingdon: Routledge.

West, S. M. (2017). Ambivalence in the (private) public sphere: How global digital activists navigate risk. In *7th USENIX Workshop on Free and Open Communications on the Internet* (FOCI 17). Retrieved from https://www.usenix.org/conference/foci17/workshop-program/presentation/west

Wulf, V., Randall, D., Aal, K., & Rohde, M. (2022). The personal is the political: Internet filtering and counter appropriation in the Islamic Republic of Iran. *Computer Supported Cooperative Work* (CSCW), 31(2), 373–409.

Yabanci, B. (2019). Turkey's tamed civil society: Containment and appropriation under a competitive authoritarian regime. *Journal of Civil Society*, 15(4), 285–306.

Yao-Huai, L. (2005). Privacy and data privacy issues in contemporary China. *Ethics and Information Technology*, 7(1), 7–15.

Yılmaz, Z. & Turner, B. S. (2019). Turkey's deepening authoritarianism and the fall of electoral democracy. *British Journal of Middle Eastern Studies, 46*(5), 691–98.

Zhang, B., Kreps, S., McMurry, N., & McCain, R. M. (2020). Americans' perceptions of privacy and surveillance in the COVID-19 pandemic. *PLoS ONE*, 15(12), 1–16.

Zhao, S., Grasmuck, S., & Martin, J. (2008). Identity construction on Facebook: Digital empowerment in anchored relationships. *Computers in Human Behavior*, 24(5), 1816–1836.

Zihnioğlu, Ö. (2019). European Union civil society support and the depoliticisation of Turkish civil society. *Third World Quarterly*, 40(3), 503–20.

Zuboff, S. (2019). *The age of surveillance capitalism: The fight for a human future at the new frontier of power*. Public Affairs.

CHAPTER 7

Detouring, Rerouting, Weaponization: The Secondary Orality of WarTok

Marloes Geboers, Elena Pilipets and Daria Delawar-Kasmai

1 Introduction

Foregrounding the audiovisual format on social media, TikTok revolutionised the dissemination of information and propaganda, especially during times of armed conflict. Following Russia's full-scale military invasion of Ukraine, a distinctive phenomenon, termed 'WarTok', emerged, engendering a new paradigm of war programming. Coined during the initial week of the invasion, WarTok refers to 'the first TikTok war' or 'the war of super-empowered individuals armed only with smartphones' (Friedman, 2022). As a portmanteau of 'war' and 'TikTok', this term, although cynical and problematic in its implications, has drawn significant attention to the platform's role in augmenting frontline reporting in Ukraine with meme-driven pro-war propaganda (Mobilio, 2022; Kiparoidze, 2022). In the internet era, where each conflict is dubbed 'unprecedented' and linked to the latest digital innovations (Tiffany, 2022), it becomes crucial to offer context-sensitive analyses of the underlying conditions that lead to such phenomena. As we demonstrate below, WarTok necessitates critical scrutiny, not only for its sensationalist stance but also for the collapse of contexts it entails.

Our contribution explores the realm of memetic TikTok performances, where war-related Ukrainian and Russian video posts amalgamate multiple media forms through remix. A focus on these expanded modalities of audiovisual 'assembly' (Parry, 2022) implies navigating the platform using networked expressive elements – sounds, effects, stickers, emojis, and hashtags. Channelling and animating engagement, such an assembly arranges content into searchable formations. Moreover, its hybrid nature allows for various trajectories of circulation that modulate affect and inscribe meaning. Our main objective is to understand how – in the context of war propaganda and its contestation on TikTok – multimodal or layered posts reroute, detour, and weaponize affective engagement by heavily relying on TikTok practices of song-appropriation and memetic cross-hashtagging. We ask how war propaganda is shaped and steered by the infrastructural and

 | DOI:10.1163/9789004711396_009

affective affordances of what we call networked soundscapes – audiovisual assemblies that emerge from the intersections of sounds and other network 'designers' such as emoji-hashtags (Highfield, 2018). We also ask how vernacular acts of sharing through hashtags and sounds establish, prolong, and disrupt the 'secondary orality' (Ong, 1982) of platform-mediated cross-cultural 'encounters.' Our focus is less on the analysis of aural memes (Abidin & Kaye, 2021) as units of cultural transmission; rather we foreground the ambient environments – or scapes – emerging from the alignments of sounds with other memetic components. The ways in which these 'soundscapes' perpetuate ideas and morph meaning require relational approaches to understanding their assembly.

In our examination of a specific soundscape that emerged during the Russian invasion of Ukraine, the phenomenon of 'secondary orality' (Ong, 1982) points to the multifaceted nature of TikTok sharing that opens up new possibilities for affect targeting, exploitation, and weaponization. To answer our research questions we adopt a platform-grounded approach, analysing posts and video metadata that were shared with either of the two TikTok songs: 'Доброго вечора Where Are You From?', commonly known by its incipit 'Good evening, we are from Ukraine' (Ukrainian: Доброго вечора, ми з України) or 'Good morning, we are from Russia' (Russian: Доброе утро, мы из России). Multimodal in that it merges textual, visual, and auditive features, TikTok engagement is networked through various digital objects, including searchable 'listed' and polyvocal 'original' sounds (Kaye, Chen & Zeng, 2021; Kaye, Zeng & Wikstrom, 2022). Cross-hashtagging practices simultaneously extend and situate these networks within the platform, resulting in a complex environment of memetic riffing (Pilipets, 2023). Drawing inspiration from Jodi Dean's (2016) interpretation of Walter Ong's secondary orality, we recognize that the online circulation of multimodal content expands the communicative possibilities of propaganda, introducing more networked, decentralised, or 'participatory' (Boler & Davis, 2021) means. This extension involves acknowledging the reciprocal relationship between users' affective responses and the media ecosystem that thrives on repetition, amplification, and attention-hijacking (see also Zulli & Zulli, 2022). As we will show, it also involves a logic of sharing according to which oral cultures become reassembled through multimodal techniques of content crafting and distribution.

In the following analysis, we explore such an ecosystem by focusing on the memetic video performances that spread horizontally through cross-hashtagging and sound-linking, specifically focusing on the antagonist encounters of 'affective publics' (Papacharissi, 2015) attuned to 'Good evening, we are from Ukraine' and 'Good morning, we are from Russia.' During Russia's full scale invasion of Ukraine, the first song, a well-known electronic track by the Ukrainian duo ProBass and Hardy, evolved into an informal Ukrainian military greeting and found widespread use on TikTok, accumulating over 803K videos. Soon enough, a similar pro-Russian counter-appropriation appeared in response, proclaiming (in Russian), 'Good morning, we are from Russia.' The former (Ukrainian) track was widely popular before the war; the latter (Russian imitation) gained visibility in March 2022 in the shape of a series of 'original sounds' distributed by multiple authors, the many of their accounts are no longer accessible.

Using multimodal network analysis, we show how these two songs, along with their accompanying video performances and hashtag publics, have become emblematic of national identity formations deeply embedded in the realm of WarTok propaganda and its contestation. In particular, we trace connections that the so-called 'hashflags' (Highfield, 2018) make visible in relation to both songs and how these connections might be subverted. Such potential (re)combinations of components open up unpredictable trajectories for targeting audiences. Arguing that hashtags, emojis, and sounds can simultaneously operate as a source of mutual amplification and a means of affective rerouting, in the next step, we zero in on a selection of ten videos shared with 'Good morning, we are from Russia' in combination with either of the two slur hashtags #путінхуйло 🔴⚫ UA (Putin is a dickhead) and #зеленскийхуйло ⚪🔵🔴 (Zelenskyy is a dickhead). These hashtags can be seen as 'hashflags' (Highfield, 2018) as the colours of the circles represent the colours of, respectively, the Bandera flag and the Russian national flag. The Bandera flag is a historical symbol of Ukrainian independence. Thinking about these entanglements in terms of their affective value, we discuss the contested logic through which Russian war propaganda on TikTok becomes a digital meme machine.

2 Participatory Propaganda: Meme Wars, Soundscapes and Secondary Orality

The participatory means of propaganda (Boler & Davis, 2021) lean into the affordances of TikTok as a space for remix, where borrowing, imitation, and adaptation of existing cultural artefacts such as song snippets allow users to forge their identity performances for broader social significance (Kanai, 2015; Lessig, 2008). In this realm, the construction and reinforcement of national stereotypes give rise to highly contested affective publics. Such publics – explicated through the use of hashtags and sounds – support both memetic riffing and attention hijacking. In our study, cross-cultural communities debunking Russian war propaganda and communities variously involved in its circulation are driven by the tactical rerouting of song-based video performances. A song, as Papacharissi suggests in her interview with Boler and Davis, is not only a story with multiple layers but a 'narrative infrastructure, a texture, a tonality' (2021, p. 60). All of these layers found on TikTok are variously embedded in video-sharing practices, creating an affective environment that fosters remix and reappropriation.

We conceptualise memetic soundscapes as emerging from the intersections of aural and (con)textual linkages forged by the use of two antagonist songs in combination with associated co-hashtags. Through these linkages, the Ukrainian song 'Good morning, we are from Ukraine' and its Russian counter-imitation shape individual and collective performances of national belonging. Specific to sound on TikTok is that it shapes engaging content formations, turning users' video performances into conduits of cultural contestation. Due to their shared expressive qualities, videos networked through the same patriotic tune, for example, may become conflictual memetic material. Attracting certain communities and repelling others, a song charged with national pride creates a 'centripetal power' (Goodman, 2010, p. 11). In its reverberation, the informal military greeting in Ukrainian spaces attracted pro-Russian counterpublics, leading to a simple modification of the lyrics. This transformed the song into a hybrid creation resembling a Frankenstein's monster adorned with Russian flag colours and other identity markers, including emojified hashtags and painted influencer cheeks. Drawing on the concept of secondary orality (see also Dean, 2016; Venturini, 2022), this dynamic can be understood through its 'fostering of a communal sense, its concentration on the present moment, and even its use of formulas' (Ong, 1982, p. 133). Such 'formulas' rely on practices of assembly, where multimodal components shapeshift meaning depending on their positionality within the larger whole.

Secondary orality encompasses the integration of literary traditions into oral storytelling practices, where repetition plays a vital role in preserving and perpetuating 'the present moment.' Primary orality results from a lack of documentation techniques, whereas the secondary orality of TikTok soundscapes arises from the reversal of this logic. Here, acts of hashtagging and sound-linking are not merely archiving communication, but they accelerate the flow of ideas and open them up for imitation and appropriation: ideas are positioned as poles within a field of oppositions (Dean, 2016) and articulated via the addition of networked elements. Allowing users to tune in and connect with like-minded others (Papacharissi, 2015), this logic provides various possibilities for content crafting and affective storytelling (Hautea et al., 2021; Geboers et al., 2022). Its workings can be understood through the lenses of affective visual rituals (Kirby & Özkula, 2023) and transmedia music (Vizcaínu-Verdu & Abidin 2022), which comprise creative configurations of visual, textual, and audio elements, opening initial content to novel appropriations. Modulated by TikTok's affordances for reworking and remixing, such multimodal affective rituals take on the shape of memes that draw strength not only from visual similarity but perhaps even more so from aural resemblances.

In the realm of cross-cultural meme wars, secondary orality, despite its perpetuating role, therefore, comes with drawbacks and can result in the distortion of intended meanings (Boudana et al., 2017; Zelizer, 2006). As content spreads and is adapted by various users, the intended message can be altered or misinterpreted, shifting and channelling collective attitudes in ambiguous ways. Similar in effect to the idea of counter-visuality and memetic weaponization (Askanius & Keller, 2021; Vicari et al., 2020), songs and speech evolve into memes and counter-memes that, over time, transform within networked arrangements of searchable hashtag publics. A national pride anthem can be subverted into a derogatory copy with an oppositional message. By foregrounding soundscapes as points of affective reverberation, we rely on their unique affordances for assembly with other memetic components. We explore how the emergent networked soundscapes of 'Good evening, we are from Ukraine' and 'Good morning, we are from Russia' create a dissonant space that hinges on cross-cultural encounters. We are particularly interested in how the alignments of the two songs with hashflags and other expressive elements develop secondary orality through content-sharing practices of those attracted and those repelled by a sound. In doing so, we elevate the dynamics of exchange across partisan networked communities. We focus less on the dissemination of memetic content; rather, we study the components of memes, and more specifically, the hybrid connections these components establish, in relation to other components with which they are strategically and tactically assembled.

3 On Studying Soundscapes: Hashtags as Connectors and Disruptors

Acknowledging the importance of platform materialities in shaping networked propaganda, we adopt a research approach that integrates the walkthrough method with online-grounded explorations of social media engagement (Rogers, 2013; Light et al., 2018). By navigating networked posts and analysing video metadata, we explore how users leverage musical artefacts to express their allegiances and engage in memetic practices of appropriation. We repurposed TikTok's search affordances to prepare a repository of TikTok posts that were networked by the use of the two aforementioned songs. Given the flexible networkedness of platform objects such as sounds and hashtags, navigating formations of related videos becomes a means of studying these objects' environments of 'expected use' (Light et al., 2018, pp. 886). Acoustic linkages between videos 'pre-scribe' these videos' memetic adaptations, each of which can be further contextualised through practices of cross-hashtagging. In order to compile a corpus of videos, we iteratively repurposed the platform's 'visibility paths' formed by the intersections of sounds and hashtags. Following these paths and their various intertwinings allows us to study networked encounters between publics and counterpublics.

We gathered metadata of 3,138 posts – videos, time of publication, engagement metrics, effects, hashtags, etc. – published with *Доброго вечора – Where Are You From* (Ukrainian sound by PROBASS ∆ HARDI) and 965 posts published with *Доброе утро! Мы из России* (one of the many pro-Russian 'original sounds'). The metadata of the posts in the search results were collected using the browser-based scraper tool Zeeschuimer (Peeters, 2022) that outputs spreadsheet files containing various data endpoints. Due to TikTok moderation measures implemented in early March 2022, many (but not all) Russian-uploaded videos were removed from global public view (Kerby, 2022). Owing to how users tweak, rework, and create music spin-offs, some of these creations continue to circulate as 'original sounds' or unlisted recordings. To attend to these practices, we perform a multi-modal network analysis of co-hashtags demarcating the soundscapes of Ukrainian resistance and Russian war propaganda.

A network analysis of sound-hashtag-relations allows us to situate video formations based on the co-hashtags that specifically connect posts with either *one* of the two songs and the hashtags that appear together with both. We identify the latter as cross-hashtagged posts, which display specific tactics of (counter)imitation and unfold their meanings through practices of attention hijacking and affective rerouting. (Tactical) cross-hashtagging and the attunements of video performances that emerge from these tactics create engaging formations that we address as memetic soundscapes. By examining the associations between sounds and written text, our exploration goes beyond focusing on content alignments with particular national affiliations. We contend that our analysis here provides valuable insights into the acoustic reverberations of war propaganda, while also highlighting how relations between sound, images, and text unfold in the transient realm of TikTok's secondary orality. In this realm, co-hashtags operate as both connectors and disruptors, shaping but also diluting (Bainotti et al., 2022) the videos' potential meanings and trajectories of circulation.

In Figure 7.1 (below), hashtags demarcate semantic neighbourhoods that arise by association with the popular Ukrainian sound 'Good evening, we are from Ukraine' and its pro-Russian appropriation 'Good morning, we are from Russia.' Colour codes indicate three co-occurring zones of engagement: hashtags in the Ukrainian soundscape are blue, hashtags in the Russian soundscape are grey, and hashtags used in combination with both sounds appear in purple. The thickness of edges and the size of nodes in the network visualisation reflect the extent of attunement between sounds and hashtags. The denser the connections, the higher the intensity of engagement. Consequently, a TikTok soundscape not only facilitates the propagation of aural memes (Abidin & Kaye, 2021) but also brings together various forms of participation and affective alignment (Papacharissi, 2015).

The means of channelling Ukrainian interventions in Russian pro-war propaganda, as suggested by the distribution of the network, encompass a diverse range of sound-hashtag combinations, which oscillate in both stance and focus. The use of TikTok-specific #fyp hashtags, along with their variations #рекомендации, #рек, #реки, #хочуврекомендации (or recommendations), holds significant importance in shaping the dynamics of engagement within both realms. TikTok #fyp hashtags hold a constitutive role in shaping users' perceptions of the algorithmic content feed, commonly known as the 'For You Page' (Schellewald, 2018; Grandinetti & Bruinsma, 2022; Kaye et al., 2022). The prevailing belief that including #fyp in a post enhances its chances of being recommended, is an integral part of the platform's vernacular knowledge shared across different linguistic spaces and (sub)cultural communities. Out of the 4,103 posts analysed, #fyp-hashtags appear in 3,908 posts, connecting Ukrainian and Russian soundscapes in tandem with more contextually relevant co-hashtags. In our network analysis, we filtered out the former to emphasise the latter, while acknowledging the significance of 'post-based virality ambitions' (Abidin, 2021) that drive the circulation of war propaganda and its contestation.

Analysing the dynamics of cross- and co-hashtagging videos networked in the use of two culturally situated songs, we can then trace how their secondary orality registers in the networks of user-generated remixed content. The following two sections map out these relations by looking closely at the

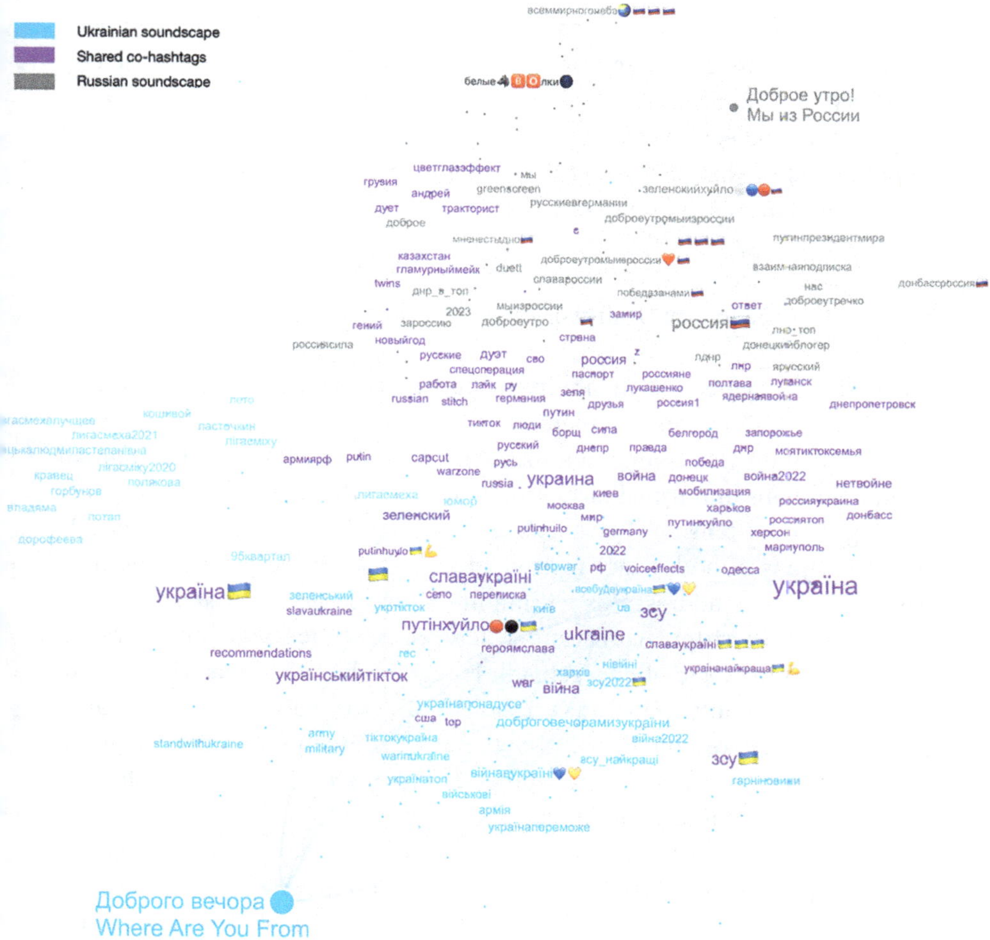

URE 7.1 A network of sound-specific and shared co-hashtags Connecting 3,138 videos published with Доброго вечора - Where Are You From ('Good evening, we are from Ukraine') and 965 videos published with Доброе утро! Мы из России ('Good morning, we are from Russia'). TikTok-specific #FYP hashtags appearing in 3,908 out of 4,103 analysed posts have been filtered out for better readability. Hashtags appearing in the network have been used in more than five posts.

cross-hashtagged intersections of sound memes and by identifying a situated formation of videos for a qualitative analysis of partisan engagement. After a discussion of cross-hashtagging in both soundscapes, we zero in on the contested connotations of the pro-Russian spin-off 'Good Morning, we are from Russia.' Demarcated by the use of hashflags, this formation exemplifies how soundscapes are always porous as meme components are used interchangeably, morphing meaning by virtue of their relational composition. By examining a selection of ten videos co-hashtagged with #зеленскийхуйло ⚪🔵🔴 (Zelenskyy is a dickhead) and #путінхуйло 🔴⚫ UA (Putin is a dickhead), we show how flags turn into antagonist symbols that simultaneously align users' identity performances and subvert the intended anti-Ukrainian stance of the song.

4 The Secondary Orality of Cross-Hashtagging

TikTok is renowned for its ability to implant short video earworms, perceived as stickier than complete songs (Vizcaíno-Verdú & Abidin, 2022). Some attribute this phenomenon to the cognitive principle that human memory retains unfinished tasks more effectively than completed ones, generating affective tension (Carson, 2022). Studying the ways in which acoustic reverberations of both songs register in cross-hashtagged video performances is one way to address this tension in online spaces where spoken words translate into written words and where evanescence and remix are ingrained into the logic of engagement. Foregrounding the memetic function of 'secondary orality' (Ong, 1982, Dean, 2016), our analysis of soundscapes sets out to explore how the ultra-nationalist landscape of Russian WarTok and the tactics of Ukrainian resistance intertwine in a complex ecology of imitation and attention hijacking.

As the analysis in Figure 7.1 shows, co-hashtags unique to the Ukrainian soundscape of 'Good evening, we are from Ukraine' revolve around two main thematic clusters: war and comedy (coloured blue). The first cluster contains patriotic hashtag-emoji combinations such as #українапонадусе (Ukraine above all), #війнавукраїні 💙💛 (war in Ukraine), and #перемогазанами 💙💛💪 (we will prevail). These hashtags are associated with a blend of videos that show events on the battlefield and eyewitness reports alongside videos of support by (predominantly) female influencers lip-syncing 'Good evening, we are from Ukraine.' Yellow and blue heart emojis, along with biceps emojis, additionally enhance the vernacular expressive force of this content formation, fostering a sense of collective support. Following Jodi Dean's discussion of Walter Ong's secondary orality (2016), hashflags accompanying musical performances of national pride can be understood as iconic elements that 'blend together speech, writing and image.' The appearance of such elements in the soundscape of 'Good Evening, we are from Ukraine' amplifies its multimodal blend through which the song assembles 'the many participating in the common practice, the many imitating each other' (Dean, 2016, p. 363).

The second cluster in the upper left part of the network comprises 37 videos that appear seemingly disconnected. These videos are dedicated to the comedy TV show *League of Laughter*, which was co-founded by the President of Ukraine Volodymyr Zelenskyy. All of these videos feature popular comedy sketches from the show's 2020 edition uploaded by a single account in April 2022, together with hashtags containing the names of the performers. Notable is also that the videos are networked by the use of the song, which, in this case, serves as a memetic linkage rather than as a soundtrack. The song itself, disappearing under the layers of editing, is not perceptible as part of video recordings. Instead, the focus is on the actors' jokes and the audience's laughter. As a humorous intervention in the flow of 'Good evening, we are from Ukraine' videos, this tactic of remediation fulfils a dual purpose. On the one hand, it allows for better searchability and amplification of otherwise unrelated content, enabling creators to reach a broader audience. On the other hand, the presence of comedy within this particular constellation may also be interpreted as a form of distraction from the atrocities of war. The juxtaposition of humour and the seriousness of the situation adds complexity to the context, inviting various interpretations and affective responses.

The co-hashtags unique to the Russian counter-appropriation of the song 'Good morning, we are from Russia' (coloured grey) also employ a structural and communicative relationship with national flag emojis. Such 'hashflags' have been previously recognized as carriers of affect and discursive flexibility (Highfield, 2018), often used in practices of demarcating in-groups and out-groups. Videos linked through #vladimirputinRU, #zанаших (for our guys), #RURURU, #мыизроссииRU (we are from Russia), and #мненестыдноRU (I am not ashamed) promote a shared stance in favour of Putin and the war. This contrasts with the use of Ukrainian flags along with yellow and blue hearts, resulting in a counter-imitative performance of 'love for the nation' (Ahmed, 2004). The juxtaposition of Ukrainian and Russian soundscapes underscores a notable division based on national belonging, while also expressing a sense of patriotic unity and a collective tendency towards othering. This juxtaposition is evident not only in the cultural connotations embedded in hashtags but also in the catchy rhythms and memetic scenarios portrayed in video performances (e.g., soldiers marching, influencers posing, tanks invading, etc.). When observed together, these performances revitalise Walter Ong's notion of technicised (or digitally mediated) orality, characterised by its ability to foster a communal sense and employ established formulas (Ong, 1982, p. 133). By blurring distinctions between speech, music, and text, 'Good evening, we are from Ukraine' and 'Good morning, we are from Russia' disseminate ambient war propaganda through reenactment of well known memetic scenarios where in- and out-groups echo each other and coalesce (see also Geboers and Pilipets, 2024).

Making use of TikTok's networked affordances, the main area of propaganda and its contestation unfolds in the shared zone of cross-hashtagging. Both songs are linked with hashtags demarcating war-associated locations – #херсон (Kherson), #киев (Kiev), #харьков (Kharkiv); national belongings – #україна UA (Ukraine), #россия (Russia); political figures – #путин (Putin); #зеленский (Zelenskyy); events such as #mobilisation; and military abbreviations such as #зсу or Зброй́ні си́ли Украї́ни (Armed Forces of Ukraine). Within this cross-hashtagged formation, a significant number of video performances revolve around the use of slurs, such as #путінхуйло 🔴⚫ UA (Putin is a dickhead). This hashflag appears alongside both songs in response to its pro-Russian counterpart #зеленскийхуйло ⚪🔵🔴 (Zelenskyy is a dickhead), exclusively associated with 'Good Morning, we are from Russia.' The pro-Russian hashtag is accentuated by three ball-shaped emojis representing the Russian flag. The red and black emojis in #путінхуйло 🔴⚫ UA also represent a flag: that of the Organisation of Ukrainian Nationalists (OUN) or the so-called Bandera Movement, which has played an important role as a symbol of anti-Soviet resistance in Ukraine. By employing the Bandera flag in combination with #путінхуйло alongside 'Good Morning, we are from Russia', users' video performances reverse the Russian appropriation of the Ukrainian song 'Good Evening, we are from Ukraine.' #путінхуйло took on the shape of a hashtag even before TikTok grew to prominence. It stems from a chorus invented and performed for the first time by fans of the Metalist football club in Kharkiv on March 30, 2014, at a joint march with Shakhtar team fans (Goncharova, 2014). During the Russian invasion of Ukraine, it spread widely, receiving many memetic variations, including those featuring the flag balls.

In the following section, we proceed by exploring the cross-cultural reverberations of 'memetic ball play' through a situated visual analysis of a montage of frames extracted from the videos shared with 'Good Morning, we are from Russia' along with #путінхуйло 🔴⚫ UA and #зеленскийхуйло ⚪🔵🔴. To exemplify how intersections of (counter)publics attuned to the song materialise through multimodal engagement, we zoom in on ten videos posted with the flag balls. All ten are aurally networked by 'Good Morning, we are from Russia', with five posts using #зеленскийхуйло ⚪🔵🔴 and five using #путінхуйло 🔴⚫ UA, the latter transforming the Russian spin-off into the object of mockery.

5 Memetic Ball Play: Cross-Cultural Reverberations

The flag ball emoji bear visual resemblance to the 'countryballs meme', originating from a comic series on the anonymous image board platform Krautchan (resembling 4Chan) (KnowYourMeme, 2023). The first countryball to appear was PolandBall in 2009, initially apolitical and intended to jest at a fellow Polish board user (Wojak). However, the meme gradually evolved, with various countryballs taking on characters embodying national stereotypes in the context of geopolitical events and international relations (Hagen, 2017). In this contested space, where the national soundscapes converge and entangle through hashtags, flag ball emojis both amplify and delineate the boundaries of national identity. They operate in ways similar to flags, symbolising territorial conquest and conveying the ideas of 'with-ness' and 'for-ness' as described by Sara Ahmed (2004, p. 74).

Our analysis (Figure 7.2, below) departs from the focus on the contested soundscape of 'Good Morning, we are from Russia' and its multimodal reverberations. The flag balls attached to #путінхуйло and #зеленскийхуйло videos are indicative of two key dynamics: 1) negotiations concerning one's alignment 'with' Ukraine or 'for' Russia, and 2) the shifts in the affective charge of these negotiations as they circulate and undergo reappropriation. Across the ten analysed video performances, the flag emojis serve as markers of collective identity, acquiring interpersonal meaning beyond their literal representations. In their role as bonding icons, they signify social belonging and hold considerable affective power (Newton et al., 2022; Stenglin, 2004; Kirby & Özkula, 2023).

The emojified layer of #putinhuilo 🔴⚫ UA, with the colours symbolising the Bandera/OUN (Organisation of Ukrainian Nationalists) flag, is linked to historical events associated with the nationalist movement for an independent Ukraine. These events have been weaponized by Russian propaganda to bolster claims of a widespread Ukrainian 'Nazism' without considering the intricate complexities of Eastern Europe's history of mass atrocities during the Soviet era (Zaitsev, 2016; Filtenborg, 2021; Goncharenko, 2022). The controversial revolutionary organisation of Ukrainian nationalists, formed in 1940 following the split of the OUN, developed its own symbols under the leadership of Stepan Bandera. The flag adopted by the revolutionary wing consists of two colours: red and black. The black colour represents the land, while the red symbolises the blood shed for Ukrainian freedom. In contrast to the yellow-blue flag that symbolises statehood and unity, the red-black flag represents the Ukrainian revolutionary spirit (Мороз, 2013).

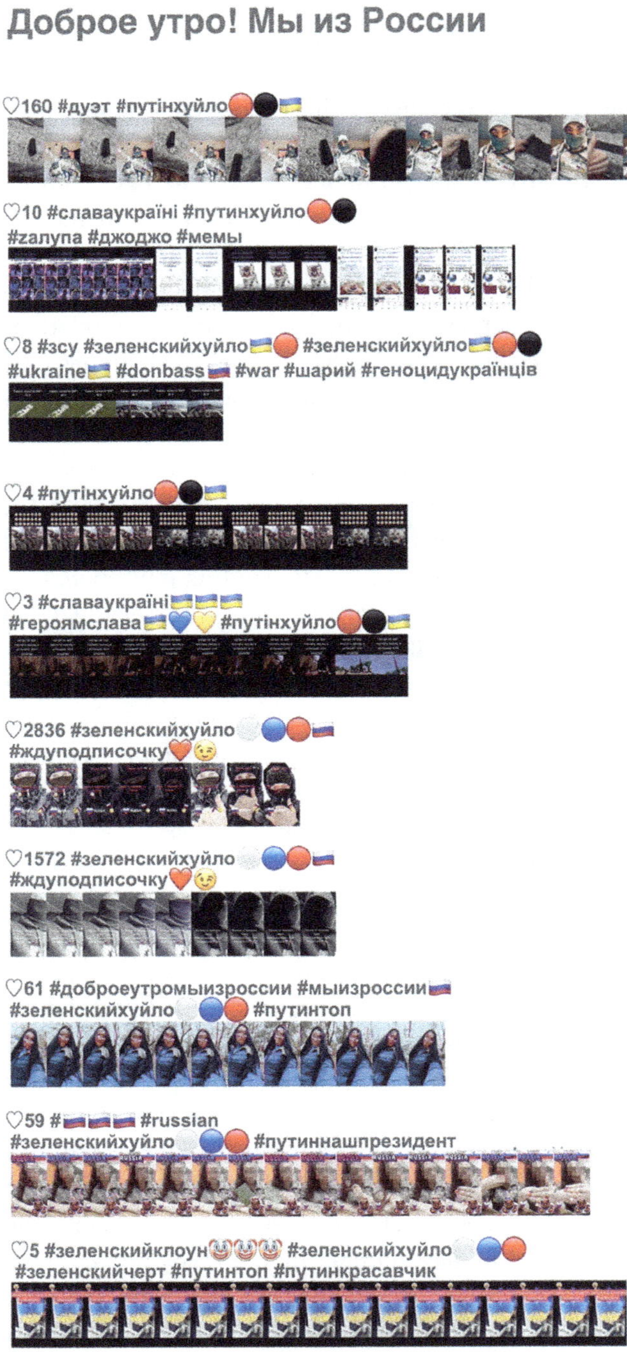

FIGURE 7.2 Montage of deidentified frames from the five most commented videos.

Notably, Russian media often employ the term 'Banderivtsi' to portray Ukrainians negatively, often associating the group with the label 'fascist' (Shevtsova, 2022). While the term originates from Bandera's surname and refers to his followers, its deliberate use by Russian propagandists aims to depict the Ukrainian government as illegitimate, aggressive, and unrepresentative of the majority (Скоркин, 2022). As a result, the colours associated with Bandera not only serve as symbols of Ukrainian struggle but also 'simultaneously' reinforce and challenge propagandist narratives, eliciting strong reactions in pro-Russian online communities. In this contested context, the slur #путінхуйло 🔴⚫ UA takes on a dual role, signifying Ukrainian national pride while being manipulated to feed propagandist agendas, provoking counter-imitations. While the videos shared with #зеленскийхуйло ⚪🔵🔴 largely follow the affirmation-seeking format that is typical of the 'influencer-aesthetic' found in other explorations of Russian war propaganda (Boichak, 2023; Geboers and Pilipets, 2024), the Ukrainian framework demarcated by путінхуйло 🔴⚫ UA, connects videos that lack distinct visual scenarios.

In the first of five 'Good Morning, we are from Russia' videos using #путінхуйло 🔴⚫ UA, the song loses its pro-Russian stance. Along with the Bandera hashflag, the video features a duet where footage of a Russian soldier boasting about his military attire is put next to someone rolling out a bin bag. The bag is a prop that alludes to a war meme about dead Russian soldiers being transported home in a trash bag (Кононенко, 2022). It can be inferred that the duet proposes an unconventional 'new uniform' for the soldier depicted in the original footage. In the second video, the sticker (overlaid text on top of the video) accompanies a random meme sequence that opens with 'when', followed up by an audiovisual performance answering the when-question. The sticker states 'Name the creepiest thing you've ever heard', followed by 'Me', all while 'Good Morning, we are from Russia' is playing. Here, seemingly unrelated images flash by in slideshow mode, indicating the absurdity of the pro-Russian song hijack vis-à-vis its origins in Ukrainian popular culture.

Combining hashtags that target both Ukrainian and Russian state leaders, the third video encapsulates ambiguous elements that further fracture the message, allowing for a multiplicity of interpretations. In this variation, #путінхуйло is used together with hashtags proclaiming 'Zelenskyy bastard' and 'genocide of Ukrainians.' The video displays carp fish either jumping on, or thrown onto, the deck of a ship. The sticker text reveals the video's political connotation: 'Karas captured an infantry fighting vehicle of the Armed Forces of Ukraine' where 'Karas' translates as both 'carp fish' as well as the name of the leader of the far right C14-movement, Yevhen Karas. The video exemplifies trollish humour by presenting a mix of nationalist ideologies – a memetic tactic of indistinction commonly used for purposes of provocation. Using the same tactic, the next performance shared with путінхуйло 🟠 ⚫ UA showcases images of zombies marked with a Z on their foreheads. These images are stitched together with images of Russian soldiers, marching to the beat of 'Good Morning, we are from Russia.' The stitch is enhanced by the repetition of laughing-with-tears emoji in the sticker text, indicating the ineptness of Putin's military attempts to conquer Ukraine. Here, the song's contrasting message actually furthers the anti-Russian stance of the post through sarcasm, tactically exposing zombie images to an antagonist audience.

The last video tagged with путінхуйло 🟠 ⚫ UA appropriates a well-known video meme called 'Shrek poop fly' that depicts Shrek being propelled into the air with great speed due to the force of the 'poop power' released (Tenor, 2020). In this humorous rendition, the overlaid text reads: 'When I couldn't sleep for a month, and I heard this masterpiece.' Through this imaginative portrayal, the video serves as a satirical critique, highlighting the perceived inadequacy of 'Good Morning, we are from Russia.' Special attention is paid to the juxtaposition that culminates from the intersection of the Putin-khuilo hashflag with the Shrek meme and the hijacked song. In the context of meme wars, Tim Highfield (2016, p. 55) describes the power of juxtaposition as a means of amplification and 'incongruity, mixing together contexts that are otherwise unrelated.' Such amplifying techniques can be seen as part and parcel of the assembling logic of secondary orality that expands the affective power of sounds into further expressive registers – video reenactments and their networked (con)textual environments. Following from the network analysis, it is clear how the hijacking of a Ukrainian song by Russian counterpublics, not merely opens the gates to pro-Russian formations of propaganda content, but also presents a path for channelling pro-Ukrainian interventions within a 'Russian soundscape', where the sounds of both parties collide.

The videos linked by the #зеленскийхуйло ⚪ 🔵 🔴 hashflag demonstrate similarities, although they do not strictly adhere to a memetic template. All four videos convey Russian identity through the use of stickers, hashtags, emojis, and embodied markers like the Russian tricolour-manicures. Creating a folkloric celebration of national identity, playful elements like flag filters and cartoon bears appear superimposed on users' bodies, along with cheerful emojis. This aesthetic aligns with what Wong et al. (2021, p. 15) term the 'cutification' of propaganda, adapting it into a fan-oriented apparatus reminiscent of the entertainment industry. While the bear motif takes on various forms in the realm of Russian WarTok, the fusion of cuteness with the bombastic sound of the 'Good Morning' song serves to detach Russian identity performances from the harsh reality of the war. Although we can only speculate about the motivations behind uploading, the use of the same propagandist sound and co-hashtags suggest a coordinated effort.

One of these videos depicts a woman displaying an expired USSR resident certificate, which she then replaces with a Russian passport, imitating an act commonly encountered in the WarTok realm. The woman's manicure features the Russian flag and the letter Z, which has acquired significance as a symbol associated with the Russian invasion of Ukraine. The video's caption aligns with typical affirmation-seeking attempts, as expressed through numerous hashtags reflecting a sense of togetherness or 'withness' (Ahmed, 2004, p. 74). The caption explicitly asks, 'Who is for Russia? Let's be friends RURURU 💪💪💪💪.' Similar examples appear in two other videos posted by the same user sporting military attire and obscuring his face. The sticker text resembles a speech template found in the wider context of Russian WarTok propaganda. The videos shout out for 'subscriptions' amplified by the hashtag #subscribe and video captions in all caps: 'LIKE and SUBSCRIBE for Russia. Russia Power RU 💪.' The underlying narrative revolves around embracing and asserting Russian identity while actively seeking communal bonds through established formulas. Such 'formulas' re-enact the secondary orality of the pro-Russian soundscape through addition and re-assembly, where multimodal components amplify each other because they repeat (Dean, 2016). As they reappear in multiple constellations, sounds, emojis, hashtags, and stickers become a form of address with 'binding effect' (see also Ahmed, 2004, p. 92).

The final video distinguishes itself from the rest in its performative and narrative elements. It features a toilet brush in the colours of the Ukrainian flag, while remixing 'Good Morning, we are from Russia' with the original track 'Good Evening, we are from Ukraine.' In this post's caption, the anti-Zelenskyy tag is accompanied by a range of slur hashtags that connect clown emojis to insults directed at Zelenskyy, Ukrainians, and Jewish people, followed by pro-Putin hashtags. The leaving-intact of the pro-Ukrainian song can be regarded as both a trace in the assembly of secondary orality and an additional layer of mockery displaying total disregard for the original cultural artefact. Important to address is how this assembly allows sound to transcend modalities of a soundscape through multimodal formulas. The evolution towards 'formulas' that create national bondings (Ong, 1982, p. 133), is contingent on assembly and reassembly, where components

– sounds, co-hashtags, video captions, emojis, etc. – remain discernible, but are always in potential conflict with each other (Parry, 2022). The pro-Russian flag ball emoji, described as sites of personal and social tension (Ahmed, 2004, p. 11), becomes saturated with affect through repeated combinations and additions. It is the accumulation of these repetitions that adds both complexity and intensity to the soundscape. By mirroring the flag balls of #путінхуйло 🔴⚫ UA and reversing the presidential insult, #зеленскийхуйло ⚪🔵🔴 establishes a link to 'the other' as a formulaic 'spell' that demarcates a pro-Russian audience through counter-imitation.

6 Conclusion: Detouring, Rerouting, Weaponization

The impact of (counter-)imitative memetic practices of war propaganda in the context of the Russian invasion of Ukraine relies on a multimodal expressive assembly, through which sounds and other elements allow for overlapping trajectories of engagement. While the platform has played an important role in boosting morale among Ukrainian troops and civilians, allowing users to share experiences and connect with others, it has also been susceptible to the strategies of weaponization. In our study, we focused on the communities involved in debunking Russian propaganda about the so-called 'special military operation' in Ukraine and those engaged in its circulation, both driven by the strategic rerouting of song-based video performances. A compelling illustration of this phenomenon involves the interweaving of the pro-Russian 'Good Morning' song with the pro-Ukrainian hashtag #путінхуйло 🔴⚫ UA. The networked connections between video performances established by soundscapes, pave the way for partisan encounters through the cross-cultural contestation of networked artefacts that are subsumed in the larger assembly of a TikTok post. While they consist of components that remain distinguishable, such content formations intertwine in an engaging, and often contradictory, manner rather than present a 'cohesive whole that supersedes its parts' (Parry, 2022, p. 4). This distinguishability affords easy exchange and subversion, giving rise to soundscapes – assemblies of engaging visual and textual elements that make the sound reverberate across different memetic communities.

Driven by the referential dynamics of memetic riffing and practices of attention hijacking, war propaganda on TikTok is formed through counter-imitative acts of sharing through combinations of hashtags, effects, duets, and sounds. Sound and its secondary orality on TikTok work as affective amplifiers. Pulling in as well as repelling people, sound animates processes of creative 'mix-and-match', attention hijacking, and in- and out-group polarisation. According to this logic, the Ukrainian song explored in our study not only exerted a binding effect for pro-Ukrainian publics; it also worked as a red flag to a bull for antagonist groups. By tracing the multimodal alignments between the Ukrainian song and its hijacked Russian version, we were able to address the interplay of mimicry and subversion rooted in the networked arrangements of audiovisual remix. The overlappings of content formations are crucial to our understanding of how engagement is maintained within such arrangements. Overlaps of oppositional soundscapes – as mediated through hashtags or, for example, shared video effects – point to the multimodal affordances of imitation and counter-imitation within TikTok's audiences. Using elements of the opponent's performance, the resulting attempts at audience coordination aim to simultaneously pre-scribe belonging and provoke a confrontation.

In the realm of WarTok, emojis, hashtags, and sounds 'condense' into reversible 'stand-ins' for affective expressions. Through imitation and repetition, networked content becomes part of a 'crowd' (Dean, 2016), turning elements into memetic formulas for communal bonding (Ong, 1982). By simultaneously preventing and provoking the sidetracking of meaning when reassembled, these formulas aid the acceleration of propagandist messages. In the soundscapes of 'Good evening, we are from Ukraine' and 'Good morning, we are from Russia' they come to serve as poles within a field of oppositions (Dean, 2016) where various elements of the assembly can be used to transform meaning. When the #путінхуйло 🔴⚫ hashflag intersects with 'Good Morning, we are from Russia,' the meaning emerging from this arrangement turns the song from a pro-Russian hijack into the object of partisan mockery. The role of sound-co-hashtag relations in shaping these dynamics is key. The design of WarTok propaganda results from a deliberate and purposeful envisioning of the potential engagement of like-minded communities, as well as of antagonists. Reverse the lyrics of a patriotic song and one has an affective component that, in assembly and reassembly with other components, shapeshifts meaning, provoking engagement on both sides.

As our analysis shows, the process of memeification through secondary orality (Ong, 1982) is, therefore, more targeted and pervasive than distinctly oral modes of propaganda. Multimodal content formations tell stories that 'blend deliberative and phatic, intentional and habitual, cognitive and affective means of expression' (Papacharissi, 2015, p. 25). This 'added deliberation' or 'level of intentionality' is relevant to the workings of WarTok, where the memeification of networked objects invites both imitative and counter-imitative action. In the propagandist realm, there is an 'added value' to the partisan other as a trigger for outrage. Owing to the high-arousal nature of polarised encounters, content that repels simultaneously provokes interactions (Berger & Milkman, 2012). Building upon the engagement of the like-minded, as well as that of the antagonists, WarTok contributions work as self-fulfilling prophecies that exploit anticipated outrage to perpetuate the flow. The absence of dialogue prioritises 'flat alignments' (Miller, 2008, p. 390) that facilitate stereotypical associations with national belonging while simultaneously expressing a collective tendency towards othering.

References

Abidin, C. (2021). Mapping internet celebrity on TikTok: Exploring attention economies and visibility labours. *Cultural Science Journal*, 12(1), 77–103.

Abidin, C. & Kaye, B. 2021. 'Audio memes, earworms, and templatability: The "Aural Turn" of memes on TikTok. In C. Arkenbout, J. Wilson & D. de Zeeuw (Eds.) *Critical Meme Reader: Global Mutations of the Viral Image* (pp. 58–68). Institute of Network Cultures: Amsterdam.

Ahmed, S. 2004. *The cultural politics of emotion*. Edinburgh University Press.

Askanius, T., & Keller, N. (2021). Murder fantasies in memes: Fascist aesthetics of death threats and the banalization of white supremacist violence. *Information, Communication & Society*, 24(16), 2522–2539.

Bainotti, L., Burkhardt, S., Cong, Y., Zhu, J., Lust, J., Babin, K., Esserghini, S., Ayaou, I., Kerboute, A., Mocombe, M., Lecat, F., Mohamed, A., Tamber, S., Mitter, D., Khosravi Ooryad, S., Leech, J., Elli, T., Zheng, K. (2022). Tracing the genealogy and change of TikTok audio memes. Available: WinterSchool2022TikTokAudioMemes < Dmi < Foswiki (digitalmethods.net)

Berger, J., & Milkman, K. L. (2012). What makes online content viral? *Journal of Marketing Research*, 49(2), 192–205. https://doi.org/10.1509/jmr.10.0353

Boichak, O. (2023) Mapping the Russian Political Influence Ecosystem: The Night Wolves Biker Gang. *Social Media + Society*, 9(2), 1–13 https://doi.org/10.1177/20563051231177.

Boler, M., & Davis, E. (2020). *Affective politics of digital media: Propaganda by other means*. New York: Routledge.

Boler, M., & Davis, E. (2020) Affect, media, movement: Interview with Susanna Paasonen and Zizi Papacharissi. In M. Boler & E. Davis (Eds.) *Affective Politics of Digital Media: Propaganda by Other Means*. (pp. 53–68). New York: Routledge.

Boudana, S., Frosh, P., & Cohen, A. A. (2017). Reviving icons to death: When historic photographs become digital memes. *Media, Culture & Society*, 39(8), 1210–1230.

Carson, E. (2022, April 6). How TikTok Songs Make the Most Intense Earworms Ever. https://www.cnet.com/culture/heres-why-tiktok-songs-are-stuck-in-your-head/

Dean, J. (2016, December 31). Faces as commons: The secondary visuality of communicative capitalism. *Open! Platform for Art, Culture, and the Public Domain*, onlineopen.org/faces-as-commons

Filtenborg, E. (2021, March 19). In Ukraine, Stepan Bandera's legacy becomes a political football … again. *Euronews*. https://www.euronews.com/my-europe/2021/03/19/in-ukraine-stepan-bandera-s-legacy-becomes-a-political-football-again

Friedman, T. (2022). We have never been here before. *The New York Times*. https://www.nytimes.com/2022/02/25/opinion/putin-russia-ukraine.html

Geboers, M., Pilipets, E., Gheli, F., Benedetti, A., Li, Y., Sathish, C., Lajosi, K., Almada, P., Xiao, F., Cano-Orón, L., Espinoza, A., E, Z., & Kleszczyńska, I. (2022). How long do we #standwithukraine? Available: https://wiki.digitalmethods.net/Dmi/Summerschool2022Howlongdowestandwithukraine

Geboers, M., & Pilipets, E. (2024). Networked masterplots: Music, pro-Russian sentiment, and participatory propaganda on TikTok. *Journal of Digital Social Research*, 6(1), 90–103.

Grandinetti, J., & Bruinsma, J. (2022). The affective algorithms of conspiracy TikTok. *Journal of Broadcasting & Electronic Media*, 1(20), 274–293

Goncharenko, R. (2022, May 22). Stepan Bandera: Hero or Nazi collaborator? *DW*. https://www.dw.com/en/stepan-bandera-ukrainian-hero-or-nazi-collaborator/a-61842720

Goncharova, O. (2014, July 11). Lifestyle blog: Kharkiv, with new anti-Russian song, becomes capital of anti-Putin music. *Kyiv Post*. https://archive.kyivpost.com/article/guide/music/lifestyle-blog-soccer-fans-chant-putin-khuilo-makes-it-to-world-fame-353315.html

Goodman, S. (2010). *Sonic Warfare: Sound, affect, and the ecology of fear*. MIT Press.

Hagen, S. (2017) Polandball is off Reddit: How r/polandball transcends memes through carefully curated geopolitical satire. Institute of Network Cultures https://networkcultures.org/longform/2017/11/15/polandball-is-of-reddit-how-rpolandball-transcends-memes-through-carefully-curated-geopolitical-satire/

Hautea, S., Parks, P., Takahashi, B., & Zeng, J. (2021). Showing they care (or don't): Affective publics and ambivalent climate activism on TikTok. *Social Media + Society*, 7(2), 1–14. 20563051211012344.

Highfield, T. (2016). *Social media and everyday politics*. Cambridge: Polity.

Highfield, T. (2018). Emoji hashtags//hashtag emoji: Of platforms, visual affect, and discursive flexibility. *First Monday*, 23(9), N/A

Kanai, A. (2015). WhatShouldWeCallMe? Self-Branding, individuality and belonging in youthful femininities on Tumblr. *M/C Journal*, 18(1). N/A

Kaye, D. B. V., Zeng, J., & Wikstrom, P. (2022). *TikTok: Creativity and culture in short video*. John Wiley & Sons.

Kaye, D. B. V., Chen, X., & Zeng, J. (2021). The co-evolution of two Chinese mobile short video apps: Parallel platformization of Douyin and TikTok. *Mobile Media & Communication*, 9(2), 229–253.

Kerby, N. (2022). TikTok, Russia & the emerging Splinternet. Available at: https://tracking.exposed/blog/tiktok-russia-the-emerging-splinternet/

Kirby, D., & Özkula, S. M. (2023). The dead pig's photo album: Affective visual rituals in collective identity formation. *International Journal of Cultural Studies*, 26 (5), 567–587. 13678779231172788.

Kiparoidze, M. (2022). TikTok influencers are dancing, lip-syncing, and posing to promote Russia's war in Ukraine. https://www.codastory.com/disinformation/tiktok-influencers-are-dancing-lip-syncing-and-posing-to-promote-russias-war-in-ukraine/

Кононенко, О. (2022) У РФ продають пакети для сміття, присвячені російським військовим https://nv.ua/ukr/lifestyle/u-rf-z-yavilisya-smittyevi-paketi-prisvyacheni-rosiyskim-viyskovim-foto-50321349.html

KnowYourMeme (2023). Polandball. https://knowyourmeme.com/memes/polandball

Lessig, L. (2008). *Remix*. London: Penguin Books.

Light, B., Burgess, J., & Duguay, S. (2018). The walkthrough method: An approach to the study of apps. *New Media & Society*, 20(3), 881–900. https://doi.org/10.1177/1461444816675438

Miller, V. (2017). Phatic culture and the status quo: Reconsidering the purpose of social media activism. *Convergence*, 23(3), 251–269.

Мороз, В. (2013). Прапор червоно-чорний – це наше знамено ...". Яким був стяг УПА https://www.istpravda.com.ua/articles/2013/10/18/138044/

Mobilio, M. (2022, February 28). Welcome to WarTok. *NewYork*. Retrieved from: https://nymag.com/intelligencer/2022/02/tiktok-ukraine-war-video.html

Newton, G., Zappavigna, M., Drysdale, K., & Newman, C. E. (2022). More than humor: Memes as bonding icons for belonging in donor-conceived people. *Social Media+ Society*, 8(1), 1–13, 20563051211069055.

Ong, W.S.J. (1982). *Orality and literacy – The technologizing of the word*. London: Methuen.

Papacharissi, Z. (2015). *Affective Publics: Sentiment, Technology, and Politics*. Oxford University Press. https://doi.org/10.1093/acprof:oso/9780199999736.001.0001

Peeters, S. (2022) Zeeschuimer. https://github.com/digitalmethodsinitiative/zeeschuimer

Pilipets, E. (2023). Hashtagging, duetting, sound-linking: TikTok gestures and methods of (In)distinction. *The Journal of Media Art Study and Theory*, 4(1), 109–135.

Pilipets, E. & Geboers, M. et al. (2023). WARTOK: Networked Soundscapes of Memetic Warfare. Panel presented at AoIR 2023: The 24th Annual Conference of the

Association of Internet Researchers. Philadelphia, PA, USA: AoIR. Retrieved from http://spir.aoir.org.

Parry, K. (2022). *A theory of assembly: From museums to memes*. Minneapolis: Minnesota Press.

Rogers, R.A. (2013). *Digital methods*. Cambridge, MA: MIT Press.

Schellewald, A. (2022). Theorizing "stories about algorithms" as a mechanism in the formation and maintenance of algorithmic imaginaries. *Social Media+Society*, 8 (1), 1–10. 20563051221077025.

Shevtsova, M. (2022) Looking for Stepan Bandera: The myth of Ukrainian nationalism and the Russian 'Special Operation.' *Central European Journal of International and Security Studies*. 16(3), 132–50.

Скоркин, К. (2022). От Бандеры до «Азова»: отвечаем на главные вопросы об украинском национализме https://meduza.io/feature/2022/04/17/ot-bandery-do-azova-otvechaem-na-glavnye-voprosy-ob-ukrainskom-natsionalizme

Sparklxxs (2020, November 12). Shrek poop fly. Available at: https://tenor.com/view/shrek-poop-fly-diarrhea-gif-19530885

Stenglin, M. K. (2004). Packaging curiosities: Towards a grammar of three-dimensional space. https://ses.library.usyd.edu.au/handle/2123/635

Tiffany, K. (2022). The Myth of the 'First TikTok War'. *The Atlantic*. https://www.theatlantic.com/technology/archive/2022/03/tiktok-war-ukraine-russia/627017/

Venturini, T. (2022). Online conspiracy theories, digital platforms and secondary orality: Toward a sociology of online monsters. *Theory, Culture & Society*, 39(5), 61–80. https://doi.org/10.1177/02632764211070962.

Vicari, S., Iannelli, L., & Zurovac, E. (2020). Political hashtag publics and counter-visuality: a case study of# fertilityday in Italy. *Information, Communication & Society*, 23(9), 1235–1254.

Vizcaíno-Verdú, A., & Abidin, C. (2022). Music challenge memes on TikTok: understanding in-group storytelling videos. *International Journal of Communication*, 16(26), 883–908.

Wong, J., Lee, C., Long, V. K., Wu, D., & Jones, G. M. (2021). "Let's go, baby forklift!": Fandom governance and the political power of cuteness in China. *Social Media+ Society*, 7(2), 1–18, 20563051211024960.

Zaitsev, O. (2016). De-mythologizing Bandera: Towards a scholarly history of the Ukrainian nationalist movement. *Journal of Soviet and Post-Soviet Politics and Society*, 1(2), 411–20.

Zelizer, B. (2006). What's untransportable about the transport of photographic images? *Popular Communication*, 4(1), 3–20.

Zulli, D, & Zulli. D. J. (2022). Extending the Internet Meme: Conceptualizing Technological Mimesis and Imitation Publics on the TikTok Platform. *New Media & Society*, 24(8), 1872–90.

CHAPTER 8

Making Sense of Post-coup Myanmar through Facebook

Paula Romero Jiménez, Ana Melchor Pérez, Siebe Dekker, Miguel Oliveira Royo and Jing Zeng

1 Introduction

Democratic societies are often confronted with internal and external challenges with disruptive potential, such as political overthrows, civil disorder, and exposure to systemic corruption (Huntington, 1996). Such events can involve diverse societal groups and, in extreme cases, test the resilience of both democratic political systems and governance structures. This is especially a problem in young or nascent democracies, such as Myanmar. Despite the country's tentative progress toward democracy in recent years, Myanmar saw a coup d'état in 2021 that disrupted its democratization process (Paddock, 2022).

This coup was planned and executed by the Tatmadaw, Myanmar's military, which has had virtually absolute political power for most of the country's history. The triggering event for the coup was the electoral defeat of the military's political party, the Union Solidarity and Development Party (USDP), in the 2020 elections (McKenna, 2024). In the aftermath of these elections, the military claimed that the victory of Aung San Suu Kyi's National League for Democracy (NLD) was fraudulent and unlawful (IDEA, 2022). However, these accusations were rejected by independent judges (IDEA, 2022). Shortly after the inauguration on February 1, 2020, the self-declared acting president, Tatmadaw General Myint Swe, announced a one-year state of emergency. As a direct response to the military coup, protesters took to the streets and laborers went on strike for much of 2021, actions which were met with a violent response by the military regime (Paddock, 2022).

As part of the military government's drastic approach to internet control, Facebook, the country's most popular online platform with millions of users, was officially banned (Schirch, 2021). However, Facebook remains the central arena where information is circulated, and collective actions are organized by activists at home or in exile (Fink, 2018). Due to a crackdown on

 | DOI:10.1163/9789004711396_010

independent journalists, including international correspondents, the role of Facebook as a central information venue in Myanmar has been further amplified during times of unrest. Journalists and activists in exile utilize the platform to inform both the Myanmar public and international audiences about the unfolding political unrest.

To critically analyze and understand how digital technology factors into political conflict and disruptions of democratization processes, the use of Facebook during the internal strife in Myanmar offers an important and insightful empirical case. Thus, this chapter sets out to explore public discussions about Myanmar, especially related to political unrest, in 2022, a period heavily affected by the earlier military coup.

The study aims to (a) outline the contexts in which post-coup Myanmar was discussed by different groups ofs on Facebook, and (b) compare how the coverage was received by the public on Facebook. To achieve this, a dataset of Facebook posts related to Myanmar from 2022 is analyzed using automated content analysis, including topic modeling and sentiment analysis.

This chapter aims to make two important contributions to the research on political conflicts in a non-Western context. First, it will provide empirical insights into the dynamics between social media discourses and political events, allowing for further theorizing of the extent to which and the contexts in which digital media reveal their 'impact' on political developments. Second, the chapter proposes a methodological approach that utilizes computational methods, considering different relevant analytical dimensions of political communication and discourse on online platforms.

This chapter begins by discussing the relevance of social media in political discourses, especially political unrest, before shifting the focus to the case of Myanmar. It then outlines the methodological approach and critically discusses the results of the social media data analysis. It concludes with suggestions for how the methodology could be further refined for future research on political unrest and the role of social media.

2 Social Media and Political Engagement

Social media platforms are increasingly significant in distributing political information and encouraging political engagement. The existing literature illuminates how politicians leverage social media to interact directly with the public across democratic and authoritarian regimes. In democratic environments, there is evidence of the effectiveness of politicians employing social media to engage with potential supporters, especially during election campaigns (Fuchs, 2018; Lu et al., 2017; Vaccari, 2017). In contrast, research on public engagement by authoritarian regimes on social media focuses on the capacity of new technologies to monitor, direct, and even manipulate public opinion (King et al., 2017; Stockmann & Luo, 2017).

From the perspective of information consumers, the central role of social media in the political information ecosystem is reflected in their reliance on these platforms for news and insights. Facebook, in particular, has become a significant source of political news in numerous countries, including Myanmar. However, this reliance comes with the burgeoning issue of unreliable political (mis)information. The decentralized model of content creation and the lack of stringent quality control substantially contribute to online users' susceptibility to disinformation and misinformation (Zeng & Brennen, 2023). Recent scholarly work has linked political misinformation to the incitement of extremism, the erosion of democratic discourse, and even the instigation of violence (Bleakley, 2023; Fuchs, 2018).

Although the role of social media in propagating political (mis)information rapidly and on a larger scale has been acknowledged, the extent to which social media fosters meaningful discussion remains less certain. Dixon (2023) found that a mere 19% of users surveyed in the United Kingdom engaged in online political discourse. Mascheroni and Murru (2017) identified in their interviews from Germany, Italy, and the United Kingdom that unpredictability and the fear of escalation are common factors leading to people's hesitation in participating in online discussions about politics. In countries with repressive political systems, there are more barriers to engaging in political discussions online. In authoritarian countries, individuals voicing critical views against authorities can face serious repercussions (Zeng et al., 2017), and even imprisonment (Bychawska-Siniarska, 2017).

Alongside propagating political information and fostering political discourse, the existing literature provides rich insights into how digital media can cultivate unique forms of collective action and social movements (Bennett & Segerberg, 2013; Earl & Kimport, 2011). Since the early 2000s, a growing body of scholarship has emerged to critically examine how new information and communication technologies bolster offline political engagement and collective action by mobilizing resources and amplifying the reach of movements (e.g., Cheung, 2019; Nip, 2004; Van Aelst & Walgrave, 2002). A significant moment in this scholarship occurred in the aftermath of the so-called Arab Spring (Lim, 2013; Meraz & Papacharissi, 2013), which some have termed "the social media revolution" (Dubose, 2011).

While many studies have acknowledged digital media's role in mobilizing and connecting supporters of political causes (Meraz & Papacharissi, 2013;

Nip, 2004; Van Aelst & Walgrave, 2002), others have urged caution regarding the presumed impacts of social media, advising against overhyping or oversimplifying these effects. For example, during the Arab Spring, social media was also exploited by authorities to conduct surveillance on activists (Markham, 2014). Another frequent critique addresses the phenomenon of "slacktivism" – a term denoting superficial engagement in activism via social media, characterized by actions requiring minimal effort, such as simply sharing, liking, or commenting (Glenn, 2015).

The extent to which social media benefits social movements is closely tied to sociopolitical and technological contexts. As Wolfsfeld et al. (2013, p. 1) contended, "One cannot understand the role of social media in collective action without first considering the political environment in which they operate." This necessitates a more comprehensive approach through in-depth case studies, particularly those involving non-English-speaking regions and non-Western democracies. The present study is centered on Myanmar in its post-coup context and pays special attention to Facebook, the country's most prominent online platform. The following section delineates Facebook's unique role in Myanmar society.

3 Myanmar and Facebook

For much of its postcolonial history, the Myanmar public has grappled with restricted access to information and stringent media control. For instance, after the military junta came to power in 1962, it immediately seized control over the media and imposed strict censorship policies (Fink, 2018). In the telecommunications sector, Myanmar has lagged significantly behind the rest of the world. Until the mid-2000s, it had the lowest mobile penetration globally; this only began to change when the sector was opened up to foreign companies offering affordable SIM cards and internet data usage (Asher, 2021).

The booming telecommunications sector in the region is most notably exemplified by the widespread proliferation of Facebook. Despite entering Myanmar in 2010, the platform did not experience significant uptake until after 2014. This surge in popularity was attributed to a particular business model adopted by local telecommunications companies that involved preinstalling the app on mobile phones and offering it without data charges (Asher, 2021). This approach facilitated Facebook's proliferation throughout the country, which in turn created a new platform for the free exchange of information and the expression of political opinions (Fink, 2018).

It is worth noting that unlike in other international markets where Facebook primarily serves as a means to connect family and friends, the platform functions as an important news and information channel in Myanmar. According to a 2016 article in *The Atlantic* (Mod, 2016), users in Myanmar tend to use Facebook for anonymously accessing information rather than for socializing. To a certain extent, Facebook has taken on the role of serving the local public as both a news portal and a search engine. Surveys conducted in 2017 and 2020 revealed that nearly all respondents relied on Facebook for news, particularly political news, and also indicated that users in Myanmar tended to trust information obtained from Facebook more than from international or official government sources (Whitten-Woodring et al., 2020). This reliance and trust on Facebook as a credible information source leave the Myanmar public especially susceptible to manipulation and harmful content on the platform.

Facebook's influence on Myanmar society came under intense international scrutiny during the 2017 Rohingya crisis. The United Nations Human Rights Council reported that Facebook had been used to host hate speech and misinformation against the Rohingya minority, inciting ethnic cleansing (Miles, 2018). Alongside its 'dark history' of inciting violence, Facebook has also been a hub for Myanmar's nascent pro-democracy movements. In February 2021, the State Administration Council (SAC), under the military government, conducted a strict crackdown on communication and information flows by issuing directives to block social media and implementing internet shutdowns (Faxon et al., 2023). Despite this drastic approach, Facebook remains a highly relevant arena for the country's activism. For instance, Faxon et al.'s (2023) study observed that after the 2021 military coup, the primary focus of social media content in Myanmar, especially on Facebook, shifted from agricultural topics to political activism. There was a surge in pro-democracy content, such as posts documenting peaceful protests and messages expressing dissent against the regime. This transformation highlights a significant evolution in the platform's use during political crises, underscoring the necessity for continued research on Myanmar-centric discourses on Facebook in the post-coup era.

4 Research Design

To study Myanmar-related discussions on Facebook, this study gathered 56,486 posts from public Facebook pages through Meta's CrowdTangle Application Programming Interface (API). The English search term 'myanmar' was used, and all available posts published in 2022 were retrieved for the study. Subsequent data cleaning eliminated non-English posts, yielding a dataset of 42,596 posts for further analysis. To address the first research objective – to compare the contexts of post-coup Myanmar on Facebook by different actor groups – this study integrates the results from topic modeling and user profile analysis. To generate a rich understanding of the Facebook-based discussion about Myanmar, the data analysis was constructed around three analytical dimensions: ***content, content creators,*** and ***audience engagement.*** Below, to answer our research questions, we elaborate on how these elements were operationalized and used.

4.1 *Content*

Two automated content analysis techniques were used: topic modeling and sentiment analysis. To identify the key contexts for discussing Myanmar on Facebook (the first research aim), we used topic modeling. Text data from Facebook posts were first preprocessed, which involved turning texts into lowercase and removing URLs, punctuation, special symbols, and stop words. K-means clustering[1] was used to reveal the key topical groups within the Facebook data. After transforming the preprocessed posts into term frequency–inverse document frequency (TF-IDF) vectors, the optimal number of clusters was determined by analyzing inertia through the elbow method, as detailed by Marutho et al. (2018). This method, which evaluates the sum of the squared distances between points and their respective cluster centroids, was used to facilitate the identification of an ideal cluster quantity. Through this approach, the number of topics was determined. Following this, the K-means algorithm was applied to categorize the posts into distinct clusters. This step allowed us to discern keywords associated with each topic.

Whereas the topic modeling results would inform us about the composition of issues associated with Myanmar, sentiment analysis would provide complementary insights into the emotional positioning taken in Facebook posts. Therefore, this study used Facebook posts' sentiment stances in tandem with their topical focus to understand how Myanmar was discussed on Facebook (the second research aim). This study employed the Valence Aware Dictionary and Sentiment Reasoner (VADER), a lexicon-based sentiment analysis. Sentiment analysis makes it possible to identify text as

---

1 The authors also conducted LDA-based topic modeling. Upon examining the results, it appeared that K-mean clustering provided more interpretable results.

negative, positive, or neutral, and established approaches can be classified into two categories: dictionary-based (lexicon) and machine learning methods (Wankhade et al., 2022). This study utilized a lexicon-based approach, as no labeled data were required. More precisely, VADER was used because it is optimized for social media texts (Hutto & Gilbert, 2014). Evidence from prior research shows that VADER performs better than other dictionary-based approaches when social media data are concerned (Al-Shabi, 2020; Srivastava et al., 2022). However, VADER is limited in how it interprets more complicated language usage, such as sarcasm (Wankhade et al., 2022).

4.2 *Content Creators*

To examine and group content creators publishing Myanmar-related content on Facebook, the study used page profile information provided by Facebook. In the original post metadata retrieved from the API, each page is assigned a predetermined label by Facebook for categorization purposes. Our database contained a variety of unique pages, totaling 593 category labels. Examples of predetermined labels include "personal_blog," "edu_site," "political_organization," and "news_site." These labels were qualitatively grouped into seven categories: media, government, politicians, education, nonprofit organizations, individual people, and others. This would enable comparisons between different content creator types to be performed in combination with the results of topic modeling and sentiment analysis.

4.3 *Audience Engagement*

To examine how audiences on Facebook responded to and engaged with different content associated with Myanmar (the second research aim), this study used two types of engagement data. The first type included standard engagement metrics, such as counts of likes, shares, and comments. The second type included users' emoji reactions, with these visual responses represented by textual labels in the structured metadata from the API – namely, "sad," "care," "angry," "love," "wow," and "haha." These parameters were used in this study to measure and compare audience reactions and sentimental responses to Myanmar-related stories published on Facebook.

The findings from three analytical dimensions – content, content creators, and audience engagement – yielded valuable quantitative insights. To further interpret and contextualize these results, a qualitative close reading was employed. In the following section, we triangulate the results of the aforementioned analysis to effectively address the primary research objectives.

5 Results

5.1 *Contextualizing Myanmar on Facebook*

As informed by the topic modeling results, there were seven key issues commonly covered in Myanmar-related discussions on Facebook: (a) commercial information; (b) Myanmar's relationship with Southeast Asia; (c) the political situation in Myanmar; (d) international news; (e) COVID-19 updates in Myanmar; (f) daily life in Myanmar; and (g) telecommunications in Myanmar.

As shown in Table 8.1, discussions related to the political situation in Myanmar encompassed a prominent topic in the dataset. The content on this subject aligned closely with major political events within the country. Figure 8.1 illustrates the number of posts on this topic throughout 2022, presented in two-week intervals. Two significant peaks could be observed in the graph. The first spike appeared on February 1, marking the one-year anniversary of the coup. Around this date, a large number of posts were shared

TABLE 8.1 Topic results from K-means

	Topic keywords	**Total number of posts**
Topic 1: Commercial information	islands, guinea, republic, congo, south, french, island, korea, sudan, united	864
Topic 2: Myanmar's relationship with Southeast Asia	thailand, vietnam, indonesia, asia, philippines, malaysia, cambodia, laos, singapore, asean	3697
Topic 3: Political situation in Myanmar	junta, military, happening, report, read, help, regime, people, coup, state	3867
Topic 4: International news	day, news, one, link, us, world, people, time, international, miss	27273
Topic 5: COVID-19 updates in Myanmar	new, year, time, cases, us, deaths, service, thailand, design, total	1320

TABLE 8.1 Topic results from K-means (*cont.*)

	Topic keywords	**Total number of posts**
Topic 6: Daily life in Myanmar	yangon, pm, please, township, job, experience, street, service, location, work	4489
Topic 7: Telecommunications in Myanmar	good, phone, pro, red, view, yangon, day, made, contact, news	1086

on Facebook to commemorate the event and express solidarity. Another event that drew attention from both national and international communities – a notable interruption – was the execution of four prodemocracy activists on July 25. The executions sparked a wave of responses from the public and human rights organizations, leading to the second major spike observed in Figure 8.1.

While the topic modeling results shed light on the range of issues related to Myanmar, sentiment analysis offered an additional perspective by revealing the emotional tone expressed in Facebook posts. This study used sentiment as the second input to examine the contexts in which Myanmar was discussed on Facebook. Table 8.2 shows the sentiment distribution across all seven topics.

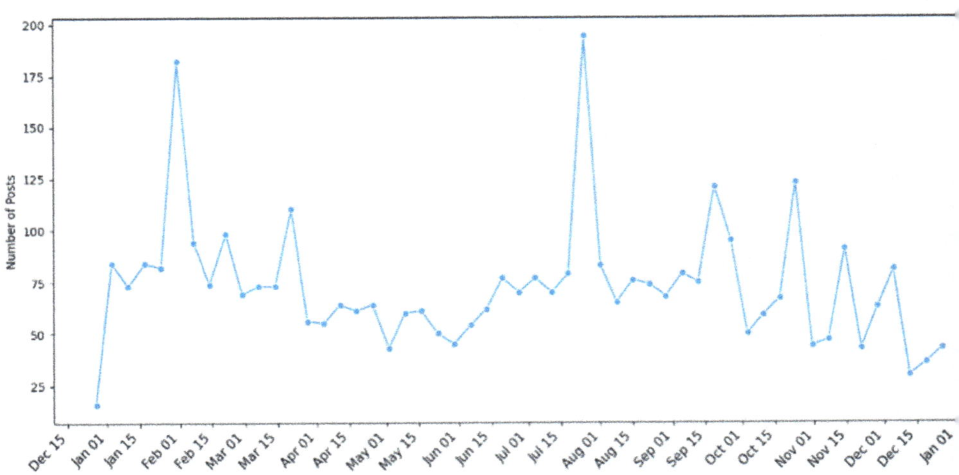

FIGURE 8.1 Political situation topic frequency over two-week intervals

TABLE 8.2 Average sentiment score per topic

	Score
Topic 1: Commercial information	0.448
Topic 2: Myanmar's relationship with Southeast Asia	0.425
Topic 3: Political situation in Myanmar	−0.184
Topic 4: International news	0.283
Topic 5: COVID-19 updates in Myanmar	0.443
Topic 6: Daily life in Myanmar	0.557
Topic 7: Telecommunications in Myanmar	0.435

As the results show, while most topics appeared to feature more positive and neutral sentiments, discussions related to the political situation were loaded with more negative sentiment. To further examine how negative sentiments arose, a time series of the sentiment score is presented in Figure 8.2.

From Figure 8.2, three significant declines in absolute sentiment could be observed. The absolute sentiment was calculated by summing the sentiment of all the posts about the political situation topic in the dataset from the corresponding week. A close examination of the posts with negative sentiment scores from the dataset revealed that these sudden shifts in sentiment

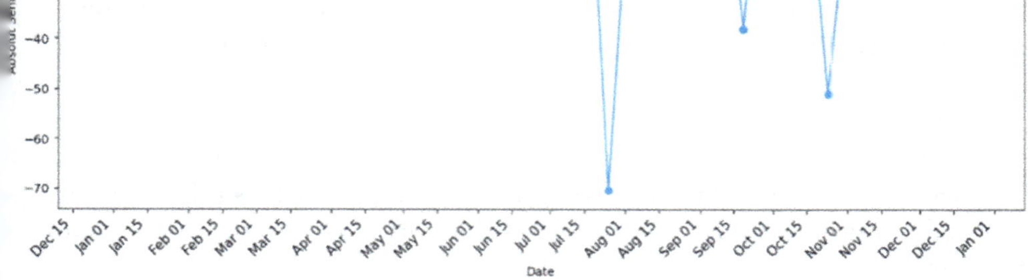

FIGURE 8.2 Absolute sentiment score over time for the political situation topic.

could be attributed to three specific events. The most pronounced drop occurred in late July, aligning with the previously discussed execution of four pro-democracy activists. The second significant dip occurred on September 16, following a school bombing that resulted in numerous dead and injured children. On October 23, another steep decline in sentiment scores could be noted. This corresponds with the Myanmar Air Force's airstrike on a concert, a deliberate attack on the anniversary of the Kachin Independence Organization that killed 80 civilians (Amnesty International, 2023).

5.2 *Whose Voice?*

The second analytical objective of this study was to examine the main content creators contributing information related to Myanmar on Facebook. As explained in the previous section, Facebook pages from the dataset were categorized into seven groups: media organizations (n = 7883), governmental institutions (n = 1039), politicians/political parties (n = 306), public figures (n = 3607), nonprofit organizations (n = 1844), education institutions (n = 1187), and others (n = 12308). Media organizations were among the most preeminent content creator groups. The most active media organizations in this dataset were The Irrawaddy and Mizzima. Both are Burmese media organizations fighting to stay independent despite the junta having revoked their licenses after the coup (Walker, 2023). In general, what stood out was that national media posts were the most dominant in the dataset. However, there were also international media organizations located, for example, in Japan (Nikkei Asia), the United States (Radio Free Asia), and Australia (Chindwin News Agency) that actively posted about Myanmar.

These content creator groups had a varied focus in their discussions about Myanmar. In Figure 8.3, which shows the topic distribution of posts published by these seven content creator groups, the media organizations' posts were predominantly about the political situation topic. This suggests the crucial role played by media organizations in informing the public of the unfolding political unrest in Myanmar. Besides the media, there were not many other content creator groups posting about the political situation.

5.3 *Audience Responses to Myanmar-Related Stories*

To identify how the Facebook audience engaged and responded to Myanmar-related stories, this study compared engagement metrics (likes, shares, and comments) and emoji responses (sad, care, angry, love, wow, and haha).

Table 8.3 provides insights into audience responses to various topics related to Myanmar. While there were variations in engagement metrics

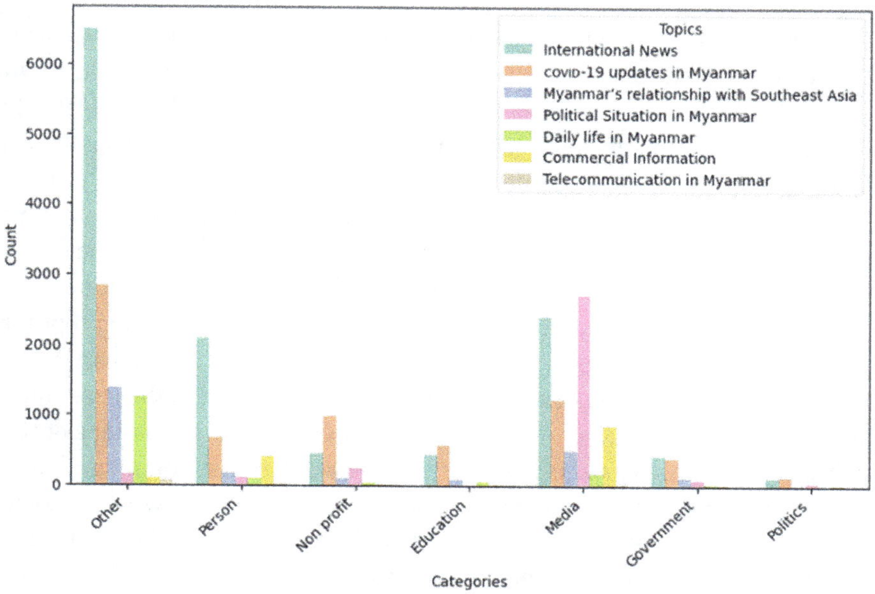

FIGURE 8.3 Bar plot with the most frequent category for each topic.

TABLE 8.3 Average engagement metrics received for each of the topics

	Average like count	Average share count	Average comment count
Topic 1: Commercial information	146.003	16.358	179.017
Topic 2: Myanmar's relationship with Southeast Asia	709.516	89.777	87.239
Topic 3: Political situation in Myanmar	431.083	61.313	21.373
Topic 4: International news	836.573	89.429	55.299
Topic 5: COVID-19 updates in Myanmar	498.924	58.458	28.176
Topic 6: Daily life in Myanmar	336.787	27.539	23.657
Topic 7: Telecommunications in Myanmar	177.840	7.413	20.678

across topics, it is noteworthy that posts concerning the political landscape commanded significant attention. Despite ranking lower on average likes compared to other topics, political posts demonstrated a notable share count (61.31). This observation suggests that while users may have hesitated to express approval through likes on politically charged posts, their inclination to share reflected their deep engagement with the subject matter. In addition, it stood out that posts about the political situation had a relatively low average comment count compared to the other topics (21.37). This could indicate that users mostly used Facebook to receive and share political news instead of engaging in discussions about politics in the comment section. This is in line with the results of Figure 8.3, which shows that most of the posts about the political situation were from media organizations.

Interestingly, the predominant reaction for all topics was love, except for the political situation topic, which primarily elicited sad reactions (Figure 8.4 and Table 8.4). In the context of Myanmar-related stories, the love emoji should be interpreted as a way of expressing solidarity. In discussions of the political situation, the sad face emoji emerged as the most dominant. These results are consistent with the findings from sentiment analysis and topic modeling. As previously noted, content under the political situation topic included distressing developments, such as the execution of prodemocracy activists and casualties among civilians, which evoked widespread negative sentiment. Given this context, it is unsurprising that sadness was the emotion most expressed by the audience on this topic.

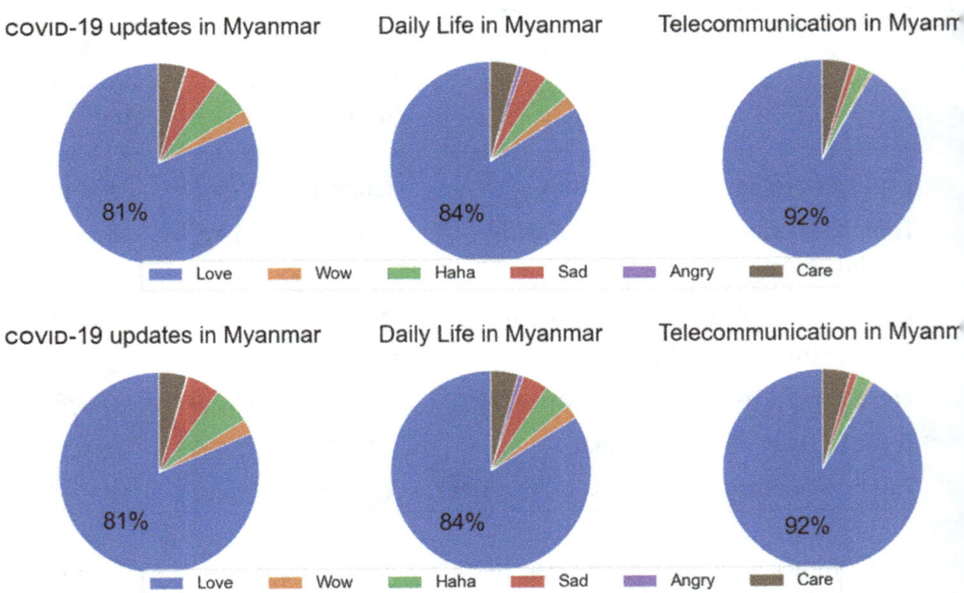

FIGURE 8.4 Percentage of Facebook emoji reactions per topic.

TABLE 8.4 Percentage of Facebook emoji reactions per topic.

	Love	Wow	Haha	Sad	Angry	Care
Topic 1: Commercial information	83.57	2.20	3.79	1.10	0.73	8.62
Topic 2: Myanmar's relationship with Southeast Asia	58.73	9.48	20.46	4.99	0.64	5.71
Topic 3: Political situation in Myanmar	27.24	1.66	11.18	44.59	10.89	4.45
Topic 4: International news	75.62	1.54	9.31	8.72	0.54	4.27
Topic 5: COVID-19 updates in Myanmar	81.31	2.40	6.16	5.36	0.21	4.55
Topic 6: Daily life in Myanmar	83.88	2.27	4.41	4.04	0.86	4.55
Topic 7: Telecommunications in Myanmar	91.58	0.35	2.34	1.11	0.06	4.55

6 Discussion and Conclusion

This study investigated how post-coup Myanmar was discussed by different actor groups on Facebook, focusing on both content and public responses to related social media posts. The analysis provides insights into the role of the platform during politically unstable times. For example, while the unfolding political situation in the region was a salient topic, our analysis also revealed that spam and other commercial posts hijacked attention on Myanmar from international audiences. At the same time, the findings of this study demonstrate that social media analytics can provide insights into the unfolding of major events and associated public responses. For instance, by combining sentiment and topic analysis, social media analytics appeared to be useful in detecting important events following the coup in Myanmar.

The analysis of actors contributing to content related to Myanmar showed that media organizations played the most important role in propagating information about the country. Although this finding is not surprising, it is important to interpret it against the media environment in post-coup Myanmar. As previously discussed, in the aftermath of the coup, news organizations, especially independent journalism organizations in Myanmar, faced severe control and sanctions. This attempt to control information flow and the narrative of the political situation could be undermined by international media and Myanmar exile journalists who contributed to publishing information from and about Myanmar. Facebook, a platform that the junta attempted to ban, remained an important medium through which an international audience was informed about the unfolding political unrest and showed support for civilians and activists.

This study has several limitations. It does not consider social media posts in Burmese, primarily due to the lack of computational tools to work with the language. Future research with access to the language and the technological expertise needed to work with Burmese should generate comparative insights into the extent to which discussions on Facebook differ between English and Burmese content. Moreover, it is worth noting that Facebook's content moderation could impact the data available for research. For instance, posts may have been removed due to hate speech or coordinated malicious campaigns by the military (Meta, 2024). Meta's internal content policy could also affect the completeness of content made available through the CrowdTangle API.

References

Al-Shabi, M. (2020). Evaluating the performance of the most important Lexicons used to Sentiment analysis and opinions Mining. *Journal of Computer Science and Network Security,20*(1), 1–7.

Amnesty International (2023). *Myanmar: 28 civilians killed in military air strike – New investigation and witness testimony.* https://www.amnesty.org/en/latest/news/2023/10/myanmar-28-civilians-killed-in-military-air-strike-new-investigation-and-witness-testimony/

Asher, S. (2021, February 4). *Myanmar coup: How Facebook became the 'digital tea shop.'* BBC News. https://www.bbc.com/news/world-asia-55929654

Bennett, W. L., & Segerberg, A. (2012). The logic of connective action: Digital media and the personalization of contentious politics. *Information, Communication & Society*, 15(5), 739–768.

Bleakley, P. (2023). Panic, pizza and mainstreaming the alt-right: A social media analysis of Pizzagate and the rise of the QAnon conspiracy. *Current Sociology*, 71(3), 509–525.

Bychawska-Siniarska, D. (2017). *Protecting the right to freedom of expression under the European Convention on Human Rights: A handbook for legal practitioners.* Council of Europe. https://edoc.coe.int/en/fundamental-freedoms/7425-protecting-the-right-to-freedom-of-expression-under-the-european-convention-on-human-rights-a-handbook-for-legal-practitioners.html

Cheung, C. H. (2019). *Social media and political participation: A case study of Facebook as a platform of communication, mobilization and action*. Lingnan University. https://commons.ln.edu.hk/cgi/viewcontent.cgi?article=1067&context=otd

Dixon, S. J. (2023, December 19). *Social media and politics in the United Kingdom (UK)*. Statista. https://www.statista.com/topics/9885/social-media-and-politics-in-the-united-kingdom-uk/#topicOverview

Dubose, C. (2011). The social media revolution. *Radiologic Technology*, 83(2), 112–119.

Earl, J., & Kimport, K. (2011). *Digitally enabled social change: Activism in the internet age*. MIT Press.

Faxon, H. O., Kintzi, K., Tran, V., Wine, K. Z., & Htut, S. Y. (2023). Organic online politics: Farmers, Facebook, and Myanmar's military coup. *Big Data & Society*, 10(1). https://doi.org/10.1177/20539517231168101

Fink, C. (2018). Dangerous speech, anti-Muslim violence, and Facebook in Myanmar. *Journal of International Affairs*, 71(1.5), 43–52. https://www.jstor.org/stable/26508117

Fuchs, C. (2018). *Digital demagogue: Authoritarian capitalism in the age of Trump and Twitter*. Pluto Press.

Glenn, C. L. (2015). Activism or "Slacktivism?": Digital media and organizing for social change. *Communication Teacher*, 29(2), 81–85.

Huntington, S. P. (1996). *Political order in changing societies*. Yale University Press.

Hutto, C. J., & Gilbert, É. (2014). VADER: A parsimonious rule-based model for sentiment analysis of social media text. *Proceedings of the International AAAI Conference on Web and Social Media*, 8(1), 216–225. https://doi.org/10.1609/icwsm.v8i1.14550

International Institute for Democracy and Electoral Assistance (IDEA). (2022). *Unconstitutionality of the 2021 military coup in Myanmar*. https://www.idea.int/sites/default/files/publications/unconstitutionality-of-the-2021-military-coup-in-myanmar.pdf

King, G., Pan, J., & Roberts, M. E. (2017). How the Chinese government fabricates social media posts for strategic distraction, not engaged argument. *American Political Science Review*, 111(3), 484–501.

Lim, M. (2013). Many clicks but little sticks: Social media activism in Indonesia. *Journal of Contemporary Asia*, 43(4), 636–657. https://doi.org/10.1080/00472336.2013.769386

Lu, Y., Lee, J. K., & Kim, E. (2018). Network characteristics matter in politics on Facebook: Evidence from a US national survey. *Online Information Review*, 42(3), 372–386. https://doi.org/10.1108/oir-09-2016-0262

Markham, T. (2014). Social media, protest cultures and political subjectivities of the Arab Spring. *Media, Culture & Society*, 36(1), 89–104.

Marutho, D., Handaka, S. H., & Wijaya, E. (2018, September). The determination of cluster number at k-mean using elbow method and purity evaluation on headline news. In *2018 international seminar on application for technology of information and communication* (pp. 533–538). IEEE.

Mascheroni, G., & Murru, M. F. (2017). "I can share politics but I don't discuss it": Everyday practices of political talk on Facebook. *Social Media + Society*, 3(4), 2056305117747849. https://doi.org/10.1177/2056305117747849

McKenna, A. (2024, January 6). *2021 Myanmar Coup d'Etat | History, facts, & explained.* Encyclopedia Britannica. https://www.britannica.com/event/2021-Myanmar-coup-d-etat

Meraz, S., & Papacharissi, Z. (2013). Networked gatekeeping and networked framing on #Egypt. *The International Journal of Press/Politics*, 18(2), 138–166. https://doi.org/10.1177/1940161212474472

Meta (2024). *Adversarial threat report.* https://scontent-zrh1-1.xx.fbcdn.net/v/t39.8562-6/428391559_372656391766973 7_4417244101689936 69_n.pdf

Miles, T. (2018) *U.N. investigators cite Facebook role in Myanmar crisis.* Reuters. https://www.reuters.com/article/idUSKCN1GO2Q4/

Mod, C. (2016). *The Facebook-loving farmers of Myanmar: A dispatch from an internet revolution in progress.* The Atlantic. https://www.theatlantic.com/technology/archive/2016/01/the-facebook-loving-farmers-of-myanmar/424812/?utm_source=copy-link&utm_medium=social&utm_campaign=share

Nip, J. Y. (2004). The queer sisters and its electronic bulletin board: A study of the internet for social movement mobilization. *Information, Communication & Society*, 7(1), 23–49.

Paddock, R. C. (2022, December 9). Myanmar coup: What to know about the protests and unrest. *The New York Times.* https://www.nytimes.com/article/myanmar-news-protests-coup.html

Schirch, L. (2021). Social media impacts on conflict and democracy: The techtonic shift. London: Routledge.

Srivastava, R. K., Bharti, P. K., & Verma, P. (2022). Comparative analysis of lexicon and machine learning approach for sentiment analysis. *International Journal of Advanced Computer Science and Applications*, 13(3). https://doi.org/10.14569/ijacsa.2022.0130312

Stockmann, D., & Luo, T. (2017). Which social media facilitate online public opinion in China? *Problems of Post-Communism*, 64(3–4), 189–202.

Vaccari, C. (2017). Online mobilization in comparative perspective: Digital appeals and political engagement in Germany, Italy, and the United Kingdom. *Political Communication*, 34(1), 69–88.

Van Aelst, P., & Walgrave, S. (2002). New media, new movements? The role of the internet in shaping the 'anti-globalization' movement. Information, *Communication & Society*, 5(4), 465–493.

Walker, T. (2023 December 23). *Covering Conflict Is Risky but Vital for Myanmar Media.* Voice of America. https://www.voanews.com/a/for-myanmar-s-media-conflict-coverage-is-risky-but-vital-assignment-/7409356.html

Wankhade, M., Rao, A. C. S., & Kulkarni, C. (2022). A survey on sentiment analysis methods, applications, and challenges. *Artificial Intelligence Review,* 55(7), 5731–5780.

Whitten-Woodring, J., Kleinberg, M. S., & Thawnghmung, A. M. (2020). Poison if you don't know how to use it: Facebook, democracy, and human rights in Myanmar. *The International Journal of Press/Politics*, 25(3), 407–425. https://doi.org/10.1177/1940161220919666

Wolfsfeld, G., Segev, E., & Sheafer, T. (2013). Social media and the Arab Spring: Politics comes first. *The International Journal of Press/Politics*, 18(2), 115–137. https://doi.org/10.1177/1940161212471716

Zeng, J., & Brennen, S. B. (2023). Misinformation. *Internet Policy Review*, 12(4).

Zeng, J., Chan, C. H., & Fu, K. W. (2017). How social media construct "truth" around crisis events: Weibo's rumor management strategies after the 2015 Tianjin blasts. *Policy & Internet*, 9(3), 297–320.

CHAPTER 9

Artificial Intelligence Governance Made in China: Negotiating Imaginaries and Power

Yishu Mao

1 Introduction

Artificial Intelligence (AI), as so-called 'disruptive technologies', will bring profound societal and economic changes. Some changes will be beneficial to certain groups of people, while some will be detrimental to others (Girasa, 2020; Anderson and Rainee, 2018). Therefore, 'AI governance' has been increasingly used globally as the framework to think about how to ensure the socially beneficial uptake of these technologies and mitigate the risks. This framework considers not only ethics, regulations, and technical requirements, but also various stakeholders' participation quintessential for the safe and beneficial uptake of AI (Bullock et al., 2024).

China appears as not only one of the most ambitious countries in pursuing the development of AI but also one of the leading actors in AI governance. Since 2015, the central government has rolled out a series of related strategies, setting goals to leverage AI innovations to construct an 'intelligent society', upgrade its economy, and become a superpower in the world. Under these policy incentives, vast amounts of public and private investments have been underway to develop and implement AI across all aspects of life in China (Ding, 2018; Zhang et al., 2022). Since 2022, the government has also stepped-up regulations with a series of policies targeting specific technologies. One of the most noteworthy regulations is the Internet Information Service Algorithm Recommendation Management Regulations (hereafter abbreviated as 'algorithm regulations') promulgated by the Chinese internet regulator, the Cybersecurity Administration of China (CAC), in March 2022. As these regulations are intended for the use of 'algorithmic recommendation services' ranging from content generation and filtering to marketing to platform work, legal professionals have interpreted them as encompassing all types of recommendation and decision-making algorithms in commercial services – the fundamental component of AI systems and one of the most widespread types of AI in use today (Ruan, 2022; Toner et al., 2021). As such, these regulations have been considered the world's first

 | DOI:10.1163/9789004711396_011

nationally binding and the most extensive set of regulations targeting algorithms and AI. They are one of the 'two major experiments in the design of AI governance' in the world,[1] offering important insights for policymakers in other parts of the world (Sheehan and O'Shaughnessy, 2023; Rolf, 2023).

However, as an 'experiment', China's AI policies are also far from complete or comprehensive, but consist of contradictions and vagueness. For example, despite the pioneering and extensive appearance of the Algorithm Regulations, a closer look may raise some questions, in particular regarding the coexistence of vagueness and specificity in some articles. Article 6 requires algorithmic recommendation service providers to 'uphold mainstream value orientations, optimise algorithmic recommendation service mechanisms, vigorously disseminate positive energy, and advance the use of algorithms upwards and in the direction of good' (Creemers et al., 2022). What are the mainstream values in question and what constitutes the direction of 'good' in China? While these big questions are not addressed in the regulation, Article 20 specifically targets the use of algorithms in 'work dispatch services to workers,' alongside articles for protecting the rights of minors and elderly. Why have these specific social groups been given special attention?

Although there are many questions around the Algorithm Regulations, there has been little doubt among scholars that China's overall AI governance approach is 'state-led' (Bradford, 2023). However, this description overlooks the more complicated picture of multiple stakeholders' roles in negotiating the future of AI development. Although China's political system is featured by growing centralization of power by the party-state, other stakeholders' roles in this context deserve increasing scholarly attention, not less. Therefore, this chapter will trace the shaping of AI and algorithm regulations in China – their existence and the inclusion as well as omission of certain issues – through the lens of 'sociotechnical imaginaries.' It will demonstrate how power between the state, industry, academia, media, and civil society is negotiated through the articulation of future visions for AI and its governance. These different actors' imaginaries and negotiation of power will elucidate the essence of an AI governance approach in China, which is a balancing act of the state's techno-developmentalist ambitions, industry's innovative capacity and profit-driven model, as well as long-due protection of Chinese labour rights and consumer rights. In delineating a

1 The European Union's efforts to regulate AI have been considered as the other major experiment.

more holistic account of how China's approach to AI governance is the result of negotiating imaginaries and power, this chapter hopes to develop a better understanding of the diversity and agency in different societies' charting of AI's path, as well as its shared features with global trends, which can contribute to ongoing debates on AI governance.

2 Negotiating Imaginaries and Power

With increasingly accelerating advancement of technoscientific projects of modern times, uncertainty and indeterminacy prevails in their paths and relations with society (Beck, 1992). At the same time, the capacity of people to imagine the future has been increasingly recognized to have important social and political functions, ranging from enabling collective interpretation of social reality to forming collective identities (Castoriadis 1987; Anderson 1991; Appadurai 1996; Taylor 2003). Imaginations of the future that contain both aspects of the social order and technological development – the 'sociotechnical imaginary' (SI) – offer considerable affordances to understand the agency in society's charting of science and technology projects' development trajectories, and to extract conscious or unconscious human preferences and values in the 'design of objects, their weighting of risks and benefits, and the behaviours they encourage, exclude, or seek to regulate' (Jasanoff & Kim, 2015, p. 3). Defined as 'collectively held, institutionally stabilised, and publicly performed visions of desirable futures (and sometimes undesirable risks) animated by shared understandings of forms of social life and social order, attainable through, and supportive of advances in science and technology', SIs were originally conceptualised to explain the difference in the US and South Korea's developmental paths of nuclear technology caused by the different visions of state and social movements deeply rooted in the two countries' different cultures and histories (Jasanoff & Kim, 2015, p6; 2009). It has been increasingly mobilised by scholars to understand the social fabrics in the weaving of various emerging scientific and technological projects (Schiølin, 2020; Bennett, 2023).

AI is both the result of and inspiration for rich SIs embedded with cultural and historical diversity around the world. Fantasies of intelligent machines capable of conducting human tasks existed in some cultures long before the 1958 Dartmouth workshop that coined the term 'Artificial Intelligence' (Cave & Dihal, 2023). But after decades of research and development, the field of AI has reached a renewed hype cycle and has caught the general public's attention, capturing imaginations around the world in recent years. Nation-states and AI companies are among the most active groups in promoting their visions (Zeng et al., 2020; Brennen et al., 2018). But more often than not, they encounter contesting visions such as those of social organisations and scholars (Mager and Katzenbach, 2021). This constitutes a historical moment when key SIs have not yet been stabilised, providing a unique vantage point in analysing different stakeholders' visions of sociotechnical futures as they manifest themselves in public communication, as well as their negotiating dynamics. Exploring this can offer important insights into how different actors and societies at large use their imaginative reservoirs to shape the paths of integrating technologies towards directions that are aligned with their interests and values.

It is important to note that SIs are not abstract ideals; they are intricately linked to practical power owing to implicit or explicit interpretations of what is good and desirable, as well as to the formulations of how science and technology can meet public needs and who the relevant publics are (Wynne 2005; Fortun and Fortun 2005, Jasanoff & Kim, 2009). Although this form of power is persuasive and soft, it shapes technological progress in terms of policies, technological design, public expenditures, and the inclusion or exclusion of citizens (Jasanoff & Kim, 2009). In short, SIs are more than fictitious fantasy; they constitute power that shapes the norms and institutions that regulate our lives.

Although scholarship analysing power in relation to SIs has focused mainly on state power and other influential actors such as corporations, the study of SIs from counter-powers is equally important. Counter-powers are those whose interests and values are excluded from the institutions of a network society, prompting them to voice their dissent and to attempt to change the institutions through their imaginaries (Castells and Cardoso, 2005). The following sections will therefore examine the imaginaries of power and counter-power on multiple levels: from state policy, corporate narratives, online discussions of tech workers, and investigative journalism to media-academia coalitions. To identify the imaginaries and counter-imaginaries about AI in China, this analysis draws from both primary sources, such as policy documents from government agencies and AI companies, collected by the author as well as extensive secondary sources analysing China's AI and overall technoscientific developments.

3 Techno-developmentalist Imaginaries of AI from National Policies

The AI imaginaries promoted by the Chinese state have deep roots in the country's modern history. AI is one among many emerging technologies that pivots a national strategy and policy package to bridge the past and future. Since China's economic reforms in the late 1970s, the country has integrated into the world economy and become a global power within four decades – many call this the 'China's economic miracle' (Lin et al., 2003; Harrison and Palumbo, 2019). This has mainly been achieved by China's reliance on export-oriented industrialization, leveraging its abundant labour force and low-cost manufacturing capabilities to become the world's factory (Naughton, 2018). However, since the 2010s, such a strategy has become increasingly unsustainable in the face of a number of challenges such as economic structural imbalance build-up, stagnating productivity, and demographic change (Magnus, 2018). Although the Chinese government has not prioritised issues related to labour conditions and the social security of the working class in the past, the so-called 'middle-income trap' and growing inequality in China have increased the urgency for solutions in recent years (Griffith, 2011).

As a response to these crises, in 2015 the central government rolled out a major industrial policy 'Made in China 2025' to move away from being a producer of cheap low-tech goods. Its aim was to upgrade the manufacturing capabilities of Chinese industries with advanced technologies such as 5G, Internet of Things, and AI (Wübbeke et al., 2016). As such, 'Made in China 2025' constitutes a perfect illustration of a state-promoted sociotechnical imaginary that outlines visions of a near future in which China not only commands the leadership in advanced technological development, but also creates a new socioeconomic order in which Chinese labour will occupy a higher position in the global value-chain. Because the label 'Made in China' has had the connotation of cheap, sweatshop-made products, some of which are counterfeits, the Chinese government consciously developed the imaginary of 'Made in China 2025' with the intention of changing the image of China's manufacturing capabilities and therefore its socioeconomic status in the world.

Due to the wide application potential of AI technologies, the Chinese government gradually incorporated them as holding a central place in the Chinese government's strategic planning. The policy narratives around AI demonstrate both a concretisation and extension of the 'Made in China 2025' imaginary. Two years after the government released the 'Made in China 2025' industrial policy, they issued a development strategy specifically for AI entitled 'the New Generation Artificial Intelligence Development Plan' (2017), which showcases the sociotechnical imaginaries of AI from the Chinese state. Not only does it chart a blueprint for technological development until 2030, but it also provides a comprehensive analysis of AI's potential applications beyond the goals of upgrading the manufacturing industry across the whole economy and society. Claiming AI as the 'core driving force of a new round of industrial transformation' that will 'reconstruct all links of economic activities' and 'trigger major changes in the economic structure', it states that AI will inject new momentum into China's economic development and become an 'intelligent economy'. In addition to the strategy's economic goals, it also outlines the various aspects that AI can contribute to the state's social goals such as coping with population ageing, environmental constraints, facilitating public services, and maintaining social stability through AI-power surveillance technologies and turning Chinese society into an 'intelligent society' (PRC State Council, 2017). This broad agenda for AI, also called 'intelligentisation' (智能化), serves as a technological solution to the nation's fundamental tension between its high material needs, comparatively lagging productivity, and ineffectual government, as seen by China's central leadership at this point in time (Webster et al., 2017). The imaginaries, which envision a cybernetic governmentality in China's AI development plan, embody a social order where socioeconomic issues will be resolved through 'a perpetual mode of citizen (self-)monitoring, adaptation, and optimization' (Bareis & Katzenbach, 2021).

Besides visions for a new economic and social order, Chinese AI strategies have included visions for a new political and world order. In both 'Made in China 2025' and the AI development plan, the government repeatedly refer to 'the Chinese dream of the great rejuvenation of the Chinese nation.' This refers to the history of the 20th century, a period in which China missed the opportunity to lead in past scientific industrial revolutions and was invaded by Western industrialised powers. 'The Chinese dream' is essentially China's return to the centre of the world through the advancement of scientific and technological capabilities enhancing China's comprehensive national power and competitiveness. In these imaginaries, AI and nationhood are closely interwoven. Here, national leadership is not just the result of AI advancement, but the very object of AI advancement. National interests are the public goods that are pursued with these policies.

In Lei's (2021, p1078) research into the perceptions of automation in China in the context of the state's economic upgrading policies, she identified this the national imaginary as 'techno-developmentalist' because it 'prioritises abstract notions of technological progress over the actual efficacy of automation, labour protection and social equality.' Although the notions of 'national competitiveness' and 'national legacy' have circulated in many countries throughout technological development as well as in the current drive for AI, Lei (2021) argues that what sets the Chinese national imaginary apart from others' is the lack of consideration for labour protection and inequality as well as the exclusion of low-skilled labours.

Although Lei (2021) did not find significant counter-imaginaries in her interviews with the factory workers that would challenge the predominant techno-developmentalist policy imaginary, as it will be shown in the following sections, contestations have emerged in several communities of tech industry white-collar workers, scholars and journalists. Although these contestations may not directly challenge the state, imaginaries from the tech industry, which further concretise the policy imaginaries, became an important site of contestations.

4 Contesting Imaginaries from the Tech Industry

Similar to findings in Western countries (Brennen et al., 2018), private companies are leading AI research and development in China and have produced a trove of media coverage promoting their imaginaries of AI. However, what is different in China is the relationship between the state and the private sector, and how they interact to shape the imaginaries and trajectories of AI development. Since 1978, China's economic reform has allowed the private sector to thrive within the limit set by the state, and a market economy provided momentum for the country's spectacular growth, with GDP per capita multiplying 28 times, making China the second biggest economy in the world within 45 years (World Bank, 2023). The ICT industry became an important sector in China's economy and the leading force in the country's pursuit of economic upgrading in the past decade. The state also had strong incentives for lax regulation of the industry and let it innovate until 2020. This co-dependent relationship between the state and the tech industry shape how imaginaries of AI have evolved in China.

However, China's tech industry is not a unitary entity and there are multiple imaginaries of AI contestation within, especially between those promoted by tech companies and those articulated by tech workers. At the 2019 World Artificial Intelligence Conference, Chinese tech titan Alibaba's CEO Jack Ma and American entrepreneur Elon Musk held a highly anticipated debate. Regarding AI's impacts on employment, while Musk expressed concerns, Ma forcefully argued that AI will enable people to enjoy their lives as human beings and focus on artistic and creative work. Moreover, due to an ageing and declining population, AI will take over jobs such as care work, so that the future will be where people can 'work three days a week and four hours a day' (*BBC News,* 2019). Jack Ma's speech represents a narrative which can be frequently found in Chinese companies' advertisements, media coverage, and corporate documents, promoting a vision of liberation, and promising an ever better future in which people will live at ease and in happiness.

An analysis of the most advanced AI companies' imaginaries reveals that the most prominent feature of industry actors' imaginaries is the sense of technological determinism and solutionism. For example, in 2019, one of the biggest Chinese tech companies, Tencent declared 'Tech for Good' (科技向善) as the company's new vision statement and launched a PR campaign under this banner. On the company's website defining this vision, information technology was compared to a tsunami, 'moving mankind from an industrial civilization to a digital society.' Internet companies were then claimed to 'play a vital role and exert huge influence' (Si, 2019) and Tencent's 'Tech for Good' vision was said to include two missions:

> on the one hand, to fully use the potential of this round of new technology, make it to the benefit of most people, solve all sorts of traditional society's inveterate diseases and problems, promote social progress, and help mankind become stronger, happier, have a better future and digital civilization. On the other hand, its mission is to ensure that new technologies are used for good, rather than abused or even for harmful purposes. In addition to commercial returns, the products and services of tech companies must have both public value and a social perspective. Due to the various negative effects brought about by new products and services, tech companies must bear corresponding responsibilities and provide solutions. In short, 'Tech for Good' means: realising technology for good use and avoiding harmful uses.
>
> SI, 2019

While the above vision statement is oriented to technology broadly, it serves as the basis for Tencent's vision 'AI for Good', which 'imagines a future-centric world that will unlock great promise for human society' (Lin, 2019). In these imaginaries, technological advancement is depicted as inevitable and society as passive receivers of technologies. Although 'AI for Good' acknowledges that technology can be used to cause harm, companies are depicted as its guardians, as opposed to actors that should be held accountable. This links to the second feature of such imaginaries – a sense of techno-solutionism, which is shown clearly by the description of a version of a 'traditional society' with 'inveterate diseases and problems' and the promotion of technologies as the solution to bringing linear societal progress and an ever better future.

Similar imaginaries permeate other companies' PR materials. For instance, China's most-valued start-up SenseTime (2020) stated its mission as 'to develop AI technologies that advance the world's economies, society, and humanity for a better tomorrow' in its *Code of Ethics for AI Sustainable Development.* Partially state-owned and partially a publicly-traded company specialising in voice recognition, iFLYTEK stated that it wants to 'build a wonderful world with AI' in its Environment, Society and Governance (ESG) report (2021). Baidu (2019) stated in the 'Self-Regulatory Principles' that the ultimate ideal for AI is to 'bring more freedom and possibilities' to humans. Huawei, in its 260-page report *Intelligent World 2030*, painstakingly painted imagined scenarios for AI's adoption across healthcare, agriculture, smart home, smart transportation, manufacturing, and energy sectors, and claimed, 'Imagination determines how far we can go in the future, and action determines how fast we can reach the future, and the best way to predict the future is to create it' (2021, p9). All of these companies employ extremely positive language to depict a utopian future created by AI.

What underlies these imaginaries is an assumption that broad scale adoption of AI will benefit the majority of people and therefore bring social good. For example, the first principle in Tencent's AI ethics framework[2] is 'available.' It implies that, to promote human development, raise well-being in society and achieve inclusive, universal, and sustainable development, more people should be able to obtain and use AI, so that people can 'share technological dividends and avoid technological gaps' (Tencent Research Institute, 2019, p12). This is in line with the main actions that Tencent has claimed to undertake to fulfil its 'Tech for Good' promises listed in the report: public service activities oriented towards the older population, people with disabilities, and missing children. This shows that its ethics and charitable actions are mostly for those who are *currently* marginalised and disadvantaged groups; its promised digital dividends are mainly directed to these groups, rather than at the social groups who *will* or *might* be disempowered and marginalised *in the future*. The negative impacts and risks of AI advancement have been mostly ignored in the corporate discourse for AI's future.

These imaginaries of enabling social well-being with wide AI adoption are shared by many tech companies, including the facial recognition company Megvii. The company's AI application principles include fairness and

2 The other three being reliable, comprehensible, and controllable.

diversity, from which it can be understood that developers should minimise systematic bias in AI and 'regardless of age, gender, race or other characteristics, everyone should be able to use AI' (Megvii, 2019). Similarly, Baidu's *Self-Regulatory Principles* stated that its vision for AI innovation is 'to promote more equal access to technology and capabilities for human beings' (2019). SenseTime listed 'shared benefits' as one of the core values to guide its development of AI and stated that AI will bridge the urban-rural divide:

> AI solutions for prevalent urbanisation issues such as pollution, security, and over-congestion can be discovered through continuous and rapid innovations. And given its efficiency, AI can now be used in rural areas to help narrow gaps amongst people, equalising opportunities for all regardless of community.
>
> SenseTime, 2020, p6

In iFLYTEK's (2021) report, the main contributions of AI to broader social issues are eradicating poverty, COVID epidemic management, application of AI in education and cultural preservation, and empowering people with disabilities.

Leading Chinese AI companies have produced a plethora of AI imaginaries in the past years that create the expectation that simple access to technology will bring such positive promises into realisation. The people who will benefit from such development are mainly the currently disadvantaged social groups who lack access to technology, not those who are disempowered now and might continue to be in the future. These imaginaries respond to those expressed in techno-developmentalist national policies which picture AI as offering solutions to the country's socioeconomic problems and as engines to transform the country's status in the global value chain.

While companies promise a utopian future, the tech industry in China has been notorious for extractive practices. The '996 working hour system' describes a prevalent working hour system where employees work from 9 a. m. to 9 p.m., 6 days per week (a 72-hour working week). Moreover, at Alibaba and many other tech companies in China, various AI applications are installed in offices and employees' computers to monitor and measure their performance (Sun, 2021). Tech giants Alibaba and ByteDance are among the clients of a 'behavioural perception system' that calculates employees' perceived 'resignation risk' as well as their efficiency based on their 'online footprints' such as browsing history and chats with co-workers (James, 2022; Sun, 2021). This high-pressure work environment was brought under public scrutiny after a string of deaths related to the excessive working hours at the internet companies[3] (Wang, 2021; James, 2021; Feng, 2022). The technology by which people are promised to work less is actually used to make some people work more directly at the site where this future is being produced. This contradiction between the promises of what AI can achieve in the future and the realities of how these technologies are developed and used now is manifested in the imaginaries of AI promoted by companies and tech workers.

Because the design of AI systems, both in the technical and social sense largely depends on the skills of individual professionals such as designers, engineers, programmers, data scientists, product managers, etc, these professionals define technologies as active creators and the face of China's techno-development regime under transformation. However, much evidence has already shown that Chinese tech workers' experience has led them to imagine a future that is far from utopian. Analysis of online discussions about the topic of AI ethics and governance reveals very different imaginaries from those promoted at the company level.

In China, social media platforms are a popular place where many tech-savvy workers turn to vent their frustrations. One social media platform – Zhihu, a forum website similar to Quora which allows anonymous posts – became a popular online forum for tech industry professionals, students of computer sciences, highly educated groups, and employees of big tech companies (Cxb168, 2019; Xueshenkeji, 2020; Zhang, 2019). As many users have been active in the discussions about AI ethics and governance, these online discussions offer valuable insights into the imaginaries at the workers' level (Mao and Shi-Kupfer, 2023). Lively discussions were held under questions such as: 'Can Super AI & Big Data replace government?' 'Suppose, in ten years, AI is capable enough to replace teachers, doctors, lawyers, manual labour, and low and middle level mental work: how should the world respond?' Many discussions were also inspired by influential figures' debates on AI, such as Stephen Hawking, Elon Musk, and Mark Zuckerberg, as well as Western science fiction productions such as *Black Mirror* and *Westworld*.

Prevalent imaginaries found on Zhihu, unlike those AI utopias constructed by company discourse, construct a dystopian future. These discussions reveal the recognition of AI as a productive technology with the potential to change social relations, power structures, and ethical norms.

3 These remind one of the heart-breaking suicides that occurred in Chinese factories in the 2010s (Johnson, 2011), which brings to the fore the irony of upgrading Chinese economy in the global value chain.

Although the imaginaries in these discussions also often appear to be technologically deterministic, they are pessimistic with an emphasis on the technology's negative impact on labour, job displacement, and increasing inequality in society. For example, one user on Zhihu wrote:

> Before the emergence of a strong AI with real self-awareness (which is almost a pipe dream at this level of technology), the issue remains an ethical one between humans. The increased productivity led by AI will further eliminate low-end labour and increase the imbalance of wealth distribution. In the future, there will be a technological oligarchy replacing the democratic government as the de facto ruler of the country, and the control of AI (which is not self-aware) is in the hands of these technological oligarchs, so this is essentially a centralised rule of a few capitalists using technological means over the majority of people at the bottom who have lost their ability to earn and consume.
>
> LAIXUCANGZHEN, 2018

This quote presents an imagination prevalent in the Zhihu discussions – a sobering anticipation of the shifting balance of power and social consequences, within a power-balance already leaning in favour of the stronger ones in society, brought by the introduction of AI technologies. Much of this is also demonstrated in the discussions regarding the labour displacement issue which directly oppose Jack Ma's argument that AI will liberate people. Instead, these discussions argue that, with increasing automation, more and more of the labour force will be squeezed out of jobs. Overall, many predictions of AI's social impacts on Zhihu paint a grim future with problems such as growing inequality, further devaluation of labour, increasing precarity in work conditions, and loss of autonomy.

What is particularly noteworthy in the discussions are the frequent references to Marxist critique of the capitalist system in these imaginaries. For example, one user wrote:

> In the past, the rich got richer and the poor got poorer because the return rate of capital was higher than the return rate of labour … now people with data and technology have more resources than those with capital. The era of AI thus brings a new concern: when data and technology are mastered by a few people, especially when they also master capital, their power will be incalculable. At the same time, when the application of AI gradually spreads, and the manufacturing and production capability of machines continues to mature, the cost

> of replacing manual labour by machines decreases, the problems of unemployment and inequality between the rich and the poor will become more prominent.
>
> FUSHOUJIUQIN, 2020

Referring to Marx's account of the process of valorisation in the circuit of capital in a profit-driven capitalist system, this user's argument also echoes what Shoshana Zuboff terms as 'surveillance capitalism', warning of the enormous power that could be amassed by the entities who own both capital and data to undermine social justice and working lives (Zuboff, 2019). These imaginaries of AI futures found among the tech workers' discussions online are frequently accompanied by critical analysis of relations of production and power. They form a sharp contrast with the corporate imaginaries' utopias where everyone is happy and free.

Therefore, the imaginaries found in China's tech industry are featured by strong contestation between those found in companies' official discourse promoting utopian and tech-solutionist imaginaries of AI, and those articulated by tech workers on social media platforms anticipating dystopian and critical imaginaries, warning of the growing concentration of power in the hands of a few.

5 Research-Based Imaginaries from Academia and Investigative Media

As illustrated in the previous sections, public imaginaries about AI in China, similar to other countries, have been largely dominated by policy and corporate narratives. However, academics and investigative journalists play important roles in revealing the realities behind the development and use of AI in China. They navigate the grand imaginaries put forward by the state, and the contrasting imaginaries from the tech industry. Due to the particularity of the media system in China, the two the groups sometimes have to collaborate to produce influence in the public sphere.

On the one hand, research related to AI's impact in China has been burgeoning in certain fields of social sciences and humanities. However, unsurprisingly, with the sweeping techno-developmentalist and utopian imaginaries from the state and tech companies, a significant amount of public research funding in China has been going towards the development of AI technologies rather than studying their societal or environmental impacts. This is evident in comparing the number of research grants awarded by the

National Natural Science Foundation of China in the STEM fields related to AI, in comparison to those awarded by the National Social Science Fund in the social sciences and humanities related to AI (Abadi et al., 2020; National Office for Philosophy and Social Sciences, 2022). Even among the latter, more research projects were oriented towards exploring how to integrate AI within specific domains or how to regulate AI from a theoretical or legal point of view.[4] Although this lack of empirical research into the impacts of AI may indicate that the latter's implementation in China's larger economy and society is still in its early phases, it is only through research that the concrete impacts of AI can be assessed. Without this evidence, the utopian imaginaries of AI's benefits for various sectors can only be misguided dreams. However, as digital platforms and embedded algorithms are major drivers for AI innovation in China, and platform companies are the so-called 'national champion' in developing AI, there has been growing empirical research into the forms of AI embedded in the platform economy and their impacts on issues such as market structure and working conditions.

On the other hand, critical journalism has played an indispensable role in the movement for the establishment of cyber laws and the dissemination of digital rights to citizens in China. Despite the tightly government-controlled public sphere, media marketization and fragmented bureaucratic structures have allowed critical news reporting to evolve, contributing to the identification and solution-searching for fundamental societal problems, in addition to dissemination of civil rights concerns (Lei, 2017; Stockmann, 2012). The ways in which investigative journalists achieve this is characterised by building extensive collaboration networks with scholars, lawyers, civil society organisations, and other social groups (Lei, 2017; Svensson, 2012), as well as continuously improvising partnership with the state to improve governance (Repnikova, 2017). With the advancement of the internet and social media in China, not only has this type of journalism quickly adapted to online formats, but journalists have also been instrumental actors in reporting issues related to digital rights, such as privacy violations. For example, one of the most well-known commercial media outlets for investigative reporting, Caixin Weekly, has engaged in unravelling the illegal market for personal data – the hackers' industrial chain (黑客产业链) since the 2010s (Zhu et al., 2012). Similarly, *Southern Metropolis Daily* even established a personal Information Security Research Centre in 2017 which is dedicated to coverage related to online personal data protection. Since 2020 they have also set up

4 Based on a review of research project titles.

an annual forum for data governance and publish reports related to data, platform and algorithm governance (Baidu Baike, 2023; Yinsihuweidui, 2022). It was with this continuous revelation of the harmful effects of data breaches and other illegal practices on the internet, that Chinese legislators promulgated a legal architecture for data protection, including the Personal Information Protection Law (PIPL) and the Data Security Law (DSL) in 2021, to address growing public concerns (Creemers, 2022).

In recent years, the collaborative networks of scholars and investigative journalists have produced some insightful reports about emerging societal issues associated with the use of AI and have successfully raised both public and policy awareness. In particular, in September 2020, one article titled 'Delivery riders, trapped in the systems' published by Chinese magazine *Portrait,* attracted widespread public attention in China. The long article exposed a problem that is prevalent in China's booming food delivery industry: food delivery platforms use algorithmic systems which allows the platform to calculate the most 'efficient' routes and estimated time for their riders to deliver an order. Although the platforms often generate demands that would be unreasonable for human riders, e.g. driving against the flow of traffic, delivery time-estimation that would only be possible by flying, and even going through walls, companies nonetheless adopted the system. They even go as far as to fine the delivery drivers if they do not meet the algorithmic demands. As a result, riders rushed through traffic and over the course of the last few years, accidents soared. However, these riders are also precarious workers with temporary contracts and no work place injury insurances. Lai asks, 'through a detailed interpretation of a system, we try to make more people think about a question: In the era of digital economy, what form of existence should algorithms take?' (2020).

The article in *Portrait* demonstrated the current social impacts of AI and provoked readers' imaginaries of future AI through its compelling stories which shed light on the individual riders' predicaments. It provided insights into the interactions of workers and algorithmic systems uncovered by social scientists' in-depth research. For example, Ping Sun is a researcher at the Chinese Academy of Social Sciences and the principal investigator behind multiple projects researching larger issues related to platforms, digital labour, technologies, social structures, and genders. The article drew from her research to explain the problems found in the sociotechnical systems in China's digital economy: algorithms developed to dispatch delivery workers are based on mathematical logic without taking ethics into account. However, algorithm developers do not have the power to change the rules of the platform, or more concretely, the platform company which aims to maximise profit has the power to evaluate the performance of the algorithmic system and its developers. Moreover, delivery workers' behaviours in which they break traffic rules to meet the increasingly strict demand of algorithms actually contribute to the 'success' of the algorithms and profits of the company. Therefore, they provide incentives for the further 'optimization' of algorithms by the developers and companies in this direction, if left without interference. As such, various actors and their logic contribute to the functioning of this larger sociotechnical system, but the majority of the workers are involved to the detriment of themselves, while only the investors benefit.

Various media outlets republished the *Portrait* article and quickly generated widespread discussions on issues of exploitative business models and algorithms' impact on workers, the monopoly practices of Chinese platform companies, and the outdated social security systems in China. 'Trapped in the system' seized public imaginaries and gained resonance within broader society. The voices of those whose interests and values have been excluded from the sociotechnical systems of AI and any broader societal support for change have directly contributed to the making of the algorithm regulations and in particular its Article 20 which stipulates that,

Where algorithmic recommendation service providers offer work dispatch services to workers, they shall protect workers' lawful rights and interests such as obtaining labour remuneration, rest and vacation, etc., and establish and perfect algorithms related to platform sign-on and allocation, remuneration composition and payment, work time, rewards, etc. (Cyberspace Administration of China, 2022).

This case of scholar and journalist collaboration to shape public imaginaries and eventually shape policy is an individual case that might not represent the overall picture of how AI governance policies are formed in China. It is therefore important to note that the opportunities and channels for such influence are not established in a structural way but are subject to fortunes and contingencies.

6 Conclusion: AI Governance Made in China – Regulation as Resolution of Multiple Conflicts

The making of AI and specific algorithm regulations in China showcases the process involving various stakeholders with multiple imaginaries, negotiating power in the process of AI governance. To successfully integrate sociotechnical systems like AI requires more than government support in science and technology policymaking, but rather governance: the participation of a wide range of actors such as industry, academia, media, civil society groups, and consumers as well as their respective formal and informal activities (Fuller, 2000; Rose, 1999; Barry, 2001; Dean, 2009). Although China's authoritarian system has drawn much attention in the policies of government agencies, Chinese academia, media, and society have shown much dynamism in negotiating the future of AI by focusing on specific applications.

However, a clear discrepancy exists in terms of power in negotiating the development trajectory of AI. The state still commands the greatest power in defining the purposes of scientific and technological advancement and what constitutes the public good, that is, upgrading China's economy and global position to achieve national 'rejuvenation' and techno developmentalist goals. When public controversies and conflicts arise, the state is able to use 'guerrilla style policy making' (Heilmann and Perry, 2011). This means that the overarching state imaginaries and development paths for AI are maintained, with details in specific regulations adjusted, responding to contestations orginated from the wider society.

The AI companies, until recently, had mostly government support and were able to promote their utopian imaginaries of AI through multiple channels that aimed to make their services and products ubiquitous. However, counter-imaginaries rooted in the lived experiences of tech workers in the industry, combined with evidence gathered by scholars and investigative journalists, were able to make themselves heard through coalitions. Nevertheless, these coalitions are very fluid and remain without supporting institutions. As the state and companies continue to harness and commercialise AI, whether the influence of counter-imaginaries can be possible for other issues related to AI governance remains unpredictable.

Therefore, to understand the emerging approach to AI governance in China, one can think of the tension between the multiple imaginaries around 'Made in China': the country's past economic success is achieved through cheap labour and the suppression of workers' rights – 'Made in China', but its aspiration future relies on upgraded technologies – 'Made in China 2025'. Will China actually achieve its dream of increased productivity through AI, or through humans who work in increasingly demanding conditions? Will the 'Chinese Dream' finally be inclusive, made of different stakeholders' aspirations and contestations? These questions need to be addressed if China is to establish effective AI governance during a period of socio-technical upheaval. At the same time, many phenomena observed in China share commonalities with global trends. The negotiated imaginaries and power involved in the development of AI and its governance in China should contribute to global debates on how to ensure inclusivity and transparency in AI governance.

References

Abadi, H. H. N., He, Z., & Pecht, M. (2020). Artificial Intelligence-related research funding by the U.S. National Science Foundation and the National Natural Science Foundation of China. *IEEE Access*, 8, 183448–183459. https://doi.org/10.1109/ACCESS.2020.3029231

Anderson, B. R. O. (1991). *Imagined communities: Reflections on the origin and spread of nationalism*. Verso.

Anderson, J., & Rainee, L. (2018). *Artificial Intelligence and the future of humans*. Pew Research Center. https://www.pewresearch.org/internet/2018/12/10/artificial-intelligence-and-the-future-of-humans/

Appadurai, A. (1996). *Modernity at large*. University of Minnesota Press. https://www.upress.umn.edu/book-division/books/modernity-at-large

Baidu (2019). *Self-regulatory principles statement* (自律性原则申明). https://ai.baidu.com/ai-doc/REFERENCE/xk3dwjgfe

Baidu Baike (2023). *Southern Metropolis Personal Information Security Research Center* 南都个人信息保护研究中心. Baidu Baike. Retrieved June 30, 2023, from https://baike.baidu.com/item/%E5%8D%97%E9%83%BD%E4%B8%AA%E4%BA%BA%E4%BF%A1%E6%81%AF%E4%BF%9D%E6%8A%A4%E7%A0%94%E7%A9%B6%E4%B8%AD%E5%BF%83/22354274?fr=ge_ala

Bareis, J., & Katzenbach, C. (2021). Talking AI into being: The narratives and imaginaries of national AI Strategies and their performative politics. *Science, Technology, & Human Values*, 47(5), 855–881. https://doi.org/10.1177/01622439211030007

Barry, A. (2001). *Political machines: Governing a technological society*. Athlone Press.

BBC News (2019, August 29). Elon Musk and Jack Ma disagree about AI's threat. *BBC News*. https://www.bbc.com/news/technology-49508091

Beck, U. (1992). *Risk society*. Sage Publications.

Bradford, A. (2023). Digital Empires: The Global Battle to Regulate Technology. Oxford University Press.

Bennett, M. M. (2023). Chinese sociotechnical imaginaries of Earth observation: From sight to foresight. Big Data & Society, 10(2), 20539517231191527. https://doi.org/10.1177/20539517231191527

Brennen, J. S., Howard, P. N., & Nielsen, R. K. (2018). *An industry-led debate: How UK media cover Artificial Intelligence*. Reuter Institute. Oxford University.

Bullock, J. B., & et al. (Eds.). (2024). *The Oxford handbook of AI governance*. Oxford Handbooks (online ed.). Oxford Academic. https://doi.org/10.1093/oxfordhb/9780197579329.001.0001

Castells, M., & Cardoso, G. (2005). *The network society: From knowledge to policy*. Johns Hopkins Center for Transatlantic Relations.

Castoriadis, C. (1987). *The imaginary institution of society*. The MIT Press. https://mitpress.mit.edu/9780262031349/the-imaginary-institution-of-society/

Cave, S., & Dihal, K. (Eds.). (2023). Imagining AI. Oxford University Press.

Creemers, R. (2022). China's emerging data protection framework. *Journal of Cybersecurity*, 8(1). https://doi.org/10.1093/cybsec/tyac011

Creemers, R., Webster, G., & Graham Toner. (2022). Translation: Internet Information Service Algorithmic Recommendation Management Provisions – Effective March 1, 2022. *DigiChina*. https://digichina.stanford.edu/work/translation-internet-information-service-algorithmic-recommendation-management-provisions-effective-march-1-2022/

Cxb168 (2019). *Thousands of Zhihu user data analysis report* (千万知乎用户数据分析报告). Jianshu. https://www.jianshu.com/p/b37d671f97e5

Cyberspace Administration of China (2022). *Internet Information Service Algorithmic Recommendation Management Provisions* (互联网信息服务算法推荐管理规定). https://www.gov.cn/zhengce/zhengceku/2022-01/04/content_5666429.htm

Dean, M. (2009). *Governmentality: Power and rule in modern society*. SAGE.

Ding, J. (2018). *Deciphering China's AI dream: The context, components, capabilities, and consequences of China's strategy to lead the world in AI*. Centre for the Governance of AI, Future of Humanity Institute, University of Oxford. https://www.fhi.ox.ac.uk/wp-content/uploads/Deciphering_Chinas_AI-Dream.pdf

Feng, C. (2022, February 23). *ByteDance worker's death after workout trends on Weibo*. South China Morning Post. https://www.scmp.com/tech/big-tech/article/3168057/bytedance-worker-dies-after-collapsing-gym-again-raising-996

Fortun, K., & Fortun, M. (2005). Scientific imaginaries and ethical plateaus in contemporary U.S. toxicology. American Anthropologist. *American Anthropologist*, 207(1), 43–54.

Fuller, S. (2000). *The Governance of science: Ideology and the future of the open society (Issues in Society)*. Open University Press.

Fushoujiuqin (2020). 我们的工作会被人工智能取代吗?. Zhihu. https://www.zhihu.com/question/58001460

Girasa, R. (2020). AI as a Disruptive Technology. In R. Girasa (Ed.), *Artificial Intelligence as a Disruptive Technology: Economic Transformation and Government Regulation* (pp. 3–21). Springer International Publishing. https://doi.org/10.1007/978-3-030-35975-1_1

Griffith, B. (2011). Middle-Income Trap. In *Frontiers in Development Policy* (pp. 39–43). The World Bank. https://doi.org/10.1596/9780821387856_CH04

Harrison, V., & Palumbo, D. (2019). *China anniversary: How the country became the world's "economic miracle."* BBC News. https://www.bbc.com/news/business-49806247

Heilmann, S., & Perry, E. J. (Eds.). (2011). *Mao's invisible hand: The political foundations of adaptive governance in China*. Harvard University Press.

Huawei (2021). *Intelligent world 2030*. https://www.huawei.com/en/giv

iFLYTEK (2021). *iFLYTEK 2021 Environmental, Social & Governance (ESG) Report* (科大讯飞 *2021* 年度 环境、社会及治理 (*ESG*) 报告 暨企业社会责任报告).

James, G. (2021, December 16). A star worker's suicide at Tencent. *The China Project*. https://thechinaproject.com/2021/12/16/a-star-workers-suicide-at-tencent/

James, G. (2022, February 21). Chinese tech workers outraged by surveillance tool that flags employees who look likely to quit. *The China Project*. https://thechinaproject.com/2022/02/21/chinese-tech-workers-outraged-by-surveillance-tool-that-flags-employees-who-look-likely-to-quit/

Jasanoff, J., & Kim, S.-H. (2009). Containing the atom: Sociotechnical imaginaries and nuclear power in the United States and South Korea. *Minerva*, 47(2), 119–146. https://www.jstor.org/stable/41821489

Jasanoff, S., & Kim, S.-H. (Eds.). (2015). *Dreamscapes of modernity: Sociotechnical imaginaries and the fabrication of power*. University of Chicago Press. https://press.uchicago.edu/ucp/books/book/chicago/D/bo20836025.html

Johnson, J. (2011). 1 Million Workers. 90 Million iPhones. 17 Suicides. Who's to Blame? *Wired*. Retrieved June 26, 2023, from https://www.wired.com/2011/02/ff-joelinchina/

Lai, Y. (2020). 外卖骑手,困在系统里. *Portait*. https://zhuanlan.zhihu.com/p/225120404

Laixucangzhen (2018). 在人工智能时代,我们创造的 AI 机器人应该满足哪些道德伦理? Zhihu. https://www.zhihu.com/question/302956493/answer/534847833

Lei, Y. (2017). *The contentious public sphere*. Princeton University Press. https://press.princeton.edu/books/hardcover/9780691166865/the-contentious-public-sphere

Lei, Y. (2022). Upgrading China through automation: manufacturers, workers and the techno-developmental state. *Work, Employment and Society*, 36(6), 1078–1096. https://doi.org/10.1177/0950017021999198

Lin, H. (2019). China's Tencent advocates "AI for Good" at AI Everything Summit in Dubai. *China Daily*. https://www.chinadaily.com.cn/a/201905/02/WS5ccaca1ba3104842260b98a0.html

Lin, J. Y., Cai, F., & Li, Z. (2003). *The China miracle: Development strategy and economic reform (Revised Edition)*. The Chinese University of Hong Kong Press. https://doi.org/10.2307/j.ctv1fj84hd

Mao, Y., & Shi-Kupfer, K. (2023). Online public discourse on artificial intelligence and ethics in China: Context, content, and implications. *AI & Society* 38(1), 373–389. https://doi.org/10.1007/s00146-021-01309-7

Mager, A., & Katzenbach, C. (2021). Future imaginaries in the making and governing of digital technology: Multiple, contested, commodified. New Media & Society, 23(2), 223–236. https://doi.org/10.1177/1461444820929321

Magnus, G. (2018). *Red flags*. Yale University Press. https://yalebooks.yale.edu/9780300246636/red-flags

Megvii (2019). *Megvii released the "Artificial Intelligence Application Guidelines" to advocate the healthy and sustainable development of AI technology* (旷视发布《人工智能应用准则》 倡导 AI 技术健康可持续发展). https://www.megvii.com/news_detail/id/60

National Office for Philosophy and Social Sciences (2022). *2022 National Social Science Fundation Project Application Announcement* (2022 年度国家社会科学基金项目申报公告). http://www.nopss.gov.cn/n1/2022/0112/c431027-32329870.html

Naughton, B. (2018). *The Chinese economy: Adaptation and growth*. MIT Press. https://mitpress.mit.edu/9780262534796/the-chinese-economy/

PRC State Council (2017). *New Generation Artificial Intelligence Development Plan* (国务院关于印发新一代人工智能发展规划的通知) http://www.gov.cn/zhengce/content/2017-07/20/content_5211996.htm

Repnikova, M. (2017). *Media politics in China*. Cambridge University Press.

Rolf, S. (2023). *China's regulations on algorithms*. https://futureofwork.fes.de/news-list/e/new-policy-briefing-chinas-regulations-on-algorithms

Rose, N. (1999). *Powers of freedom: Reframing political thought*. Cambridge University Press. https://doi.org/10.1017/CBO9780511488856

Ruan, D. (2022, June 13). Mainland China adopts new regulation on algorithmic recommendation technologies. https://www.dentons.com/en/insights/articles/2022/june/13/mainland-china-adopts-new-regulation-on-algorithmic-recommendation-technologies

Schiølin, K. (2020). Revolutionary dreams: Future essentialism and the sociotechnical imaginary of the fourth industrial revolution in Denmark. Social Studies of Science, 50(4), 542–566. https://doi.org/10.1177/0306312719867768

SenseTime (2020). *Code of ethics for AI sustainable development*. https://oss.sensetime.com/20210305/a75c053e36e0f93607a769e2e098bee2/Code%20of%20Ethics%20for%20AI%20Sustainable%20Development.pdf

Sheehan, M., & O'Shaughnessy, M. (2023). *Lessons from the world's two experiments in AI governance*. Carnegie Endowment for International Peace. Retrieved March

15, 2023, from https://carnegieendowment.org/2023/02/14/lessons-from-world-s-two-experiments-in-ai-governance-pub-89035

Si, X. (2019). Five core questions about Tencent's "Technology for Good" (关于腾讯 "科技向善" 的五个核心问题). *Tencent Research Institute.* https://www.tisi.org/15076

State Council of the People's Republic of China (2017). New Generation Artificial Intelligence Development Plan (新一代人工智能发展规划). https://www.gov.cn/zhengce/content/2017-07/20/content_5211996.htm

Stockmann, D. (2012). *Media commercialization and authoritarian rule in China.* Cambridge University Press. https://doi.org/10.1017/CBO9781139087742

Sun, N. (2021). China's tech workers pushed to limits by surveillance software. *Nikkei Asia.* https://asia.nikkei.com/Spotlight/The-Big-Story/China-s-tech-workers-pushed-to-limits-by-surveillance-software

Svensson, M. (2012). Media and civil society in China. *China Perspectives,* 2012(3), 19–28. https://doi.org/10.4000/chinaperspectives.5934

Taylor, C. (2003). *Modern Social Imaginaries.* Duke University Press. https://www.dukeupress.edu/modern-social-imaginaries

Tencent Research Institute (2019). *Technology ethics in the Intelligent Era – Reshaping trust in digital society* (智能时代的技术伦理观——重塑数字社会的信任). https://cloud.tencent.com/developer/news/411837

Toner, H., Triolo, P., & Creemers, R. (2021, August 27). Experts examine China's pioneering draft algorithm regulations. *DigiChina.* https://digichina.stanford.edu/work/experts-examine-chinas-pioneering-draft-algorithm-regulations/

Wang, V. (2021, February 1). Worker deaths put big tech in China under scrutiny. *The New York Times.* https://www.nytimes.com/2021/02/01/business/china-technology-worker-deaths.html

Webster, G., Creemers, R., Kania, E., & Triolo, P. (2017). Full translation: China's "New Generation Artificial Intelligence Development Plan" (2017). *DigiChina.* https://digichina.stanford.edu/work/full-translation-chinas-new-generation-artificial-intelligence-development-plan-2017/

World Bank (2023). GDP per capita (current US$) – China. Retrieved July 31, 2024, from https://data.worldbank.org/indicator/NY.GDP.PCAP.CD?locations=CN

Wübbeke, J., Meissner, M., Max, Z., Ives, J., & Conrad, B. (2016). *Made in China 2025: The making of a high-tech superpower and consequences for industrial countries.* MERICS. https://merics.org/en/report/made-china-2025

Wynne, B. (2005). Reflexing complexity: Post-genomic knowledge and reductionist returns in public science. *Theory Culture & Society,* 22(5), 67–94.

Xueshenkeji (2020). *Average Zhihu users are 985? Python-based crawler gives you the answer* (知乎人均 *985?* 用 *Python* 爬虫告诉你答案). http://www.bilibili.com/read/cv4317550/

Yinsihuweidui (2022). *Southern Metropolis Daily's first algorithm governance report: Over 60% of respondents believe that companies should be forced to disclose algorithms* (南都首发算法治理报告:超六成受访者认为应强制企业公开算法_. https://www.sohu.com/a/www.sohu.com/a/620403827_121258695

Zeng, J., Chan, C., & Schäfer, M. S. (2020). Contested Chinese Dreams of AI? Public discourse about Artificial intelligence on WeChat and People's Daily Online: Information, Communication & Society. Information, Communication & Society. https://www.tandfonline.com/doi/full/10.1080/1369118X.2020.1776372

Zhang, D., Maslej, N., Brynjolfsson, E., Etchemendy, J., Lyons, T., Manyika, J., Ngo, H., Niebles, J. C., Sellitto, M., Sakhaee, E., Shoham, Y., Clark, J., & Perrault, R. (2022). The AI Index 2022 annual report. AI Index Steering Committee, Stanford Institute for Human-Centered AI, StanfordUniversity. https://aiindex.stanford.edu/wp-content/uploads/2022/03/2022-AI-Index-Report_Master.pdf

Zhang, S. (2019). *Zhihu user data analysis report* (知乎用户数据分析报告). 知乎专栏. https://zhuanlan.zhihu.com/p/79521617

Zhu, Y., Yu, D., Ye, D., Xu, C., Wang, S., Yu, S., & Jin, Q. (2012). [*Cover Report*] *Internet Leaks* 【封面报道】网络大泄密. Caixin Weekly. Retrieved June 30, 2023, from http://magazine.caixin.com/2012-01-06/100346334.html

Zuboff, S. (2019). *The Age of surveillance capitalism: The fight for a human future at the new frontier of power*. Profile Books.

PART 3

Inequalities, Resistance and Alternatives

CHAPTER 10

"With great power comes great responsibility": Lending Visibility to Risky Political Content

Özlem Demirkol Tønnesen

1 Introduction

All content on social media is produced with an audience in mind (Senft, 2008) and negotiating the boundaries between catering to audience expectations and the need for self-expression is a common dynamic for content creators. This dynamic is particularly essential when analysing the political content produced by microcelebrities, which refers to the accounts that cultivate a large audience and attention on online platforms via their strategic self-presentation and skilful content production. In settings where freedom of speech is protected, getting involved in politics may put microcelebrities at risk of alienating sponsors and audiences (Goodwin et al., 2020) which may result in loss of income and status. However, in the high-stakes environment of repressive regimes, these risks of expressive political participation are also accompanied by the risk of prosecution for those who hold opposing views to the government. As the risks are staggering for microcelebrities who wish to amplify and express these opposing views, a fine balancing act is needed to maintain status and lend visibility to marginalised discourses on a platform where fame is currency. In this chapter, I explore how Twitter (now rebranded as X) microcelebrity accounts in Turkey negotiate their roles within the anti-government political discourse and justify their expressive political participation. In Turkey and many other repressive regimes, Twitter users are more scared than ever to see, interact with, or produce political content (Tanash et al., 2017), which necessitates careful approaches for those who want to express political opinions to acclimate their audiences into accepting the incoming political content.

2 Twitter under Repression

The Justice and Development Party (AKP) in Turkey has been in power for the last 21 years, a period which has been marked by a significant

 | DOI:10.1163/9789004711396_012

authoritarian turn. Currently, over 90% of the mass media in Turkey is controlled by the government (Ahval, 2018). When the mass media is stripped of its duties as a watchdog and systematically erases dissident viewpoints from the public arena, those who are in opposition rely on social media to find political allies, express grievances (Tufekci & Wilson, 2012), take on citizen journalism (Hassid, 2012), and police government's narratives (Kavanaugh et al., 2016). Much like in other repressive settings with similar challenges, Twitter has been hailed as a critical tool for dissent and a platform 'reflecting everyday politics in Turkey' (Dogu & Mat, 2019, p. 232).

In an effort to tame Twitter, in February 2014, the AKP government introduced an Internet law, which gives the government absolute authority to block entire websites and target individual social media accounts without needing to go through the judiciary process (Letsch, 2014). In 2020, another law was introduced to target posts that disturb public order, which includes criticisms of the government and the president. As a result of these interventions, studies note an increase in self-censorship in online spaces due to the fear of being penalised (Wilson & Hahn, 2021; Tanash et al., 2017). Dissidents interviewed by Wilson and Hahn (2021) consistently report the difficulty of criticising the government without attracting a surge of abuse from trolls and pro-government accounts alike. Political views on Twitter get policed regularly by ordinary users and state employed social media moderators. A report by Freedom House (2018) showed that 65% of content removal requests to Twitter have been issued by Turkey and in 2017, Turkey was classified for the first time as 'not free' in terms of Internet freedom.

In this media landscape, the two most recent election periods have shown the extent to which those in opposition to the government, including opposition parties, are dependent on the imperfect freedom of speech social media provides (Lerner, 2018). However, while having high visibility and popularity on social media is necessary to mitigate the erasure of dissent from other media platforms, it also increases vulnerability to hostile surveillance. Therefore, researching how microcelebrities navigate this playing field is essential to understanding the ways subordinated groups can still claim limited agency in a repressive setting.

3 Why Does Expressive Political Participation of Microcelebrities Matter?

Before discussing their role in political discourse, I would first like to clarify my use of the term 'microcelebrity'. While there is a growing trend of studies using the term political influencers in reference to similar accounts (see. the special issue on Political Influencers in *Social Media and Society* (2023)), I refrain from using this term because the accounts I study are not primarily political; therefore, their political content is not an exclusive signifier of their microcelebrity status. Some studies that explore political content by microcelebrities recognise this distinction and refer to these accounts as non-political influencers (Tang, 2023). However, such a distinction encourages a binary that assumes accounts are either exclusively political or entirely non-political, while in reality, most accounts on social media fall somewhere in between. This is why I opted for using the term 'Twitter microcelebrity' without additional descriptors.

On the other hand, microcelebrity is also a contested term as there is a lack of consensus in academic research on whether microcelebrity refers to what someone 'is' or what someone 'does'. Pioneers of this field, Senft (2008) and Marwick (2015), both defined microcelebrity as a set of practices, a mind-set of carefully constructing the online self as a consumable good with the expectation that others do the same. In this sense, they consider microcelebrity not as a status but as a demand of social media architecture that encourages users to conduct themselves as celebrities competing for visibility. Whereas in other academic works microcelebrity is mostly used as a status one obtains by successfully maintaining a large audience (e.g. Tufekci, 2013; Abidin, 2015; Jerslev, 2016, etc.). In eliminating any confusion that may arise from this dichotomy, I find the need to separate microcelebrity practice from microcelebrity: the former refers to the practices that users engage with to gain visibility and fame in online spaces, and the latter refers to the accounts that successfully employ these practices.

4 Casually Political: Microcelebrities as Intermittent Political Actors

To date, most of the studies that explore political content produced by influencers or microcelebrities focus on accounts that exclusively produce political or advocacy content. However, as I argued above, the online world does not only consist of accounts that exclusively discuss politics and those who never do; there is a spectrum of political engagement that applies to microcelebrities and ordinary users alike. Microcelebrities who primarily use their social media accounts to promote political events and ideologies attract a limited but politically active audience (Tufekci, 2013), whereas microcelebrities who intermittently engage with politics can attract more politically diverse audiences. Thus, more attention needs to be paid to microcelebrities who do not define themselves as activists but create political content as a part of their larger narratives (Pande, 2018).

For instance, studies by Gaddini (2021), Abidin (2019), and Abidin and Cover (2019), show how the decentralization of attention via social media allows those who are marginalized within their own community, society, or other media to claim political agency if they can command the scattered attention. Gaddini (2021) explores the claiming of religious and political authority by white evangelical women microcelebrities, while Abidin (2019) and Abidin and Cover (2019) explore the advocacy work of LGBTQ microcelebrities. These studies engage with microcelebrities who claim agency by intermittently making their politics explicit and at other times create political agency through embodied representation of marginalised lifestyles. They show that there is much more to microcelebrity politics than the limited domain of activist/political microcelebrity.

I focus on Turkish-speaking/posting microcelebrities whose audience consists of those within a national Twittersphere, which creates further nuances in the ways they partake in political discourse in comparison to microcelebrities that exist in a multinational network. Evidently, microcelebrity theories are predominantly based on the experiences and practices of English-speaking/posting microcelebrities (Abidin, 2021). Acknowledging the caveats that come with focusing on microcelebrity practice in a repressive setting, Pande (2018) points to a lack of attention paid to the characteristics of local Internet cultures in microcelebrity research. They argue that local cybercultures make microcelebrities act differently from their Western counterparts due to issues such as censorship, lack of freedom of speech, and different cultural norms.

In repressive settings like Turkey where online dissent is criminalised, maintaining an exclusively activist/political account with a large following is incredibly risky. Research shows that, alongside self-censorship, Turkish Twitter users avoid engaging with activist accounts due to the fear that association with these profiles may put them at risk of losing their jobs (Yanatma, 2018). Therefore, the type of microcelebrity accounts that I discuss here are not exclusively activist or political accounts even though they mostly do generally narrate life in Turkey from an anti-government standpoint. Their lack of apparent ties to specific political parties or NGOs helps them in reaching a more diverse audience (Özdemir, 2019) and helps them to be perceived as 'people like us' rather than institutionalised political actors with a specific agenda.

When those who do not use social media to exclusively engage with politics express political opinions, due to the diversity of their audience they need to employ impression management strategies to avoid alienating audiences and attracting backlash from those who hold opposing views (Pearce et al., 2018; Pearce & Vitak, 2016; Mor et al., 2015). Particularly for microcelebrity accounts that are not exclusively political, engaging with politics may mean losing industry contacts or audiences (Goodwin et al., 2020), which is why impression management is essential to mitigate these risks. Impression management on social media has long been the subject of scholarly attention in which studies have shown how people employ different impression management strategies to overcome context collapse (Marwick & boyd, 2014); avoid disclosing political alignment (Marder et al., 2016); and circumvent unwanted audiences by manipulating the structural affordances of social media (Abidin, 2021a).

Those who studied impression management in politically risky settings focused mainly on strategies that allow for favourable self-presentation to family, friends, colleagues, and others with whom the content creator is connected in offline settings (Pearce and Vitak, 2016; Mor et al., 2015). These studies reference self-censorship as the most common impression management tool; Pearce et al. (2018) add that managing one political and one non-political profile is also increasingly common. However, these strategies are not accessible for microcelebrities in repressive settings whose expressive political participation is needed to mitigate the erasure of dissent from traditional media (Tufekci, 2013); using an alternative account for politics or self-censorship would defeat this purpose. Therefore, how microcelebrities who intermittently engage with politics manage these tensions needs further attention. This chapter contributes to this field by exploring how microcelebrities negotiate and justify their oppositional political content to their audiences, in a setting where political dissent creates fear and anxiety.

5 Methods and Ethics

Findings presented here are a part of a larger project in which I use interpretive content analysis to explore the content shared by 97 Twitter microcelebrity accounts over the three months leading up to the 2018 parliamentary and presidential elections in Turkey. Throughout the election period, I observed these accounts by using a Twitter list I created. I scraped their tweets to an Excel spreadsheet on a daily basis using the Web Data Research Assistant plug-in on Google Chrome. The final dataset included approximately 21,000 tweets. Interpretive content analysis was used to explore latent meanings reflexively and qualitatively by operationalising my insider knowledge into the conventions of anti-government political discourse on Turkish Twitter. During this time, I also tracked important events, rallies, and speeches that took place to provide the necessary context for these tweets for later analysis.

Exploring political talk in a repressive setting and on a platform rife with context collapse demands that researchers pay attention to latent and polysemic meanings that may resonate with different audiences in different ways. Due to the complexity of this data, my approach to interpretive content analysis differs from the common applications of this methodology. I use the coding principles of interpretive content analysis to create coding categories that enable me to categorise this large dataset of 20,981 tweets into manageable subsets that I then further analyse using a textual analysis approach. In this chapter, I mainly draw from a subset of the tweets that describe microcelebrities' justification of and motivation for sharing political content.

As the tweets I analyse were sent at a politically turbulent time in an oppressive regime, I pseudonymize them by using the row numbers in which they appear in the Excel spreadsheet to avoid re-identification and to give some context to their chronological occurrence. I have also ensured that the tweets shared here are translated in a manner that protects the meaning but does not result in a true match when reverse translated using an online translation tool or service.

Below, I identify three interconnected strategies that microcelebrities put forward to justify their political engagement to their audiences, which are priming, retargeting, and redefining roles. I trace how these strategies support and build on each other throughout the election period and culminate in the framing of political participation as a duty for all microcelebrity accounts. I conclude by arguing that as 'focusers of public attention' (Tufekci, 2013), the forced politicisation of microcelebrity accounts may have significant consequences for the way political opinions are discussed, disseminated, and popularised on social media.

6 Priming the Audience: Negotiating the Twitter Cool

Most of the impression management work of microcelebrities was focused on the negotiations around the unusual increase in their political content. In the early days of the election campaign period, these negotiations were managed by priming the audience for the likely increase in political content.

Priming is a strategic communication tool, and it refers to emphasizing certain messages or topics in earlier content to influence how the subsequent messages on the topic should be viewed (Van Duyn & Collier, 2018). Even though priming is often discussed in relation to mass media, research has shown that this strategy is also useful for influencers when presenting sponsored content, as it helps them to sustain their authenticity (Luoma-aho et al., 2019). I propose that priming as a microcelebrity tactic has uses beyond the limited scope of creating a more amicable outlook for sponsorships. It can be applied to all self-preservation efforts to control audience perceptions about an upcoming change or disruptions in the topics, alignment, frequency, and intensity of content. Therefore, if consistency is a key tenet of sustaining authenticity and fame on social media (Marwick & boyd, 2011), any disruption to coherence – including new sponsorships – will require a similar amount of 'warm-up' to ready the audiences for what is to come.

Even though priming has a close relationship to impression management, it is not widely discussed in the influencer or microcelebrity literature. Particularly in the context of social movements and political discourse where microcelebrities are considered the 'focusers of attention' (Tufekci, 2013), priming, which allows for selectively directing the attention of audiences to a desired interpretation of content, should be further investigated as a core microcelebrity strategy. In April 2018, the first reactions to the announcement of snap elections in Turkey reflect how this strategy is essential in politically turbulent settings. I will now present how priming is used through four tweets that were sent within the first few days after the announcement of snap elections:

> Elections being scheduled around the same time as the World Cup is really bad. We were going to follow websites for scores but now we will chase ballot boxes on Twitter instead, damn!
>
> ROW 50

> Wow, this is my last day at work, I quit, and I wanted to come here to talk about holidays and not having to go to work tomorrow but once again certain political things have happened so we will need to talk about those
>
> ROW 146

> Ramadan, Elections, World Cup; look,these three are the periods that Twitter is most fun: we will have a lot of fun together you will see!
>
> ROW 92

> These snap elections are seriously like a lifeline to many. People were chasing cats on streets to increase their interactions on Twitter. We have clearly become addicted to this, let's have a referendum after the elections too!
>
> ROW 1469

The primary use of priming is justifying topical shifts, which can be clearly seen in the first two tweets. The first tweet is sent by an account that often provides football commentaries. For these accounts, not focusing entirely on the World Cup is unusual; thus, amplifying the importance of the content that will replace the World Cup commentaries is essential to maintain their audiences' attention. Similarly, the second tweet is also addressing an upcoming diversion from planned content. On the other hand, the final two tweets add to this sentiment by pointing to Twitter's reputation as a platform for political discussions and leveraging this aspect of platform culture, which helps build an expectation that this abrupt change in content is normal and positive.

What makes these tweets priming statements is that even though the topic of concern (i.e., elections) is introduced, it is not discussed. The focus is rather on preparing the audience for the upcoming shift in the frequency of election related content. In this sense priming is different from framing; where framing concerns how a topic is evaluated and understood, priming focuses on the salience of the content (Van Duyn & Collier, 2018). All these approaches to preparing the audience for a topical change further exemplify how content production is a two-way process between audiences and microcelebrities. A common aspect of these priming statements is that they help sustain coherence despite upcoming change which is essential to maintaining microcelebrities' personal brand (Marwick and boyd, 2011). The expected outcome is that audiences primed in this way will be more forgiving about the focus on elections, even if they do not want to see more political content.

Priming can also be used to sustain attention rather than to channel or encourage it. Later during the election period, priming statements repeatedly appeared to refocus attention. For instance, in the week leading up to election day when the frequency of political tweets had reached overwhelming levels, statements such as the one below were used to reassure audiences and remind them why their sustained attention is important, Here, priming is used mainly as a tool for negotiation by openly positioning the content as a mutual commodity of audiences and the content creator even though the content creator is in full control:

> There are only a few days left until the election, so where there is a need for a ballot box observer, what the opposition candidate has said etc., whatever crosses my way I will RT. If this ends in the way we want, I promise there will only be jokes and comedy here. This account will be like a carnival, so please be patient.
>
> ROW, 13970

A priming strategy implies that involvement in political discourse for popular accounts has multiple dimensions that are not all at the discretion of the content creator. While previous research suggests that an increase in political content is more commonly observed when the content creators get positive feedback from audiences (Luoma-aho, et al., 2019), it can be seen here that microcelebrities also pre-emptively try to mitigate negative reactions they may receive for their future content.

7 Retargeting and Keeping Up Appearances

As the election related tweets started appearing with greater frequency, other techniques of impression management also started emerging. These were bolder and riskier than priming. Despite the priming efforts, due to the light-hearted and humorous tone of their usual content, these accounts faced backlash from some of their followers who demanded to see fewer political tweets. Even though the narrative established by priming may push the blame away from the content creators, audience backlash redirects the responsibility back to them. When audiences point to a lack of coherence, it can be addressed in one of two ways: either by defending this change, which may lead to loss of audiences, or by succumbing to audience pressure, which may lead to loss of trust by the audiences who saw the microcelebrity as an independent voice, thus decreasing the audiences' belief in the microcelebrity's authenticity.

Much of the research on audience backlash focuses on content creators in visual social media platforms and hostilities based on race, gender, and sexual orientation (Abidin, 2019; Glatt 2023). As Glatt (2023) observes, on visual social media such as Instagram and TikTok there is a blurring of lines between making and being content, which speaks to a different playing field than Twitter where embodied aspects of identity are much less front and centre. These studies suggest that in dealing with hostility, influencers directly shut down opposing views only as a last resort and often either ignore or use these instances as teachable moments to start a constructive conversation (Abidin, 2019). In addressing audience backlash, I have observed two strategies that go hand in hand and have certain similarities with the ones observed by previous research.

First, the overwhelming majority have opted for ascertaining confidence in their authenticity and right to self-expression by aligning themselves with the expectations of a politically minded audience while rejecting others. Secondly, by creating a narrative of responsibility for people who command public attention, they have rendered the option of refraining from politics on Twitter altogether or due to audience demands intolerable and cowardly. I see this as akin to using hostility as a teachable moment (Abidin, 2019) in which microcelebrities were trying to convince their audience that partaking in politics is a moral imperative. In doing so, they deferred responsibility for choosing to engage with politics by indicating that this was a requirement of their visibility on Twitter. I will unpack how these strategies have unfolded respectively.

The first narrative was actualised by publicly rejecting the audience demand for less political content and thus retargeting content to those who appreciated its political substance. Some examples of how Turkish microcelebrities responded to audience backlash can be seen below:

> Quote RT response to a tweet saying 'don't do politics':
> Do you think you own me? If you don't like it, there is the door!
>
> ROW 13210

> I write as I please, dear friends. I am not your clown; all I do is to permit you to watch me and that's all.
> If you can't stand what I say, you can just f*** off!
>
> ROW 1401

These tweets aim to undermine the importance of audience demands for these accounts' status on the platform by ascribing the locus of their success to the quality of their content. This may seem contradictory to the priming practices I have discussed earlier, but it complements them by drawing up boundaries to reinforce a level of control that these accounts have over their own narrative. That is, this allows these accounts to reinforce trustworthiness and authenticity by creating a perception that they do not care about the commercial interests and fame large audiences bring as much as they care about staying true to their opinions. Similarly, Glatt (2023) observes how some content creators turn to creating content they can be proud of over content that brings the most commercial gain by aligning with audience demands. However, retargeting can still mitigate the loss of audiences by appealing to the audiences who are disgruntled by the lack of tenacity and commitment shown by other microcelebrities who do not partake in political discourse. In doing so, these accounts deflect the criticism about their content back towards the audiences who wish to see fewer political tweets. This is also supplemented by retweeting supporting content by audiences such as the one below:

> I follow you for your political tweets otherwise your jokes are bad.
>
> ROW 13216, RT

Which audience responses microcelebrities choose to engage with and which of them they ignore is a performative act, especially when it comes to those they choose to retweet or quote retweet. By publicly rejecting certain audience demands and amplifying others that align with the accounts' agenda, these tweets first communicate a pledge, a reassurance to those who demand more political content during elections. Secondly, these tweets serve priority-setting and boundary-forming functions between audiences and microcelebrities when it comes to editorial independence, which reinforces the authentic image of microcelebrity.

8 A Twitter State of Emergency: Shifting Roles and the Responsibility of Being Seen

Even though the tweets shared in relation to retargeting so far have shown a very bold and unyielding stance against audience backlash, this was justified through the adoption of a second strategy that I call redefining roles. Retargeting concerned microcelebrity accounts' right to self-expression whereas the redefining roles strategy framed political participation as a responsibility of those commanding public attention. Thus, by framing the creation of political content as a duty and not just a personal decision, these accounts have strengthened the underlying morality of partaking in political discussions at the expense of audience demands. Through this they assigned a grander purpose to their right to self-expression and created an understanding that their refusal to succumb to audience demands was not just self-indulgence. For example, through the tweets such as the ones below they signal the importance of lending visibility to political content on Twitter in a repressive regime and reinforce the framing of this as a duty:

> These unfollows do not hurt me! Tag and send me anything that needs to be disseminated. I just want this to be a fair election.
>
> ROW 15594

> Dear microcelebrities, [...] many thanks to you for choosing to remain silent. I kiss you on your non-existent honour.
>
> ROW 16091, RT

This duty narrative was most prevalent towards the final days of the election period when attention was on the outcome and mobilisation against electoral fraud increased. During this time, much of the discontent was directed towards those microcelebrity accounts who avoided speaking about politics, which further enabled the politically minded microcelebrities to push back against audience demands by framing expressive political participation as a key part of Twitter culture and microcelebrities as key actors within the political discourse (Özdemir, 2019). These tweets were presented as if they were calls for a state of emergency on Twitter that demanded all to channel their efforts towards a unified purpose.

These demands echoed both in retweets and microcelebrity tweets were not just addressing the accounts whose usual content includes or can be easily adapted to accommodate political discourse. They were indiscriminately extended to all microcelebrities. Those who stepped outside of their usual narrative to partake in the political discourse were praised for their involvement. Most notably, a well-known beauty vlogger's unexpected involvement in addressing electoral fraud and encouraging their followers to go to rallies on Twitter attracted much attention and support for its potential to mobilising a young, beauty vlogger audience that is often considered apolitical.

This sense of duty towards amplifying anti-government voices and aiding the mobilisation against electoral fraud was most felt on June 24th, the election day, when 97% of the tweets were about the elections. As highlighted in the case of the beauty vlogger, the high visibility of microcelebrity accounts was seen as essential in the mobilisation efforts to prevent electoral fraud and to encourage reluctant voters to exercise their democratic rights. The emphasis was put on the microcelebrities' responsibility to disseminate information and aid collective mobilisation by leveraging their visibility. This resonates with how within social movements social media savvy participants see disseminating information on social media as a moral responsibility (Mercea and Levy, 2020). In this case, such an emphasis was also conveyed by separating Twitter from other social media platforms that are seen as more frivolous. Twitter was seen as a more 'serious' platform that deals with bigger issues as opposed to other platforms; the blame was placed on microcelebrities on other platforms who do not understand the importance of political participation. This was also why the political engagement of the beauty vlogger was reciprocated by such enthusiastic support.

> Where are those YouTubers who are constantly screaming, he stole my video, she stole my joke? I have no words for you! Do you understand what they are "stealing" from you right now?
>
> ROW 15523

> Let's not forget those bloggers, Instagrammers and Youtubers who post photos from holiday resorts on the day of the elections rather than helping us [in preventing electoral fraud]. You can't earn millions that comfortably ...
>
> ROW 14452, brackets added for clarity

Such tweets directly criticise the content creators of other platforms for their indifference. This hostility towards those who refrain from expressive political participation and the separation of Twitter from other platforms are also key to the pushback against audiences who demand fewer political tweets. Framing Twitter as a space for expressive political participation and emphasising the importance of microcelebrity involvement allow these accounts to justify their commitment to political content and supplement their priming and retargeting efforts.

9 Fear and Loathing: Mitigating the Consequences of Visibility

Assigning key roles to microcelebrity accounts during elections exacerbates the risks these accounts need to take to fulfil these responsibilities. So, beyond the negotiations with audiences, these accounts have to grapple with the fear of prosecution and the fear of losing industry contacts. Many Turkish social media users are afraid of expressive political participation on Twitter due to the criminalisation of dissent and so they resort to self-censorship (Tanash et al., 2017). Even though fear due to the risk of arrest and harassment is common on Turkish Twitter for all account holders, for microcelebrities whose income is reliant on good industry relationships, expressive political participation carries additional risks. One of the accounts for instance shared this tweet when they noticed that they were left out of a viral campaign:

> Am I not included in this [viral campaign for a brand] because of my support tweets for the opposition's presidential candidate?
>
> ROW 17460

Similarly, another account got into trouble with the brand they were advertising as the company did not like the fact that the sponsored tweet was indicating that their products make people want to vote for the opposition party (Row 2596, 2600). This is because anti-government political engagement is controversial, hence strictly policed on Turkish Twitter not just by the government but by audiences, companies, and sponsors alike (Wilson & Hahn, 2021). Much like their counterparts who work in other creative industries, microcelebrities in Turkey also battle with precarity and a lack of legal support systems that would protect them when they conflict with employers or sponsors. Not aligning with the government can lead to loss of contacts, as companies need to be on the good side of the government to stay afloat. Indeed, sharing these sentiments was also a part of impression management: making these sacrifices visible strengthens authenticity, as accounts distance themselves from the commercial aspects of the microcelebrity culture. Similarly, others also found the fear of losing income and followers as a weak excuse to neglect citizenship duties:

> We are seeing the fake micro-celebrities who don't tweet about the most important elections to ever happen in the country out of fear of losing their industry contacts and viral campaigns, you cannot hide. When the times change, we will know who to trust.
>
> ROW 14703

> Fear is like this. Those micro-celebrities who try to get girls on Twitter 24/7 remain silent during the elections because they are afraid of losing followers or being arrested. […] they act dead because they are afraid.
>
> ROW 16089

The priorities of the popular accounts were expected to shift during important political events from merely holding on to their microcelebrity status by producing their usual content, to actively engaging in the political discourse. Therefore, these accounts were expected to get involved in the anti-government political discourse not only by taking on the risk of arrest but also by accepting the long-lasting risk of losing income and industry contacts. Furthermore, the trivialisation of fear and the frequent reminders by the microcelebrities about these risks they took on created pressure for involvement in anti-government political discourse to sustain authenticity.

These pressures also showed how the topical configuration of Twitter is policed by the politically engaged:

> When there are election topics to discuss no one would care about this aid campaign. You should bring this up again after the elections so that we can do something.
>
> ROW 12352

> Didn't wanna say it during the election period but our baby food stocks are gone, and I need your help. If you can't send money or food, please help by RT. Please use this link [...]
>
> ROW 16587

Similar to these tweets, this pressure to partake in political discourse was demonstrated even a few weeks after the elections when many were humorously questioning whether it was too soon to speak about matters other than politics without a backlash from their followers.

After the general and presidential elections were won by the AKP and Recep Tayyip Erdogan, the incumbent President gained additional powers which meant that the fear of prosecution in the Twittersphere has increased. In this climate, even though deleting past critical tweets may attract negative attention and backlash, it was a valid self-preservation tactic and more common than publicised. Through the course of this project, I had to delete and replace the tweets I present in this chapter as examples of these practices because they were gradually deleted by the account owners. In fact, four days after the elections a police investigation started regarding a well-known Twitter account and the fear of sharing a similar experience was present in other post-election tweets:

> You won't see as many microcelebrities [...] who talk about the irregularities and reveal the misconduct by politicians anymore ... everyone has a family, and everyone is scared
>
> ROW 16156

Evidently, the demand for weaponising visibility by stepping outside of established microcelebrity personas and the expectation to overcome fear come at a great personal risk for these microcelebrity account holders who need to maintain followers, industry relations, and their authenticity while avoiding arrest. Even though the strategies of priming, retargeting, and redefining roles are useful in encouraging and normalising expressive political participation, they are not enough to protect these highly visible accounts from the risks that come with creating oppositional content. Navigating the dangers of visibility that makes these accounts susceptible to scrutiny is a difficult task, especially when they are given the responsibility to partake in risky discussions which, as I have shown throughout this chapter, is a responsibility these accounts have helped to create. In fact, by pointing out the risks they took and the fears they had to overcome when partaking in political discourse, they strengthened the message that microcelebrities have a duty to their audience and country that goes beyond the need for self-preservation.

10 Conclusion

Furthering others' work on the strategies for online impression management in politically risky settings (Pearce & Vitak, 2016; Mor et al., 2015), this chapter introduced three strategies that work together to increase the acceptance of expressive political participation by microcelebrities in the Twittersphere. I have shown that while the narratives introduced through priming aim to sustain attention and normalise the increase in political content, the retargeting strategy is used to present microcelebrities' expressive political participation as a testament to their authenticity and to cast their political participation as a condition of microcelebrity status. These culminated in the framing of lending visibility to oppositional content as a duty for all accounts that command public attention. Therefore, the increase in political content was justified as fulfilling a responsibility rather than an outright rejection of audience demands to see fewer political tweets.

Personal preferences, platform culture, and audience expectations all play a role in shaping these accounts' content production, perhaps at a much higher level than other accounts with less public presence. Here, I used Turkish Twitter microcelebrities as a case study to first highlight the importance of the political, systemic, and cultural conditions under which microcelebrities create content, which affects how they conduct themselves as public figures. Secondly, I argued that these complexities of content production lend importance to not just political discourse but also the choreographies around politically charged content that are designed to increase its acceptance and the attention paid to it. This means that the impression management techniques employed by these accounts when engaging in political discourse are as important as the actual political commentaries they tweet.

The moralisation of political participation is of particular importance within and beyond the Turkish case, as it creates a pressure that all microcelebrities are expected to channel all their efforts to amplify the reach of political tweets. These pressures arise from the idea that visibility is fundamental for the dissemination of information online which naturally puts the onus on those that command the most attention. This shift in the roles of microcelebrities corresponds with a trend that is not isolated to Turkey. Recent research shows an increasing demand by audiences for microcelebrities to engage in expressive political participation in the United States, which accounts for a much larger and globally known market than Turkey (Goodwin et al., 2020). This means that impression management of microcelebrities in catering to this demand while maintaining audiences with varying degrees of interest in politics needs to be researched in other contexts as well.

References

Abidin, C. (2015). Micro-microcelebrity: Branding babies on the internet. *M/C Journal*, 18(5).

Abidin, C., & Cover, R. (2018). Gay, famous and working hard on YouTube Influencers, queer microcelebrity publics and discursive activism. In P. Aggleton, R. Cover, D. Leahy, D. Marshall, & M. Lou Rasmussen (Eds.), *Youth, Sexuality and Sexual Citizenship* (1st ed., pp. 1–23). London: Routledge.

Abidin, C. (2019). Yes Homo: Gay influencers, homonormativity, and queerbaiting on YouTube. *Continuum,* 33(5), pp. 614–629.

Abidin, C. (2021). Activism in Singapore in the Digital Age: Influencer cultures, meme factories, and networked virality. In P. Natalie, & Z. Shamil (Eds.), *Singapore Perspectives: Politics.* (pp. 49–55). Institute of Policy Studies.

Abidin, C., (2021a). From "Networked Publics" to "Refracted Publics": A Companion Framework for Researching "Below the Radar" Studies. Social Media + Society, 7(1).

Ahval, 2018. How social media influenced Turkey's election campaigns: Looking back. [online] Ahval. Available at: https://ahvalnews.com/turkey-elections/how-social-media-influenced-turkeys-election-campaigns-looking-back

Dogu, B., & Mat, H. (2019). Who sets the agenda? Polarization and issue ownership in Turkey's political Twittersphere. *International Journal of Communication,* 13(22), pp. 229–250.

Freedom House (2018). Freedom on the Net – Turkey. Freedom on the Net. Available at: https://freedomhouse.org/report/freedom-net/2018/turkey

Gaddini, K. (2021). 'Wife, mommy, pastor and friend': The rise of female Evangelical microcelebrities. *Religions,* 12(9), p. 758.

Glatt, Z. (2023). The intimacy triple bind: Structural inequalities and relational labour in the influencer industry. *European Journal of Cultural Studies,* 27(3), pp. 424–440.

Goodwin, A., Joseff, K., & Woolley, S. (2020). Social media influencers and the 2020 U.S. election: Paying 'regular people' for digital campaign communication. Austin: *University of Texas at Austin Center for Media Engagement.*

Hassid, J. (2012). Safety valve or pressure cooker? Blogs in Chinese political life. *Journal of Communication,* 62(2), 212–230.

Jerslev, A. (2016). Media Times| In The Time of the Microcelebrity: Celebrification and the YouTuber Zoella. *International Journal of Communication,* 10, pp. 5233–5251.

Kavanaugh, A. L., Sheetz, S. D., Sandoval-Almazan, R., Tedesco, J. C., & Fox, E. A. (2016). Media use during conflicts: Information seeking and political efficacy during the 2012 Mexican elections. *Government Information Quarterly,* 33(3), pp. 595–602.

Lerner, D. (2018). Turkish opposition gets creative as Erdogan tightens grip on media ahead of elections. Available at: https://www.haaretz.com/middle-east-news/turkey/.premium-erdogan-elections-turkish-opposition-gets-creative-1.6166971

Letsch, C. (2014). Turkey pushes through new raft of 'draconian' internet restrictions. *The Guardian, 06 February.* Available at: https://www.theguardian.com/world/2014/feb/06/turkey-internet-law-censorship-democracy-threat-opposition

Luoma-aho, V., Pirttimäki, T., Maity, D., Munnukka, J., & Reinikainen, H. (2019). Primed authenticity: How priming impacts authenticity perception of social media influencers. *International Journal of Strategic Communication,* 13(4), pp. 352–365.

Marder, B., Slade, E., Houghton, D., & Archer-Brown, C. (2016). "I like them, but won't 'like' them": An examination of impression management associated with visible political party affiliation on Facebook. *Computers in Human Behavior, 61,* pp. 280–287.

Marwick, A. (2015). You May Know Me From YouTube: (Micro)-Celebrity in Social Media. In D. Marshall & S. Redmond (Eds.), *A companion to celebrity.* Wiley.

Marwick, A., & boyd, d. (2014). Networked privacy: How teenagers negotiate context in social media. *New Media & Society,* 16(7), pp. 1051–1067.

Marwick, A., & boyd, d. (2011). To see and be seen: Celebrity practice on Twitter. *Convergence: The International Journal of Research into New Media Technologies,* 17(2), pp. 139–158.

Mercea, D. and Levy, H., (2020). The activist chroniclers of Occupy Gezi: Counterposing visibility to injustice. In: A. McGarry, I. Erhart, H. Eslen-Ziya, O. Jenzen & U. Korkut, (Eds.), *The Aesthetics of Global Protest: Visual Culture and Communication,* (1st ed., pp. 234–246). Amsterdam University Press.

Mor, Y., Kligler-Vilenchik, N., & Maoz, I. (2015). Political expression on Facebook in a context of conflict: Dilemmas and coping strategies of Jewish-Israeli youth. *Social Media + Society*, 1(2), pp. 1–10.

Özdemir, S. (2019). Sosyal Medya Ünlüleri Üzerine Bir İnceleme: Türkiye'de Twitter Mikro Ünlüsü Olmak. *Moment Journal*, 6(2), pp. 406–427.

Pande, R. (2018). It's just a joke! The payoffs and perils of microcelebrity in India. In C. Abidin & M. Brown (Eds.), *Microcelebrity Around the Globe* (1st ed., pp. 1–17). Bingley: Emerald Publishing Limited.

Pearce, K. E., & Vitak, J. (2016). Performing honor online: The affordances of social media for surveillance and Impression Management in an honor culture. *New Media & Society, 18*(11), pp. 2595–2612.

Pearce, K., Vitak, J., & Barta, K. (2018). Socially mediated visibility: Friendship and dissent in authoritarian Azerbaijan. *International Journal of Communication,* 12, pp. 1310–1331.

Senft, T. (2008). *Camgirls: Celebrity and community in the age of social networks* (1st ed.). New York: Lang. pp. 1–25.

Tanash, R., Chen, Z., Wallach, D., & Marschall, M. (2017). The decline of social media censorship and the rise of self-censorship after the 2016 failed Turkish coup. In 7th USENIX Workshop on Free and Open Communications on the Internet.

Tang, J. L. (2023). Issue communication network dynamics in connective action: The role of non-political influencers and regular users. *Social Media+ Society,* 9(2), pp. 1–13.

Tufekci, Z. (2013). 'Not this one': Social movements, the attention economy, and microcelebrity networked activism. *American Behavioral Scientist,* 57(7), pp. 848–870.

Tufekci, Z., & Wilson, C. (2012). Social media and the decision to participate in political protest: Observations from Tahrir Square. *Journal of Communication,* 62(2), pp. 363–379.

Van Duyn, E., & Collier, J. (2018). Priming and fake news: The effects of elite discourse on evaluations of news media. *Mass Communication and Society,* 22(1), pp. 29–48.

Wilson, J., & Hahn, A. (2021). Twitter and Turkey: Social media surveillance at the intersection of corporate ethics and international policy. *Journal of Information Policy,* 11(1), pp. 444–477.

Yanatma, S. (2018). *Reuters Institute Digital News Report – Turkey Supplementary Report.* Reuters Institute for the Study of Journalism. Available at: https://reutersinstitute.politics.ox.ac.uk/sites/default/files/2018-11/Digital%20News%20Report%20-%20Turkey%20Supplement%202018%20FINAL.pdf

CHAPTER 11

Artificial Intelligence's Sexual Politics: Three Modes and the Case of Japan

Hiromi Tanaka and Michelle H. S. Ho

1 Introduction

Since the late 2000s, artificial intelligence (AI) has attracted much attention as a herald of emerging digital technologies. The current debate about AI began in the 2000s, following a prolonged 'AI winter' (Matsuo, 2020) that came after the first and second AI boom in the 1950s to 1960s and the 1980s to early 1990s, respectively. This winter ended with the advent of deep learning, a new type of machine learning technology gaining traction in 2006. The advancement of internet-related technologies from the 1990s to 2000s played a pivotal role in this development. New information and communication technologies and the Internet of Things (IOT) were instrumental in generating big data. A significant increase in computing power also enabled the collection, processing, accumulation, and analysis of huge amount of data at an unprecedented scale and speed.

Further technological progress and the increasing versatility of AI have sustained interest in the technology to this day, as evidenced by the more recent buzz generated by ChatGPT, an AI chatbot developed by OpenAI and released in November 2022. Their large language models, known as GPT (general pre-trained transformer), were developed in 2018 and updated three times since then (2019, 2020, and 2023). As of late 2023, their most recent model GPT-4, which can process both text and images with an unprecedented level of precision and complexity, has become a global sensation.

Although AI technologies are promoted as innovative and beneficial to our lives, their applications have been criticized, too. Various risks have been identified, such as the potential spread of disinformation, violations of copyright and privacy, and the concentration of key technologies and data in the hands of a few dominant companies. Furthermore, scholars have become increasingly concerned about the reproduction of technology-facilitated inequalities, social biases, and injustice, particularly those related to gender and race (Benjamin, 2019; Buolamwini & Gebru, 2018; Crawford &

DOI:10.1163/9789004711396_013

Calo, 2016; Noble, 2018; O'Neil, 2016; West et al., 2019; Zou & Shiebinger, 2018). However, these problems remain underexplored despite their relevance and urgency. While critical debates about AI ethics are expanding, they remain insufficient, particularly in addressing broader social aggregations of inequalities and power dynamics (Munn, 2022).

We address these recurrent issues of hierarchies and power through a critical feminist lens. Reviewing critical discourses about AI related to gender and, to a lesser extent, race, and ethnicity, over the past two decades, we propose and discuss three modes of AI's contemporary sexual politics, namely a) discrimination caused by non-organic actors or 'actants' (Latour, 2005) such as algorithms, b) AI-facilitated sexual violence, and c) gendered embodiment of machines serving human needs. Since most existing research has been conducted in the West, we apply this framework to a non-Western case study to illustrate how AI's sexual politics unfold in a different sociocultural context.

For the present study we focus on Japan, a leading non-Western country in technology with a relatively long history of technological development characterised by a culturally unique fascination with nonhuman objects, which is distinct from the West. In Japan, nonhuman objects are not necessarily 'othered' by humans. While advanced automation and robotics research and development began in the 1950s, following technological modernisation between the 19th and the mid-20th century, a unique intimate relation between humans and nonhuman entities has manifested culturally. Unlike in the West, machines have been imagined, designed and perceived as friendly beings to humans.[1] Such 'techno-animism' (Allison 2006, see also Jensen & Blok, 2013) has shaped Japan's contemporary ubiquitous digital culture as well as popular culture, through media output such as manga, anime, games and science fiction (e.g., Katsuno & White, 2022; Petman, 2009; Robertson, 2017; Roquet, 2022; Steinberg, 2019; Tomita, 2005; Taylor, 2007).

1 Some examples include contemporary robots such as Sony's Aibo (1998–2006) and Softbank's Pepper (2014–2021) as well as Gakutensoku (1928), the first robot invented in Japan/Asia (Frummer, 2020). Some trace this lineage back to *karakuri ningyo* or mechanised puppets which were produced to entertain people and popular in the Edo period (1603–1867). Such strong affinity with nonhuman machines can be seen in many later popular cultural narratives (e.g., Tezuka Osamu's *Austro Boy,* 1952–1968; Fujiko F. Fujio's *Doraemon,* 1969–1997, etc.).

Importantly, Japanese digital- and pop cultures are gendered. Japan is known for rigid gender relations, being ranked 125th among 146 countries in the Global Gender Gap Report 2023. The nation is thus characterised by a stark ambivalence, in that its economic, technological and sociocultural development is not in tandem with a democratisation of traditional gender relations. This makes Japan an insightful case study for exploring how emerging technologies such as AI are socio-culturally constructed and located. It also underlines the importance of critically considering culture and the embeddedness of gender therein.

In what follows, we will first review critical discourses about AI related to gender and, to a lesser extent, race and ethnicity, from the 2010s to date. We demonstrate three major modes of contemporary sexual politics related to AI. Drawing on existing research, which is mostly available via Anglophone literature, we will illuminate various ways in which AI operates as a present-day actant and device of discrimination, violence, and oppression (Section 2, 3 and 4). We will then use this framework to examine the Japanese case (Section 5). To outline the current AI landscape in Japan and analyse how its sexual politics is played out therein, we will conduct discourse analyses of various textual and visual data, including media reports, promotional materials of certain products or services as well as governmental documents. Supplementing this data are other materials, such as interview data we have collected for our study of AI companions (e.g., see Tanaka & Ho, 2022). We conclude by discussing their implications for governance and future research.

2 Algorithms as Emerging Actants of Discrimination

Contemporary discussions about AI typically concern machine learning, whereby other terms such as 'algorithm' and 'deep learning' are frequently used together. They are, however, only parts of AI. Technically, AI refers to 'a class of computer system designed to model some aspect of human intelligence' (Adam, 1998, p. 1), such as learning, speaking, moving (e.g., automated cars) and making decisions. Machine learning is one component of such a system that uses data and algorithms – a set of mathematical rules given to a computer or procedure used for solving a computational problem – to achieve allegedly human-like learning. It enables computers to identify patterns in data and make a prediction or judgement about unknown data or the future. Deep learning, which is based on a four-layer network to mimic the structure of human brain synapses, significantly improved the accuracy of machine learning systems and enabled more complicated modelling and application for a wide variety of objectives.

These technologies, however, entail various gendered and racialised problems. They have emerged as a new agent of discrimination against certain groups of people, especially women and people of colour (Benjamin 2019; Buolamwini & Gebru, 2018; Noble, 2018; West et al., 2019; Zou & Shiebinger, 2018). Such technologies exacerbate already existing sociocultural discrimination in at least three ways we highlight below: symbolically annihilating, trivialising, and condemning women and other marginalised groups.[2]

Symbolic annihilation can be seen in Google Search, one of the most widely used services on the internet. In its search results, women are often erased from predominantly male occupations such as engineering (Gaskell, 2022). Similarly, Google Translate has transformed women historians or doctors into their male counterparts in a translation of gender-neutral words for these occupations from European languages into English (Kayser-Bril, 2020). That said, more recently, Google has begun addressing these problems in a bid to better their image (e.g., Kuczmarski, 2018), but critics have read such efforts as tantamount to "window dressing" (Kayser-Bril, 2020).

Women and minorities are trivialised in various systems using machine learning algorithms including image recognition, which turns out to be less accurate in identifying women and people of colour than White men. As early as 2015, this issue was highlighted in a controversial Google photo app that inaccurately labelled a Black woman as a gorilla. Similarly, the Gender Shades project at MIT (http://gendershades.org/) found that the accuracy of different systems such as Microsoft, IBM and FACE++ depended on a person's gender and skin tone. An error rate of up to 34.7 percent was recorded, whereby White men and darker-skinned women were the most and least accurately recognized, respectively (Buolamwini & Gebru, 2018; for similar studies, see e.g., Dooley et al., 2021). Other studies revealed that facial recognition systems do not perform well on transgender and non-binary people (Urbi, 2018; Sheuerman, Paul & Braubaker, 2019).

Condemnation refers to false accusations and defamatory statements against women. Misogyny is not a new phenomenon but it intensified with the spread of new information and communication technologies. The internet has become a place where anti-feminist and misogynist sentiments are widely shared, which is reflected in the big data retrieved from the web

2 These concepts were originally elaborated by Gaye Tuchman (1978) in her analysis of gender representation in the mass media.

used for machine learning. Accordingly, Google's search algorithms, for example, learn from prejudiced views about feminists, people of colour, and environmentalists available on the internet. As a result, search results can be offensive to them, both in text and image (Vlasceanu & Amodio, 2022) as well as in search recommendations (Cadwalladr, 2016; Noble, 2012; for a similar case on transgender people, see Shadel, 2022). In short, search engines can contribute to further spreading hate towards them (Solomon & Levin, 2016).

While algorithms play a significant role, humans can amplify the problems of condemnation, as in the case of Tay.AI, a Microsoft chatbot released on Twitter in 2016 (Hunt, 2016). Designed as a 19-year-old (White) girl and targeted at young people aged 18 to 24, Tay.AI quickly learned Twitter users' languages and started making anti-feminist and racist statements. Although Tay.AI shows its own 'symbiotic agency' (Neff & Naggy, 2016) through its metamorphosis, some human users allegedly misused her ability to learn language and fed her offensive words. On one hand, there was a technical deficit that the bot was not equipped with a filtering system. On the other hand, Tay.AI's identity as a young (White) woman likely emboldened misogynist-racist harassment (Vorsino, 2021) as is often the case with online misogyny and gender-based violence (Henry & Powell ([2016]2018).

All these cases of discrimination require accountability. However, it is not always clear why computers produce certain outputs. The so-called 'black box' problem does not mean that machines or computers are sole bearers of responsibility. The case of Tay.AI shows how machine learning systems can be exploited by humans with malicious intent. In fact, humans are engaging in algorithmic discrimination. Firstly, data are biased because they reflect prevailing social inequalities which have long accumulated through humans' practices and related processes in society. Secondly, those who have access to the data often prioritise capitalist interests and are thus far quicker in advancing technologies than working on the problems that emerge. This indicates their limited interest in or awareness of such problems. It is no wonder that a handful of leading (mostly U.S.) companies in the technological fields are dominated by White men who are least affected by these problems.

3 Generative AI and New Forms of Technology-Facilitated Sexual Violence

Many of the problems described above result from discriminative models of machine learning. These models are used to predict labelling or classifications, whereby a machine is trained on a dataset so that it can 'learn', i.e., recognize statistically, what the boundaries between different kinds or classes are and predict what e.g., an image possibly visualises. Major criticisms against these models concern bias in training data in which women, people of colour and other minorities are underrepresented. Recently, generative models have attained popularity since the mid-2010s. While discriminative models are mainly for classification, labelling and prediction, generative models can create new data samples which may not exist in the training set through an observation of training data. Although capable of creating seemingly new output, these models have also caused problems, such as abusive usage by humans and inaccurate or even harmful outputs, especially for women.

There are many frameworks for generative AI, such as the popular GANs (generative adversarial networks) and GPTs. GANs were introduced in 2014 (Goodfellow et al., 2014). Compared to earlier models such as VAE (variational autoencoder), GANs excel in producing high quality images from low quality images. The term 'deepfake' refers to employing GANs (i.e., deep learning) to synthesise false human images that look highly realistic. Initially, deepfakes appeared to be entertaining but soon their role in deceit, manipulation and disinformation surfaced. For example, deepfakes of influential politicians can pose a real threat to democracy. Another abusive use of deepfakes is fake pornography. In 2019, an app called DeepNude caused public uproar. Released in 2018, DeepNude enabled users to undress and produce a realistic nude image of women. DeepNude targeted women because its dataset contained only images of naked women (Schick, 2020). In the same year, Telegram, a Russian messenger app, came under fire after its bot service was used to create fake nude images, victimising at least 100,000 women (Hao, 2020). AI-generated pornography has grown by more than 290 percent since 2018 (Verma, 2023).

Arguably, generative AI not only reinforces objectification and commodification of female bodies, forms of sexism deeply rooted in society and culture, but its rise has also added a new dimension to technology-facilitated sexual violence (TFSV), which is a term Henry & Powell ([2016]2018) coined to index harms related to modern digital technology use, such as sexting, online misogyny, and non-consensual sharing of intimate images. Unlike other types of TFSV, AI-facilitated sexual violence is difficult to regulate because the images are of non-existing humans and thus there are no victims. Even if someone creates a fake nude image of a real person using a regular photo, the perpetrator is unlikely to be prosecuted unless the image is publicly distributed.

Having said that, AI technologies have been used to combat sexual violence, too. Computer vision-based pornography filtering, for example, can recognise the difference between pornographic and non-pornographic images. This approach, however, identifies only 'standard' naked women, thus excluding other types of bodies (e.g., gay, trans, fat, and hairy etc.; see Gehl et al., 2017). It thus reinforces existing cultural expectations about the human body, namely, which bodies are supposed to be 'standard' and which ones are not. Another example is using AI chatbots to counter sexual violence. While enabling users to report sexual violence cases, these chatbots are often exploited under 'surveillance capitalism' (Zuboff, 2019), a new global economic order which has emerged in the current digital age, characterised by power relations between technological companies exceling in advanced technologies and other actors, whereby people's behaviour is observed and their experiences are transformed into and exploited as free raw material for capitalist practices. Tech companies' interest does not primarily lie in protecting users from various forms of harm, because they prioritise maximising profits through accumulation of users' data as capital (Henne et al., 2021). These cases show how AI functions as a double-edged sword to humans' wellbeing.

4 Gendered Embodiment of Intelligent Machines

The controversies about gendered AI discussed so far are related to supposedly intangible algorithms. Gendered ideologies, however, can be materialized on another dimension, through an embodiment of intelligent machines in concrete material forms. Such gendered embodiment has a longer history than the current AI technologies, having emerged from robotics research and industry since the late 1940s. It is therefore important to look at how chatbots or text- and voice-based conversational agents have developed since the 1950s. Though it was not until the 2010s that contemporary chatbots using AI emerged, there has been a consistent trend of gendered anthropomorphisation in chatbot development, strongly characterized by a preference for chatbots gendered as 'female' over 'male' (Feine et al., 2020).

Many chatbots, voice assistants, and smart agents are given feminine names, voices, appearances (e.g., avatars),[3] and personality traits (e.g., friendly, supportive, caring, passive) and assigned to perform tasks traditionally

3 Vorsino (2021) notes that they are mostly styled as young White women.

associated with women's labour, such as secretarial duties and care work or emotional labour (Chin & Robison, 2020; Costa & Ribas, 2019; Feine et al., 2020; Vorsino, 2021; Strengers & Kennedy, 2020). Aside from Tay.AI, other well-known examples of such feminised chatbots include ELIZA (1960s), A.L.I.C.E. (1990s–) and more recent ones such as Apple's Siri (2011–), Amazon's Alexa (2014–), and Microsoft's Cortana (2014–). Often seen as a forerunner of chatbot technology, ELIZA played a central role in the formation of modern AI narratives (Natale, 2019). Considering this, the fact that it was given a White woman's identity was crucial for chatbots' future feminisation and racialization.

Recently, as a response to mounting criticism of gender stereotypes ingrained in chatbot design, technology companies have tried to make them more gender-neutral. For example, Apple updated Siri's voice setting to offer a variety of voices in different genders and regional accents, even including a gender-neutral voice (Perez, 2022). Still, many others remain feminized, notably Amazon's Alexa (Costa & Ribas, 2019). This might be justified based on studies that found a feminised design is more suitable for robots which substitute traditional feminised labour (see Eyssel & Hegel, 2012), but as McDonnell and Baxter (2019) ask, what do we want to do with these findings? Do we want to reinforce stereotypes? From a different angle, the very effort to increase diversity by adding different types of voice ironically reveals how obsessed designers are with human gender, which machines actually do not have (Phan, 2017).

Furthermore, machines have been developed explicitly as a 'surrogate' (Atanasoski & Vora, 2019) for human women and thus feminised: women companion agents for male heterosexual desires. Some agents such as Replika developed by Luka, Inc. in 2017 (Luka, Inc., 2024) are flexible in its gender setting but others are 'female' as their default. Many websites and apps now allow users to easily create virtual girlfriends. Also, sex robots using AI technologies (e.g., speech recognition) are now available on the market. One of the well-known examples is Realbotix's Harmony, which is both an avatar and a humanoid robot developed since 2017 (https://realbotix.com). It has a realistic human woman's physical body with synthetic soft skin, a feminine-sounding voice and moving body parts (e.g., eyes, mouth). Hailed as 'the world's first talking sex robot' (Kragen, 2017) using AI, Harmony is equipped with a virtual reality camera, speaker, and microphone (Coursey et al., 2019). Though Harmony is marketed as bisexual and her appearance is customizable to a certain degree, sex robots are generally female and a majority of their owners and users are men who often have a high socioeconomic status and a 'robot fetish' (Döring, 2017; see also Döring &

Pöschl, 2018). Sex robots typically represent slender women with thick lips, heavy eye make-up, and big breasts and bottom, resembling 'sexy', normatively feminine bodies that are prevalent in popular culture.

Since the late 2000s, sex robots have attracted much attention with the growing digital romance and sex market. Some argue sex robots can enhance users' satisfaction and mitigate sex workers' vulnerability (Levy, 2007), but they have been designed based on a narrow understanding of gender, including how women behave and what they desire. In fact, diverse and complex interaction and communication are not fully reflected in designing such robots. Many scholars associate this problem with a legacy of gendered robots represented in popular culture, especially science fiction (Chin & Robinson, 2020; Costa & Ribas, 2019; Devlin, 2018; Devlin & Belton, 2020). Little empirical research has been conducted on human-sex robot interaction, not to mention cultural differences. One exception is a survey which found that more men than women and more people in the U.S. than in Indonesia tend to accept sex robots (e.g., see Döring, 2017).

Advocates against sex robots (e.g., the campaign against sex or porn robots, https://campaignagainstsexrobots.org) have expressed ethical concerns about their role in reinforcing the sexual objectification of women and facilitating violence against them. Ethics scholars have raised critical questions, such as whether sex robots should be produced at all (Nyholm & Frank, 2019; Sterri & Earp, 2021; see also Danaher, 2019), whether they can be subjects in ethical discussions, and if so, how their 'consent' can be considered in their interactions with human users (Kaufman, 2022). While sex robots cannot hastily be associated with real-life occurrences of sexual violence against women without any evidence (Devlin, 2018), media and communication scholars have long pointed out that humans tend to equate computers with humans and treat them as if they were interacting with other humans (Reeves & Nass, 1996). This demands more investigation into the relation between humans' mistreatment of machines and their actual interaction with other humans.

5 Gendered (and Racialised) AI in Japan

Japan's advanced robotics and automation research and development started in the 1950s soon after the Second World War. Since then, the country has been a leading non-Western country in the technological field for more than half a century. In the past decade, other countries such as China and South Korea have emerged as strong players in the changing technological landscape characterised by emerging technologies such as AI. Despite this competition, there are hundreds of AI companies in Japan and they are slowly growing in number (Bochev, 2021). Large Japanese companies such as NEC, Fujitsu, NTT and Hitachi as well as new venture businesses domestically develop AI systems; some of these systems are exported. Moreover, the Japanese government regards AI as a key for national development, having established various AI-focused bodies (e.g., strategic councils focusing on AI) and drafted many AI-related programs (e.g., AI Technology Strategy, 2016; AI Technology Strategy Action Plan, 2018; the Fifth and Sixth Science and Technology Basic Plans, 2016–2020 and 2021–2025).

Japan's AI policy has been characterised by strong emphasis on industrialisation and high expectations in promoting innovation, enhancing mobility (e.g., development of automated cars) and solving social problems, particularly those related to the declining population and aging society, i.e., anticipated labour shortage and increased needs for elderly/health care (Government of Japan, 2015, 2016, 2021; Shibata & Watanabe, 2023). In fact, the automotive and robotic sectors have large shares in Japan's AI related research and development. Companies such as Toyota, Honda, and Nissan invested four to eight billion euros in AI-related research and development (Greimel, 2019; Ishii, 2018, cited in Dirksen & Takahashi, 2020). To implement this, the government formed 'networks' (Dirksen & Takahashi, 2020) with private companies, their research institutes, public research and funding institutions and universities to promote AI research and development.

Alongside these industrial and governmental developments, Japanese popular culture has been increasingly digitalised over the last several decades, leading to new types of products and services. However, the official AI-related documents rarely refer to culture, much less problems such as biases, stereotypes, and violence. The latest discussion at the newly created AI Strategy Council (2023-) mentioned only education as an important cultural area for AI application, but in other cultural areas such as media, communication, and entertainment, AI is already heavily used, though no discussion about ethical implications takes place. Similar to other countries, AI is used for advertisement delivery and feed recommendation on social media. Facial recognition technologies are increasingly used not only for policing and immigration control, but also for leisure purposes such as games. Various entertainment devices using AI such as companion robots (e.g., Aibo, LOVOT) have become commercialised, too. More recently, ChatGPT has received much attention in Japan as elsewhere and became a popular tool for various purposes, ranging from content creation to intimate therapeutic conversations about serious personal matters (Tohata, 2023).

Despite a relative disinterest in culture among officials and experts who advise them, it is an important domain where the sexual politics of AI is played out. Examining Japan's 'robo-sexism' in the 2000s and the early 2010s, Robertson (2017) analysed various robots made in Japan such as gynoids (e.g., Repliee Q2 and HRP-4C) and mechanical looking cute robots (e.g., Pino and Posy) and showed how gender was a powerful force in guiding Japanese robotics research and development. While useful, Robertson's work mostly focused on the gendered embodiment of robots and was too early to cover more recent AI developments. Considering this, we ask the following questions to be further addressed in the present chapter: what does Japan's robo-sexism look like now? Are there intersections of gender with race and ethnicity? How might we incorporate the other two modes of AI sexual politics, algorithmic discrimination and abusive use of AI technologies?

5.1 *Algorithmic Discrimination*

Algorithmic biases manifesting in gendered and racialised AI have mostly been framed in Japanese media discourses as problems that occur abroad, that is, outside Japan. However, this is changing with ChatGPT. A recent study conducted on ChatGPT by *Asahi Shimbun*, a major nationwide newspaper, revealed that its chatbot produces gender biases and stereotypes in Japanese (Shino, Yamazaki, & Niizuma, 2023). Being the first study as such conducted on outputs in Japanese language, both GPT-3.5 and -4 were asked about occupation and gender 3000 times each, resulting in biased answers accounting for 41.5% (GPT-3.5) or 22.9% (GPT-4). These findings were considered so significant by the newspaper's editors that *Asahi Shimbun* reported them on its front page.

Another critical gender (and race and ethnicity) issue related to machine learning algorithms concerns facial recognition systems. In Japan, critical debates about facial recognition typically concern surveillance without referring to gender, race or ethnicity. This view comes with two pitfalls. First, they do not consider the sociocultural implications of these technologies for women and darker-skinned people. Apart from foreigners who temporarily reside in Japan and migrants from other countries, there are darker-skinned Japanese who are of mixed race and have partial roots or backgrounds in South Asia, Africa, and other parts of the world. Racial hierarchies exist based on physical traits such as skin and hair. This has led to numerous public controversies in Japanese society (see e.g., Ho, 2023; Ho & Tanaka, 2022). Using facial recognition technologies without adequate reflection and discussion therefore exacerbates algorithmic discrimination in Japan.

Moreover, the increasing popularity of facial recognition for other purposes, such as skin beauty assessment, must also be viewed from a gendered and racialised lens. In Japanese culture, beauty ideals are strongly associated with whiteness. Skin colour is understood in a dichotomy of 'white' and 'black', whereby the 'white skin' (*shiroi hada, irojiro* in Japanese) is embraced as a beauty standard (Ashikari, 2005), particularly for women. To maintain or acquire fair skin constitutes a significant part of feminine aesthetic labour (Elias et al., 2017). There are clear gender and racial hierarchies based on skin tone, and this can be seen in various practices of feminine aesthetic labour offered by cosmetics companies, such as *bihaku* (beautiful white) products and, most recently, *bihada* (skin beauty) AI, a scoring system for skin beauty assessment. These companies typically use lighter-skinned East Asian, Western, or Eurasian models for advertising such services. Since darker-skinned people are rendered invisible, the new technologies reinforce their exclusion in Japanese society.

5.2 *AI-Facilitated Sexual Violence*

AI-facilitated sexual violence in Japan resembles that of other countries. Various tools available on the internet such as Stable Diffusion are popular and used for the production of fake pornography. As mentioned earlier, if fake pornographic images of real humans are identified, they will likely be regulated; however, even such images remain difficult to crack down on. This applies to Japan as well, which has no law for regulating fake pornography. Perpetrators who privately create and possess sexual images cannot be prosecuted even if these images are taken non-consensually from real humans.

Japan has one of the largest and most diverse sex markets in the world (Koch, 2021). Though prostitution and pornography are legally prohibited in Japan, they have existed for centuries. Various pornographic contents are available in live-action (e.g., AV or adult videos) and manga/comic format (e.g., adult manga). These contents are consumed both domestically and internationally. Unsurprisingly, it did not take long until AI was adopted to produce fake pornographic contents of women and children, particularly girls. In fact, AI-generated pornographic images made in Japan are already popular beyond its borders because they are readily available and 'more realistic' (Kuwabara, 2023).

Since 2020, the Japanese police started regulating deepfake nude contents, leading to several arrests for defamation against female celebrities and copyright violation against a company which produced the original adult video (Kakizaki & Suzuki, 2020). According to current law, however, completely fake content is beyond police jurisdiction, even if AI-generated characters resemble real persons. The onus, it would seem, falls on large publishing companies who remain cautious about copyright violation. For example, in May 2023, a weekly magazine, *Shukan Purei Boi* (Weekly Playboy) published a photobook of Ai Satsuki, an AI-generated *gurabia aidoru* (gravieur idol)[4]. Soon after public opinion criticised this AI-generated model for resembling a real person, Shueisha, the publisher and a major publishing company in Japan, decided to pull the photobook. Still, many pornographic contents including fake ones circulating on the internet are created by individuals and small publishing companies who do not necessarily care about copyright violations as long as they are not legally prohibited (Kuwabara, 2023). It remains to be seen what measures will be implemented in Japan in the future. So far, the Japanese government has been observing discussions about and regulations of AI in other countries, particularly the European Union and the United States, without taking action – a typical cautious stance in Japan's policy making.

5.3 *Gendered Embodiment of Machines*

Gendered embodiment continues to operate with the rise of contemporary AI technologies, which are deeply entangled with Japanese popular culture. While some non-gendered robots such as LOVOT (https://lovot.life/; released in 2018) are also available, the overwhelming majority of AI technologies are gendered. Examples include feminine-presenting AI receptionists, AI moderators, and AI companions who are typically designed as avatars in a manga/anime style appealing to men and who carry out traditional women's labour.

For instance, AI Sakura-san (https://www.tifana.ai) is an AI receptionist who has appeared in some 300 companies (Osaka, 2020) and public institutions (e.g., the Supreme Court; see Watanabe, 2022) since its introduction in 2021. Designed as a twenty-one-year-old woman named Shibuya Sakura, she has her own social media accounts on X and Instagram (@sakurasan_ai).

AI newscasters are also becoming popular. Sony developed Araki Yui, a twenty-seven-year-old virtual woman who reads texts by learning news speech data (https://www.sony.jp/professional/ai-announcer/). As an AI-generated feminine-presenting newscaster, she has been introduced in

---

4 *Gurabia aidoru* resemble pin-up girls who pose in a sexually provocative way. They often reveal much skin, wearing a bikini or other 'fetish' outfits (e.g., uniforms). Their images are not defined as pornography because their sexual organs (and nipples) are covered. Their contents constitute a popular genre in Japan's publishing market targeted at cisgender heterosexual men.

many TV and radio programs. NHK, Japan's national broadcasting company, also developed its own AI newscaster called *Nyusu no yomiko* or News' Yomiko, who made her first appearance on TV in 2018 (Sony, n.d.). AI newscasters have been introduced in many countries but in the Japanese cases, the avatars are all young anime-style women characters.

Ubiquitous in Japanese popular culture, this particular anime-style aesthetic of women characters is known as *bishōjo* (beautiful girl), which stimulate *moe* or emotional closeness or affection among a certain group of male manga/anime fans (Galbraith, 2014). Exemplifying this aesthetic is an AI companion and 3-D holographic chatbot with which 'her' users can interact called 'Healing Bride, Azuma Hikari.' This product was launched by Vinclu/Gatebox in 2016 (https://www.gatebox.ai/gatebox). As a virtual wife or girlfriend who addresses 'her' users as '*masuta-san*' (my master), Azuma Hikari's primary role is to cheer them up. Up until March 2023, Azuma Hikari was using an AI platform called CLOVA and had various functions as a smart assistant. After CLOVA was terminated, Gatebox started testing GPT-4 and in late 2023, announced plans to incorporate it into their services.

Compared to robots which were developed before 2020 and analysed in Robertson's (2017) work, more advanced technologies are used for the recent forms of AI receptionists, moderators, and companions, but their embodiment is still gendered within the same logic: being given a normatively feminine name, voice, appearance, and personality traits. Furthermore, they are all designed as feminised anime-style characters (e.g., *bishōjo*) who are subordinate to human users. Hence, they replicate traditional gender relations prevailing in Japanese society.

Few empirical studies have been conducted so far about gendered interactions between these AI receptionists, moderators, and companions and their human users. On the one hand, an intimate relationship with AI technologies appears to be possible. In our study of Azuma Hikari, one of our interviewees maintained that perceived 'consent' was important for him in his interaction with 'her'.[5] On the other hand, however, human users have abused AI agents. For example, when AI Sakura-san was installed at a train station in Tokyo, some human users asked 'her' questions such as whether she has a lover and what her breast, waist, and hip sizes are. These are questions that are tantamount to sexual harassment and objectification of women. AI Sakura-san responded to these questions with 'I don't know' or 'I'm sorry, I can't hear well' (Kokumai & Kuribayashi, 2020). Such reactions

5 The first author conducted this interview on 25 October 2022 in Tokyo.

further reflect dominant cultural norms and expectations of women as obedient in Japanese society. They ultimately reinforce how difficult it is for victim-survivors to reject, protest, or report sexual harassment and violence.

6 Conclusion: Future Interventions

Throughout this chapter, we have identified and developed three modes of AI's contemporary sexual politics – algorithmic discrimination, AI-facilitated sexual violence, and gendered embodiment of machines – arguing that they are characterized by gendered entanglements of human bodies and non-human entities such as computers, technologies, cultural ideas, and human emotions. Applying this framework to Japan, an under-researched non-Western case in critical AI studies, we also demonstrated that while AI technologies appear to be neutral and objective, they nevertheless encompass specific cultural and symbolic meanings in their development, deployment, and diffusion. We contend it is important to pay more attention to this techno-social-cultural configuration and deploy a critical feminist lens so that we can better understand contemporary technologies from a culturally sensitive approach.

We have shown in the Japanese case that a culturally grounded demand for entertainment technologies related to local popular culture as well as the government's strong preference for industrialisation, economic, and technological development and conservatism in adopting new radical measures promote gendered creation, uses, and perception of AI technologies. Misogyny is not a Japan-specific problem, but its androcentric economy, politics, and culture characterised by deep-rooted sexism encourage such an environment for adopting new technologies while reinforcing existing gender-based discrimination. Moreover, changes in policy making and public opinions often occur due to *gaiatsu* (pressure from abroad). This has been especially true of gender-related issues,[6] which would mean that to bring about change, interventions need to come both from within and outside.

6 Laws against gender and sexual discrimination such as Equal Employment Opportunity Law (1985/1986), Basic Law for a Gender-equal Society (1999) and Law for Promoting Understanding LGBT People (2023) were all drafted and passed by the Diet under great external pressure. The global #MeToo movement also significantly changed public discourses on sexual violence.

Possible interventions are threefold. First, a dialogue between different stakeholders including businesses, the government, and researchers is important. Although this is regularly mentioned and recognised by AI scholars in Japan (e.g., Japanese Society for Artificial Intelligence, 2020), not much has happened yet. To achieve this, all parties must pay heed to the complexities of gender, inequality and power.

In 2020, computer programming was introduced as a compulsory subject in all Japanese primary schools. However, students should not only learn technical skills, but also acquire a deep understanding that such technologies do not exist in a vacuum. In higher education, this is particularly important for engineering and science university students who will most likely end up developing and designing AI in the future, but most of them fail to learn about critical perspectives of such technologies. Hence, for the second intervention, we recommend that any curriculum involving AI requires inter- and/or transdisciplinary learning across STEM, humanities, and the social sciences. Even for older generations who do not attend school anymore, they can learn or brush up on their knowledge through courses offered by both private and public organisations including companies, universities and local municipalities.

Third, media organisations and journalists could also play a key role in building awareness for the public to be more critical of AI (see Nguyen & Hekman, 2024). This is a major challenge in Japan and other capitalist economies, where women and minorities are marginalised in economy, politics, education, and media and powerful actors, such as businesses and the government, often pursue economic profits over cultural values. Collectively, realising these three interventions necessitate more critical research into AI and its contemporary sexual politics in Japan and beyond.

Acknowledgement

This work was supported by JST-RISTEX Japan Grant Number JPMJRX19H6 and Meiji University Institute of Humanities Grant 2024–2025.

References

Allison, A. (2006). *Millennial monsters: Japanese toys and the global imagination.* University of California Press.

Adam, A. (1998). *Artificial knowing: Gender and the thinking machine.* Routledge.

Ashikari, M. (2005). Cultivating Japanese whiteness: The 'whitening' cosmetics boom and the Japanese identity. *Journal of Material Culture*, 10(1), 73–91.

Atanasoski, N., & Vora, K. (2019). *Surrogate humanity: Race, robots, and the politics of technological futures*. Duke University Press.

Benjamin, R. (2019). *Race after technology: Abolitionist tools for the new Jim Code*. Polity.

Bochev, S. (2021, February 9). Trailing behind: The 2020–2021 Japanese AI landscape. *Medium* Retrieved 6 July 2023 from https://sbochev.medium.com/trailing-behind-the-2020-2021-japanese-ai-landscape-2568a03b8522

Buolamwini, J., & Gebru, T. (2018). Gender shades: Intersectional accuracy disparities in commercial gender classification. *Conference on Fairness, Accountability, and Transparency, Proceedings of Machine Learning Research*, 81, 1–15.

Cadwalladr, C. (2016, December 4). Google, democracy and the truth about internet search. *The Guardian*. Retrieved from https://www.theguardian.com/technology/2016/dec/04/google-democracy-truth-internet-search-facebook

Chin, C., & Robinson, M. (2020, November 23). How AI bots and voice assistants reinforce gender bias. The Brookings Institution. Report. Retrieved 19 May 2023 from https://www.brookings.edu/research/how-ai-bots-and-voice-assistants-reinforce-gender-bias/

Costa, P., & Ribas, L. (2019). AI becomes her: Discussing gender and artificial intelligence. *Technoetic Arts*, 17(1–2), 171–193. https://doi.org/10.1386/tear_00014_1

Coursey, K., Pirzchalski, S., McMullen, M., Lindroth, G., & Furuushi, Y. (2019). Living with Harmonay: A personal companion system by Realbotix™. In Y. Zhou & M.H. Fisher (Eds.), *AI Love you* (pp. 77–95). Springer.

Crawford, K. & Calo, R. (2016). There is a blind spot in AI Research. *Nature*, 538, 311–313.

Danaher, J. (2019). Building better sex robots: Lessons from feminist pornography. In Y. Zhou & M.H. Fisher (Eds.), *AI Love you* (pp. 133–147). Springer.

Devlin, K. (2018). *Turned on: Sciences, sex and robots*. Bloomsbury Sigma.

Devlin, K., & Belton, O. (2020). The measure of a woman: Fembots, fact and fiction. In S. Cave, K. Dihal & S. Dillon (Eds.), *AI narratives: A history of imaginative thinking about intelligent machines* (pp. 357–381). Oxford University Press.

Dirksen, N., & Takahashi, S. (2020, October 5). *Artificial intelligence in Japan 2020: Actors, market, opportunities and digital solutions in a newly transformed world*. Embassy of the Kingdom of the Netherlands, Innovation, Science and Economic Affairs and Climate Policy. Retrieved 27 June 2023 from https://www.rvo.nl/sites/default/files/2020/12/Artificial-Intelligence-in-Japan-final-IAN.pdf

Dooley, S., Downing, R., Wei, G., Shankar, N., Thymes, B., Thorkelsdottir, G., Kurtz-Miott, T., Mattson, R., Obiwumi, O., Cherepanova, V., Goldblum, M., Dickerson, J.P., Goldstein, T. (2021, October 21). Comparing human and machine bias in face recognition. arXiv preprint arXiv:2110.08396. https://doi.org/10.48550/arXiv.2110.08396

Döring, N. (2017). Vom Internet yum Robotersexä Forschungsstand und Herausforderungen für die Sexualwisenschaft. *Zeitschrift für Sex-Forschung*, 30(1), 35–57. https://doi.org/10.1055/s-0043-101471

Döring, N., & Pöschl, S. (2018). Sex toys, sex doll, sex robots: Our under-researched bed-fellows. *Sexologies*, 27(3), e51–e55. https://doi.org/10.1016/j.sexol.2018.05.009

Elias, A., Gill, R., & Scharff, C. (2017). Aesthetic Labour: Beauty Politics in Neoliberalism. In A. Elias, R. Gill, & C. Scharff (Eds.), *Aesthetic Labour: Rethinking Beauty Politics in Neoliberalism* (pp. 3–50). Palgrave Macmillan.

Eyssel, F., & Hegel, F. (2012). (S)he's got the look: Gender stereotyping of robots. *Journal of Applied Social Psychology*, 42(9), 2213–2230. https://doi.org/10.1111/j.1559-1816.2012.00937.x

Feine, J., Gnewuch, U., Morana, S., & Maedche, A. (2020). Gender bias in chatbot design. In A. Følstad et al. (Eds.), *Chatbot research and design. Paper presented at CONVERSATIONS 2019* (Lecture notes in computer science, vol. 11970). Springer, 79–93. https://doi.org/10.1007/978-3-030-39540-7_6

Frummer, Y. (2020). The short, strange life of the first friendly robot. *IEEE Spectrum*. 21 May 1010. Retrieved 16 November 2023 from https://spectrum.ieee.org/the-short-strange-life-of-the-first-friendly-robot

Galbraith, P. W. (2014). *The moe manifesto: An insider's look at the worlds of manga, anime, and gaming*. Tuttle.

Gaskell, A. (2022, 6 September). How biased Google Search results affect hiring decisions. *Forbes* (online). Retrieved 9 May 2023 from https://www.forbes.com/sites/adigaskell/2022/09/06/how-biased-google-search-results-affect-hiring-decisions/

Gehl, R.W., Moyer-Horner, L. & Yeo, S.K. (2017). Training computers to see internet pornography: Gender and sexual discrimination in computer vision science. *Television & New Media*, 18(6): 529–547.

Goodfellow, I. J., Pouget-Abadie, J., Mizra, M., Xu, B., Warde-Farley, D., Ozair, S., Courville, A., Genjio, Y. (2014). Generative adversarial networks. *NeurIPS Processing: Advances in Neural Information Processing Systems* 27 (NIPS 2014). Retrieved 18 May 2023 from https://doi.org/10.48550/arXiv.1406.2661

Government of Japan (2015). 日本再興戦略 改訂 2015 [Japan Revitalisation Strategy. Revised in 2015].

Government of Japan (2016). 科学技術基本計画 [The Fifth Science and Technology Basic Plan].

Government of Japan (2021). 科学技術・イノベーション基本計画 [The Sixth Science, Technology and Innovation Basic Plan].

Greimel, H. (2019, August 8). Automotive sector tops r&d spending in Japan. *Automotive News Europe*. Retrieved 27 June 2023 from https://europe.autonews.com/automakers/automotive-sector-tops-rd-spending-japan

Hao, K. (2020, October 20). A deepfake bot is being used to "undress" underage girls. *MIT Technology Review*. Retrieved 31 May 2023 from https://www.technologyreview.com/2020/10/20/1010789/ai-deepfake-bot-undresses-women-and-underage-girls/

Henne, K., Shelby, R., & Harb, J. (2021). The datafication of #MeToo: Whiteness, racial capitalism, and anti-violence technologies. *Big Data & Society*, 8(2). Retrieved 18 May 2023 from https://journals.sagepub.com/doi/10.1177/20539517211055898

Henry, N., & Powell, A. ([2016]2018). Technology-facilitated sexual violence: A literature review of empirical research. *Trauma, Violence, & Abuse*, 19(2): 195–208.

Hunt, E. (2016, March 24). Tay, Microsoft's AI chatbot, gets a crash course in racism from Twitter. *The Guardian*. Retrieved 18 May 2023 from https://www.theguardian.com/technology/2016/mar/24/tay-microsofts-ai-chatbot-gets-a-crash-course-in-racism-from-twitter

Japanese Society for Artificial Intelligence, The (2020). 人工知能 [Artificial Intelligence]. Vol. 35, No. 5.

Jensen, C. B., & Blok, A. (2013). Techno-animism in Japan: Shinto Cosmograms, actor-network theory and the enabling powers of non-human agencies. *Theory, Culture & Society*, 30(2), 84–115. https://doi.org/10.1177/0263276412456564

Kakizaki, M., & Suzuki, T. (2020, October 2). フェイクポルノ：フェイクポルノ初摘発 AIで作成、公開 2容疑者逮捕 [Fakeporunography, uncovered for the first time. Two suspects arrested for creating it with AI and releaesing it online]. *The Mainichi Newspaper*, Tokyo Evening Edition, 7.

Kaufman, E. M. (2022). Reprogramming consent: Implications of sexual relationships with artificially intelligent partners. In E. M. Kaufman, *Nuances of sexuacl consent*. Routledge. https://doi.org/10.4324/9781003276180

Katsuno, H., & White, D. (2022). Haptic creatures: Tactile affect and human-robot intimacy in Japan. In Y. Minowa & R. Belk (Eds.), *Consumer culture theory in Asia: History and contemporary issues* (pp. 242–262). Routledge.

Kayser-Bril, N. (2020, September 11). Female historians and male nurses do not exist, Google Translate tells its European users. *Algorithm Watch*. Retrieved 2 August 2021 from https://algorithmwatch.org/en/google-translate-gender-bias/

Kokumai, A., & Kuribayashi, F. (2020, March 19). 高輪新駅の AI「さくらさん」、セクハラ受け流しで物議 [AI Sakura-san at a new station in Takanawa, dodging sexual harassment brings about a controversy]. *The Asahi Shimbun Digital*. Retrieved 3 July 2013 from https://www.asahi.com/articles/ASN3M6RPHN3MUTIL044.html

Kragen, P. (2017, September 22). The world's first talking sex robot, Harmony, is ready to make her debut. *Australian Financial Review*. Retrieved 21 June 2023 from https://www.afr.com/technology/the-worlds-first-talking-sex-robot-harmony-is-ready-to-make-her-debut-20170922-gymism

Kuwabara, T. (2023, November 12). AI 生成の児童性的画像、サイト運営者「削除するつもりない」… 専門家から対策求める声 [AI generated child pornographic images, website owner 'we're not going to delete them' … experts ask for counter-measures]. *Yomiuri Shinbun Online*. Retrieved 25 November 2023 from https://www.yomiuri.co.jp/national/20231112-OYT1T50002/

Kuczmarski, J. (2018, December 6). Reducing gender bias in Google Translate. Retrieved 26 May 2023 from https://blog.google/products/translate/reducing-gender-bias-google-translate/

Latour, B. (2005). *Reassembling the social: An introduction to Actor-Network-Theory*. Oxford University Press.

Levy, D. (2007). *Love and sex with robots: The evolution of human-robot relationships*. HarperCollins.

Luka, Inc. (2024). Replika. Retrieved August 16, 2024 from https://replika.com

Matsuo, Y. (2020). 人工知能関連技術の歴史と技術動向 [History and trend of artificial intelligence technologies]. *Denki Joho Tsushin Gakkaishi* [Journal of IEICE], 103(5), 450–455.

McDonnell, M., & Baxter, D. (2019). Chatbots and gender stereotyping. *Interacting with Computers*, 31(2), 116–121.

Munn, L. (2022). The uselessness of AI ethics. *AI and Ethics*, 3, 869–877. https://doi.org/10.1007/s43681-022-00209-w

Natale, S. (2019). If software is narrative: Joseph Weizenbaum, artificial intelligence and the biographies of ELIZA. *New Media & Society*, 21(3), 712–728. https://doi.org/10.1177/1461444818804980 49

Neff, G., & Nagy, P. (2016). Talking to bots: Symbiotic agency and the case of Tay. *International Journal of Communication*, 10, 4915–4931

Nguyen, D., & Hekman, E. (2024). The news framing of artificial intelligence: a critical exploration of how media discourses make sense of automation. *AI and Society*, 39(2), 437–451. https://doi.org/10.1007/s00146-022-01511-1

NHK Science & Technology Research Laboratories (2018, May). AI キャスターがニュースチェック11 に登場 [AI newscaster appeared in News Check 11]. *Giken Dayori* [Newsletter of Science & Technology Research Lab]. Retrieved 23 November 2023 from https://www.nhk.or.jp/strl/publica/giken_dayori/158/3.html

Noble, S. U. (2012). Missed connections: What search engines say about women. *Bitch*, 12(54), 37–41. Retrieved 30 May 2023 from https://safiyaunoble.files.wordpress.com/2012/03/54_search_engines.pdf

Noble, S. U. (2018). *Algorithms of oppression: How search engines reinforce racism.* New York University Press.

Nyholm, S., & Frank, L. (2019). It loves me, it loves me not: Is it morally problematic to design sex robots that appear to "love" their owners? *Techné*, 23(3), 402–424. https://doi.org/10.5840/techne2019122110

O'Neil, C. (2016). *Weapons of math destruction: How big data increases inequality and threatens democracy*. Crown Publishers.

Osaka, N. (2020, May 25). 駅の案内で大活躍「AI さくらさんの別の顔 社員向けヘルプデスクなど導入企業は300社 [AI Sakura-san, popular at a train station's support desks, also introduced for employees' helpdesk in 300 companies]. *Toyo Keizai Online.* Retrieved 7 July 2023 from https://toyokeizai.net/articles/-/352077?page=2

Perez, S. (2022, February 24). Siri gains a new gender-neutral voice option in latest iOS update. *TechCrunch.* Retrieved 16 June 2023 from https://techcrunch.com/2022/02/24/siri-gains-a-new-gender-neutral-voice-option-in-latest-ios-update/?guccounter=1

Petman, D. (2019). Love in the time of Tamagotchi. *Theory, Culture & Society*, 26(2–3), 189–208. https://doi.org/10.1177/0263276409103117.

Phan, T. (2017). The materiality of the digital and the gendered voice of Siri. *Transformations*, 29(11), p. 1–24. Retrieved 30 May 2023 from https://www.semanticscholar.org/paper/The-Materiality-of-the-Digital-and-the-Gendered-of-Phan/f1b11ccf3e30632b65e6b781dbf2d0e3013568c7

Reeves, B., & Nass, C. I. (1996). *The media equation: How people treat computers, television, and new media like real people and places*. Cambridge University Press.

Robertson, J. (2017) *Robo sapiens japonicus: Robots, gender, family, and the Japanese nation*. University of California Press.

Roquet, P. (2022). *The immersive enclosure: Virtual reality in Japan*. Columbia University Press.

Schick, N. (2020). *Deepfakes: The coming Infocalypse*. Twelve.

Shadel, J. D. (2022, April 7). How Google's autocomplete predictions encouraged transphobic searches. *Them.* Retrieved 23 May 2023 from https://www.them.us/story/google-autocomplete-suggestions-transphobia-celebrities

Sheuerman, M. K., Paul, J. M., & Brubaker, J. R. (2019). How computers see gender: an evaluation of gender classification in commercial facial analysis and image labelling services. *Proceedings of the ACM on Human-Computer Interaction*, 3 (CSCW, Article 144).

Shibata, H., & Watanabe, A. (2023, May 27). AI 政策、推進色強く 戦略会議提言、リスクは列挙 [AI policy, its strategic council promotes AI, with its risks mentioned]. *Asahi Shimbun.* Morning Edition (p. 3).

Shino, K., Yamazaki, K., & Niitsuma, T. (2023, November 21). チャットGPT 職業に性別偏見 [ChatGPT, gender biases in occupation]. Asashi Shmbun. Tokyo, Morning Edition (p. 1).

Sodhi, G. S., & Kaur, J. (2022). Scientific racism faced by Indian fingerprint scientists during colonial rule: Need to correct a historical wrong. *Journal of Scientific Temper*, 10(1&2), 52–69.

Solomon, O., & Levi, S. (2016, December 16). How Google's search algorithm spreads false information with a rightwing bias. *The Guardian*. Retrieved 26 May 2023 from https://www.theguardian.com/technology/2016/dec/16/google-autocomplete-rightwing-bias-algorithm-political-propaganda

Sony (n.d.) AI アナウンサー 人工知能でより人に近い音声読み上げを実現 [Achieving more human-like voice reading using artificial intelligence]. Retrieved 23 November 2023 from https://www.sony.jp/professional/ai-announcer/

Steinberg, M. (2019). *The platform economy: How Japan transformed the consumer internet*. University of Minnesota Press.

Sterri, A. B., & Earp, B. D. (2021, November 10). The ethics of sex robots. In Carissa Véliz (ed.), *The Oxford handbook of digital ethics* (online edition, Oxford Academic). https://doi.org/10.1093/oxfordhb/9780198857815.013.13

Strengers Y. & Kennedy, J. (2020). *The smart wife: Why Siri, Alexa, and other smart home devices need a feminist reboot*. MIT Press.

Tanaka, H., & Ho, M. H. S. (2022). Romancing AI? Gender and new digital intimacies in contemporary Japan. Paper presented at the Conference "AI and the human: Cross-cultural perspectives on science and fiction." Humboldt Institute for Internet and Society, The Japanese-German Center Berlin, 11–13 May 2022.

Taylor, E. (2007). Dating-simulation games: Leisure and gaming of Japanese youth culture. *Southeast Review of Asian Studies*, 29, 192–208.

Tohata, K. (2023, June 22). AI に心の相談 弱さが生む人間の役割 [Mental health counseling with AI. Humans' role emerging from their weakness]. *Asahi Shimbun*. Morning Edition (p. 13).

Tomita, H. (2005). Keitai and the intimate stranger. In Ito, M., Okabe, D., & Matsuda, M. (Eds.), *Personal, portable, pedestrian: Mobile phones in Japanese life* (pp. 183–204). MIT Press.

Tuchman, G. (1978). Introduction: The symbolic annihilation of women by the mass media. In Tuchman, G., Daniels, A.K., & Benét, J. (Eds.), *Hearth and home: Images of women in the mass media* (pp. 3–38). Oxford University Press.

Urbi, J. (2018, August 8). Some transgender drivers are being kicked off Uber's app. *CNBC*. Retrieved 28 August 2021 from https://www.cnbc.com/2018/08/08/transgender-uber-driver-suspended-tech-oversight-facial-recognition.html

Verma, P. (2023, November 5). AI fake nudes are booming. It's ruining real teens' lives. *The Washington Post*. Retrieved 22 November 2023 from https://www.washingtonpost.com/technology/2023/11/05/ai-deepfake-porn-teens-women-impact/

Vlasceanu, M., & Amodio, D. M. (2022, July 12). Propagation of societal gender inequality by internet search algorithms. *PNAS*, 119(29) e2204529119. https://doi.org/10.1073/pnas.2204529119

Vorsino, Z. (2021). Chatbots, gender, and race on Web2.0 platforms: Tay.AI as monstrous femininity and abject whiteness. *Signs: Journal of Women in Culture and Society*, 47(1), 105–127.

Watanabe, R. (2022, June 15). 最高裁判所、AI アバター接客「AI さくらさん」を導入 [AI avater receptionist, AI Sakura-san introduced at the Supreme Court]. *Digital Gyosei/Digital Administration*. Retrieved 3 July 2023 from https://www.digital-gyosei.com/post/2022-06-15-news-supremecourt-ai/

West, S. M., Whittaker, M., & Crawford, K. (2019). *Discriminating Systems: Gender, Race and Power in AI*. New York: AI Now Institute. Retrieved 5 April 2019 from https://ainowinstitute.org/discriminatingsystems.html.

Yomiuri Shimbun, The (2023, June 7). AI が生み出した女性モデルのグラビア写真集、集英社が販売中止 … 著作権を問題視する声 [AI female model's gurabia photobook withdrawn by Shueisha. Public opinions question its copyright]. *Yomiuri Shimbun*. Retrieved from 3 July 2023 from https://www.yomiuri.co.jp/culture/20230607-OYT1T50185/

Zou, J., & Schiebinger, L. (2018). Design AI so that it's fair. *Nature*, 559 (12 July 2018), 324–326.

Zuboff, S. (2019). *The age of surveillance capitalism: The fight for a human future at the new frontier of power*. PublicAffairs.

CHAPTER 12

The Tech Gender Gap: A Closer Look at Women's Experiences in the Technology Industry

Julia Luteijn and Rhied Al-Othmani

1 Introduction

The Netherlands – renowned for its top-tier technology companies and its robust knowledge-based economy – is not only aspiring to advance in the realm of technological innovation, but also actively pursues progressive social policies. The country is home to prominent global players such as Philips and chip producer ASML, underscoring its status as a leader in technological developments. Parallel to this, Dutch society takes pride in its commitment to gender equality. Most notably, the Netherlands introduced policies and action plans such as the Emancipation Note (Ministry of General Affairs, 2022) and a target of 30% women in the technology sector specifically (FME, nd). Dutch technology companies themselves strive to adopt progressive-inclusive policies internally as well, illustrating a considerable step forward in social equity within the industry. However, despite these efforts, the workplace culture in the technology sector appears still marred by a lack of gender diversity, ultimately resulting in an imbalanced work environment. The male-dominated technology world is often associated with a 'macho' and 'brogrammer' culture, and it sometimes resembles an exclusive 'boys' club' that can be unwelcoming to women and other gender orientations (Cauterucci, 2017; Edgar, 2021). Policies alone may not suffice to address this imbalance if women do not feel compelled to enter the tech industry and to stay there.

Gender disparity in the technology sector is particularly evident in three key areas: educational enrolment, workforce participation, and leadership positions. In the Netherlands, the representation of women in STEM fields (science, technology, engineering and mathematics) is notably low, with less than a quarter of students being female (CBS, 2023). Globally, there is a rise in the percentage of women pursuing careers in STEM fields after graduation, but a clear decline is observed after an average of only one year in the workforce, possibly related to the influence of traditional gender norms (Zacharia et al., 2020). To understand, develop, and work with digital

 | DOI:10.1163/9789004711396_014

technologies, knowledge associated with STEM fields is often needed for potential employees to be considered viable in the labour market (Zollman, 2012). Despite its policies, the situation in the Netherlands can be considered concerning, with women holding only 16.5% of jobs in the IT sector, as reported by the Dutch Central Bureau for Statistics (2024). These jobs include roles such as software developers, engineers, data and network specialists, and IT specialists.

While there has been an increase in female leadership positions across various sectors over the past decade, women's representation in executive tech positions still remains disproportionately low. Currently, women hold 29.4% of entry-level positions, but only 17.8% are in roles at the director or vice president level (Global Gender Gap Report, 2023). This disparity highlights the ongoing challenges in achieving gender parity in leadership within the technology industry. For instance, the largest and most well-known Dutch technology firms, ASML and Atos, which specialise in system development and machine supply, respectively, both have male CEOs. Another well-known payment platform, Adyen, also has a male CEO at the helm. The leaders of Dutch-based Booking.com and WeTransfer are men. These prominent examples illustrate the persisting overrepresentation of men in decision-making, policymaking, cultural influence, and hiring positions within the technology industry even in a relatively progressive Western country.

The present chapter investigates the potential obstacles that women may face when entering the technology sector by presenting the results of a series of in-depth interviews with female tech professionals. The exploratory analysis uncovers how personal motivations linked to individual upbringings play an important role and what challenges mark women's career journeys as well as how they successfully navigated them. Our critical exploration sheds further light on lingering issues pertaining to gender inequality in allegedly open and inclusive tech labour markets from the perspective of those who are directly affected by prevailing imbalances. The observations are linked to a critical discussion of the risks that the gender gap carries for the future of more representative and inclusive technology. It is important to critically explore the interconnectedness of preconceptions and experiences that define what it means to be a woman in the technology field, and how these factors affect women differently than men. A combination of different factors related to implicit gender biases in the Dutch technology sector seems to contribute to women being less likely to work in male-dominated sectors. It is necessary to understand women's experiences and perspectives in the technology sector, with particular attention to the role of implicit bias in shaping their professional journeys. The argumentation and analysis are thus anchored around the research question: *How does implicit gender bias influence women's experiences and perspectives in the technology industry?*

The present chapter begins with a comprehensive literature review on biases, the gender gap, representations and environmental support, followed by an outline of the methodology which includes interviews and their results. It closes with a discussion of findings and practical recommendations specifically for the Dutch tech sector.

2 The Risks Associated with Gender Gaps in Technology

The underrepresentation of women in the technology industry can have profound consequences, one of which is the absence of diverse perspectives through the obvious omission of female viewpoints. This scarcity of alternative perspectives often leads to a male-centric approach in the design of products and services, limiting the scope of their applicability and relevance. Perez (2019) emphasizes that this one-sided design process neglects the diverse needs and experiences of a wider user base. An illustrative example of this can be found in Apple's health app, which was primarily designed by male developers. When the app was initially launched, it lacked a feature to track menstrual cycles, despite the obvious importance of menstruation in affecting women's physical and mental well-being (Duhaime-Ross, 2014). Compounding this issue, companies such as Google and Microsoft, whose digital products for self-monitoring have become integral to many users' daily routines, show a gender disparity within their technical workforce. Google reports a gender ratio of 74.7% men to 25.3% women, and Microsoft has 73.4% male and 26.6.% female employees (Statista, 2023; Microsoft Diversity & Inclusion Report, 2023). This imbalance is concerning, as it suggests a substantial exclusion of women from shaping the technological developments entering society through everyday products and services. Big Tech companies wield unprecedented power here through their monopoly over the data sphere, granting them control over categorising individuals into specific boxes based on data profiling and influence on user behaviour, ultimately leading to a possible abuse of power (Nguyen & Beijnon, 2023). In view of the exponential growth and transformative influence of technology on society, particularly in domains such as Engineering, Data, and Artificial Intelligence, it becomes imperative to prioritize inclusive representation of different gender orientation and sexes to increase equality. In addition to the potential rise in the gender pay gap, technology jobs are known for their high salaries and high positions and are situated in a rapidly expanding and dynamic sector that requires a growing workforce.

Fostering greater participation of women in this sector not only addresses this increasing demand but also plays a crucial role in avoiding a widening of the gender pay gap. Importantly, gender is only one aspect of individual's positionality in society and any disadvantages and forms of exclusion that women experience are likely to be exacerbated by additional factors such as race and sexual orientation, while it is also important to acknowledge that social groups self-identifying as nonbinary face even greater challenges (e.g., related to LGBTQ+ people). Addressing the gender gap in tech with an eye on women provides a critical lens through which broader systemic issues within the sector can be understood.

3 Perceived Barriers among Women in Technology

There are several social, cultural, and economic factors that can contribute to a wider gender gap in technology. During childhood, parents often exert significant influence on their children's views on viable careers and hence their eventual choices. Dutch research (De Vleeschouwer et al., 2020) shows that parents tend to believe technology careers are unsuitable for girls. During formal education, it appears that secondary-school girls receive less support from their parents and instructors than boys, and girls are more frequently told that they should not pursue a career in STEM. The presence of role models and representation is often seen as a motivating factor in initiating a career in technology. It appears that familiarity with the tech sector, whether through parents or teachers at school, is lacking, which may have an impact on girls opting to study technology (Besecke & Reilly, 2017; Papastergiou, 2008; Schorr, 2019). Research (Besecke & Reilly, 2017) indicates that even if parents of women in STEM may express support for their daughters' careers, they may not take concrete actions to positively influence their career choices. Furthermore, young women often report feeling misunderstood by their parents, who may not fully understand their work. If young women receive adequate support in this regard, they are more likely to enrol in studies focused on technology in higher education (Tandrayen-Ragoobur & Gokulsing, 2022). Increased exposure to science and technology also fosters encouragement and interest in learning science, technology, engineering and mathematics among young women (Idris et al., 2023).

Consequently, parental support is considered a crucial factor in promoting young women's enrolment in STEM tertiary education (Tandrayen-Ragoobur & Gokulsing, 2022; Makarem & Wang, 2020). Parents' beliefs, expectations, and behaviours can have an impact on their daughters' interest in math, for example, and relatedly, their developing sense of gender identity as of early childhood. For instance, fathers can serve as role models and actively encourage their daughters to pursue careers in STEM. This indicates that parents should not rely solely on teachers or schools to motivate their children to choose STEM subjects (Tandrayen-Ragoobur & Gokulsing, 2022), but they should rather actively encourage and promote their daughters' interests in STEM disciplines early on (Wang & Degol, 2017). However, the fact that there are not many female STEM role models may also be a contributing factor to women's underrepresentation in the industry (Beede et al, 2011). This illustrates that women's social environment, particularly in terms of support and education, may have a substantial impact on whether they pursue a career in technology. During STEM education, other critical negative factors that have been documented are a lack of support and even forms of (sexual) harassment, often enacted in subtle yet deliberate ways (Zacharia et al., 2020).

These different biases already affect young females at an earlier age. Consequently, it has been argued that strong gender preconceptions hinder women from entering the industry (Beede et al, 2011).

Even after choosing to pursue a STEM or computer science degree, women are less likely than men to work in STEM occupations (Beede et al, 2011; Sassler et al, 2017). This is also evident in STEM higher education (Tandrayen-Ragoobur & Gokulsing, 2022; Papastergiou, 2008). 'Social norms still play a significant role in the choices people make in our country,' says former Minister Dijkgraaf of Dutch Education, Culture, and Science. 'This maintains financial inequality between men and women and perpetuates persistent traditional imbalances in terms of work or childcare. Accessible childcare, good leave arrangements, and breaking gender-stereotypical role patterns should contribute to this. Above all, a cultural shift is necessary' (NOS, 2023). Cycling in and out of the workforce to raise a family is an important factor that can put women in tech at a disadvantage (Beede et al, 2011; Sassler et al, 2017). Tech jobs are typically given as full-time positions, whereas women often choose part-time positions, partly due to maintaining a family-work balance (De Wit & Kalkhoven, 2019). Although the Dutch government has taken steps to encourage women to work longer hours by improving access to childcare services, leave options are limited in terms of flexibility, affordability, and quality (Yerkes & Hewitt, 2019).

4 Understanding (Implicit) Gender Bias and Its Implications for Women in Tech

Forms of bias against women are not always explicit and proactively articulated but often have more implicit manifestations. Implicit gender bias refers to the unconscious attitudes or stereotypes that affect people's understanding, actions, and decisions in a gendered manner. Implicit gender bias operates subtly, often without conscious awareness, influencing behaviour and decision-making processes in the workplace and beyond (Ellemers, 2014; Friedmann & Efrat-Treister, 2023). Individuals may consciously believe that they harbour no intention to discriminate against others, yet they can still hold unconscious biases that subtly influence their actions. These implicit biases, as described by Banaji and Greenwald (2016), Baron and Banaji (2006), and Boysen and Vogel (2008), are fragments of knowledge about social groups that accumulate through repeated exposure to our cultural environment. Biases can gradually shape behaviour towards specific social groups including during decision-making processes that ultimately create unequal opportunities. For example, if a person has an unconscious bias that associates certain traits or abilities with a particular gender, they may be more likely to hire or promote individuals who fit that stereotype, even if other candidates are equally qualified.

This can result in fewer opportunities for individuals who do not fit the stereotype, leading to a gender gap in the workplace (Chadwick & Baruah, 2019). A study by Wang and Redmiles (2019) shows implicit gender biases among software engineers. This research reveals that software engineers associate technical leadership roles more with men than women. These biases can influence professional decisions in the technology sector, impacting the assessment of job candidates and the evaluation of contributions. Despite efforts to combat explicit biases, implicit biases remain a critical, yet often unnoticed, barrier affecting women's career paths in software development and other tech jobs. Implicit gender bias and the gender gap in technology can be entwined with the experiences that women have in this sector, from the choice of going into technology during university and throughout their careers in the industry.

Arguably, the underrepresentation of women in the tech sector leads to a predominantly male-dominated work environment. Male-dominated workplaces can produce inequalities and power imbalances that affect women's experiences in various ways. To begin with, women may face explicit discrimination, harassment, and exclusion from informal networks and decision-making processes (Wright, 2016). In particular, women in IT face numerous structural barriers throughout their careers, including gender discrimination, the glass ceiling, stereotypes, a lack of representation, and advancement barriers (Kenny & Donnelly, 2019; Amon, 2017) including stereotypical images of e.g. female scientists, such as 'frumpy or sexy, over-emotional or cold, victimised or 'bitchy'' (Makarem & Wang, 2020; Kitzinger et al., 2008, p. 20). Women also face intimidation, lack of support and encouragement, and 'the boys club', which keeps power where it has always been (Tandrayen-Ragoobur & Gokulsing, 2022; Wentling & Thomas, 2009; Besecke & Reilly, 2017). Likewise, women's limited participation in the decision-making process within these fields is problematic (Wentling & Thomas, 2009). STEM careers are often characterised by a gendered organisational culture that reproduces traditional gender roles. Persistent (implicit or explicit) assumptions, that women are less capable in IT careers contribute to this situation. In male-dominated workplaces, women often lack contact with other female professionals, which diminishes gender-based support and networking opportunities, perpetuating gender-based discrimination and power imbalances (Wright, 2016).

In contrast, diverse teams and companies with more women in leadership positions perform better financially and make more informed decisions (Dixon-Fyle et al., 2020; Ellemers, 2014). The scarcity of role models and mentors is also a challenge (Tandrayen-Ragoobur & Gokulsing, 2022; Makarem & Wang, 2020; Kitzinger et al., 2008). Role models and mentors play a crucial role in broadening girls' perspectives on the achievements of science and technology in the modern world. Finally, women in STEM face difficulties in achieving work-life balance, particularly when it comes to managing family and children. Struggles with multiple roles, such as being a mother versus pursuing a career in science, present ongoing challenges (Makarem & Wang, 2020; Wentling & Thomas, 2009; Besecke & Reilly, 2017). Finding a balance between these competing demands can be a source of stress and strain.

5 Methodology

The present study aims to investigate the personal experiences of women in the tech industry, with a focus on understanding specific examples of bias encountered in their professional journey. The study likewise explores how women navigate and respond to biases in the male-dominated sector and work environment. Additionally, this chapter considers the male/female ratio within departments or teams to better comprehend the gender dynamics. It also explores women's perspectives on the current gender ratio in their department and their preferences regarding potential changes. By delving into these areas, the critical analysis seeks to unearth the complex relationships between various factors and implicit bias.

For this study, 35 interviews with women in technology-centric positions were conducted in 2021 and 2022; respondents work in the areas of architecture, engineering, education, consulting, and management. These female professionals were approached via professional networking platforms such as LinkedIn. Through snowballing, further respondents fitting the target population were recruited. The interviews followed a structured format, ensuring that the questions posed were predetermined and provided to the participants in advance. A comprehensive document containing the interview questions and background information on the study was shared prior to the scheduled interview session. The participants consisted of individuals of different ages, including both juniors and seniors. The interview questions are related to the literature review, from which themes were extracted (see Table 12.1). The interview questions focused on the origin of their careers, perspectives on the gender gap in the workplace, experiences with bias, role models, involvement from their social environment, and

TABLE 12.1 Themes and related interview questions

Themes from theory	Interview questions
Path to technology: This theme examined how the participants ended up pursuing careers in the field of technology. It aimed to uncover the diverse backgrounds, motivations, and pathways that led them to their current positions.	*How did you get into the tech industry?*
Role models: The analysis explored the presence and influence of role models in the participants' tech journeys. It sought to understand whether they had role models who inspired them and the identity of these role models.	*Who is your role model?* Can you explain why this person inspires you?

TABLE 12.1 Themes and related interview questions (*cont.*)

Themes from theory	Interview questions
Family support: This theme delved into the participants' experiences of family support and understanding.	*How does your family react to you working in tech?* Do your family members also work in tech? And are they women or men?
Prejudices: The analysis investigated whether and what types of prejudices the participants encountered as women working in the tech industry.	*Have you experienced prejudice as a woman in the tech industry?* If so: Were those underestimations? Were those overestimations? Can you name examples? How often do you experience that? How do you deal with these situations? How are you approached or addressed by your male and female colleagues?
Gender ratio at work: The analysis examined the observations and perspectives on the representation of women at their jobs and the potential implications it had on their experiences.	*What is the gender ratio in the department?* What do you think of the current ratio? Would you see it differently? If so, why?
Women's hiring quota: This theme explored the participants' opinions and attitudes towards women's hiring quota. It aimed to understand their perceptions of the effectiveness and impact of such measures in promoting gender diversity and equality in tech.	*What do you think of the quota scheme?*

perspectives on implementing a gender quota as a potential solution. After transcribing the interviews, a thematic text analysis uncovered patterns within the themes. Thematic text analysis involves systematically coding and categorising qualitative data based on recurring themes or patterns within the text (Kuckartz, 2013). The analysis focused on several key themes and corresponding questions that aimed to explore various aspects of the women's experiences, challenges, and perspectives in the tech industry.

The research involved multiple participants who had diverse work experiences. Some individuals transitioned from their initial roles as developers to their current positions in management. During the interviews, the participants held diverse positions, as outlined in Table 12.2.

Most interviews were conducted in Dutch and some in English, with durations ranging from 30 to 50 minutes. All interviews, except for one, were conducted by both researchers. A total of two interviews were conducted face-to-face and recorded. Additionally, one interview was conducted using an online form, and 32 interviews took place through video calls, which were recorded for audio transcription purposes. Subsequently, the recorded files were transcribed using an online AI tool. However, to ensure accuracy, the transcriptions were manually edited to eliminate errors. Additionally, names of people and companies were removed from the transcripts to keep the data anonymous. Once the transcripts were finalised, they were shared with the respective participants for review and verification.

TABLE 12.2 Participant's professions

Development, architecture and engineering	Data Engineer, 2 × Software Engineer, Front-end Campagne Creator, 2 × Software Tester, Software Architect, Cloud Engineer, Mechanical Engineer, Data Trainee, 4 × Developer, UX/Web Designer
Education	Lecturer Data Engineering, Lecturer Interaction Design/Visual Artist, Makerspace Creator/Lecturer, Lecturer Programming and Media Design, Lecturer Communication Multimedia Design in Tech
Consultation	3 × SAP Consultant, Business Information Analyst, Business Intelligence Consultant, Consultant Road Construction Materials
Management	Product Manager Autonomous Vehicles and Robotics, Business Analyst Manager, Service Delivery Manager, 2 × Project Manager, SAP Manager, Traffic Resource Manager, Manager Consultant, Technology Innovation Manager

6 Results

The results are categorised into three themes: 1) Path to Technology & Role Models 2) Biases and 3) Gender Gap & Power Dynamics. Some participants provided multiple distinct responses to a single question.

6.1 *Path to Technology and Role Models*

The majority of the participants indicated that they were inspired by their environment, or that they had intrinsic motivations to start a career in technology. First, it appears that often, a young woman's job path is influenced by her father. Participants indicated that the father 'opened their eyes', introduced technology into their lives, or encouraged them to seek a profession in technology because they believed it to be a good fit for their skills. The father is mentioned multiple times as being supportive in pursuing a career in tech. Not only the father, but also other family members and the environment played a significant role in influencing the decision to pursue a career in tech. Factors such as having a friend who was already working in the tech industry, or a sibling employed in tech, contributed to the motivation to enter the field. Second, certain participants expressed a more intrinsic sense of motivation, which was evident in their expressed sentiments such as: 'I started studying physics because I liked it in high school' or 'So then I started focusing more on coding and learning there because I found it was more creative, more challenging.' Some participants shared their experiences of unintentionally finding themselves in the field of technology, with answers such as: 'And then I kind of just happened to get into tech, just fell into it, really.' Some participants initially pursued different paths, such as studying cultural sciences, political science, or pedagogy, before eventually transitioning into the field of technology. The majority of the participants answered that their role models were family members or individuals within their immediate environment. Among them, some participants specifically mentioned their father as a role model, stating, 'And yes, I also think my father who first inspired me to go into technology ... he was always busy with cameras and videos and yes had a technical profession with printers.' Some participants responded with statements indicating that they do not have a role model, such as: 'So it's not just that you're a woman in tech, but you actually don't have that role model' or 'I never really had a role model.'

Participants brought attention to the limited presence of women serving as role models within the tech industry, expressing their thoughts through statements such as: 'I don't really have a role model in the sense of a woman in technology at all' and 'I can imagine that if there had been more female teachers, it would indeed be purely a role model and someone who could empathise a little better with how you would feel as a woman in tech where the male-to-female ratio is still very uneven. That in retrospect it might have been nice to have had more female teachers who had such experience.' Furthermore, some participants cited renowned figures such as Buddha or Angela Merkel as their role models. Other participants expressed that they do not have role models in the sense of specific individuals, but rather look up to certain character traits or qualities. Next, family members are mostly viewed as supportive. However, participants also encountered unsupportive family members who struggled to grasp the nature of their work, as reflected in statements such as: 'My family doesn't understand what I'm doing at all.' and 'It really is a kind of black hole for them.' Some family members reacted with surprise, according to some participants as they held a different perception of work in technology, 'imagining it as a place where one would work in a dark basement.' The father figure was again mentioned, as fathers who work in tech passed on the intention to work in technology: 'My dad, he works in tech too, so I kind of got it from him I guess.' Also, 'And my father is a carpenter, so a contractor. And it is quite technical. So, I think [that is where] I got it [from] too.'

These findings show that their family, especially fathers, and the broader social environment significantly influence women's entry into the tech field. This highlights the role of implicit gender bias where girls may unintentionally receive less encouragement to pursue tech careers compared to boys. The unexpected prominence of fathers in the themes of the path to technology, role models, and family support contrasts with previous research that limited the father's role primarily to the path to technology. The influence of peers and acquaintances in shaping the tech journey underlines the importance of social networks and knowledge-sharing.

6.2 *Biases*

The results indicate that most of the participants face both implicit and explicit gender bias in various aspects of their lives. Bias is identified in relation to family planning, with women encountering remarks and judgments regarding their intentions to have children or not, as well as concerns about how their careers would be affected by such decisions. One participant shared her personal experience, stating, 'I also experienced a disadvantage, because I have also worked in a company where it was all men who were heavily reformed [who held very traditional views] and thought that a woman should sit at home with the children.' Furthermore, another participant shared an incident during her studies, saying, 'And then there was also a tour, where a number of guys bet with beer crates how quickly I would quit [the study programme].'

Another recurring theme was that women are being underestimated and need to prove their capabilities, or even have their abilities questioned altogether. Half of the participants shared that they had felt underestimated specifically because of their gender while working in the field of technology. One participant highlighted this issue, stating, 'Yes very implicitly, where you are sometimes heard less in meetings. That communication between man and woman sometimes really differs from getting the impression that someone else is overvalued despite the fact that he is probably less able, but perhaps because of his masculinity he has a more self-assured attitude or brings things differently, more dominant.' Some of the interviewees also experienced 'mansplaining.' One participant shared her experience of underestimation, saying, 'When I went somewhere, I was always asked where the (technical) expert was, and usually, I was the expert.' Furthermore, another participant mentioned, 'The developers often don't expect that I could actually code and or understand how code works. But they know that I did ICT and data visualisation study. They expect that I can only make beautiful things while I can also do more than that.'

The results also shed light on a persistently male-dominated culture that can exclude and marginalise women. One participant described this dynamic, saying, 'What often happens is that guys build their own little boys' club and then you're a little bit you know, the woman is a bit left out.' Multiple participants experienced prejudice in which a position of power is abused by touching and talking improperly by a male co-worker or boss. This sentiment was articulated through statements such as: 'I was touched improperly and he is talking to me improperly. [I asked HR to] please do something about that. Two weeks later, they made him my manager.' Finally, the participants highlighted instances of preferential treatment and experiences of being stereotyped. A few participants remained neutral on the biases topic. Women in technology often find themselves underestimated and needing to prove their capabilities before being recognized. This pattern of bias, both implicit and explicit, affects women variably, depending on individual responses and mindsets. This tendency reflects broader societal biases where women's abilities in tech are often questioned, leading to a need for continuous validation of their skills.

6.3 *Gender Gap and Power Dynamics*

Working in a diverse environment, according to most respondents, is the 'best' since it changes the 'dynamic,' 'atmosphere,' and 'culture,' and adds a 'creative approach.' Some participants never had the opportunity to work in a diverse team. Furthermore, some participants expressed a perspective that working in a diverse team does not make a significant difference, or in some cases, being the only woman in a team could be advantageous for them. This was seen in statements such as: 'Now at work I've been, as I said, more than 10 years in this industry and a lot of situations when I was alone in a meeting, for example, with five or ten men and I didn't feel anything like if they are looking at me somehow different or if I don't have the chance to talk or to speak up.' or 'Although I must say that I also think I have an advantage to be a woman because I am already positively discriminated against.' Finally, a few participants remained neutral and did not have a particular opinion regarding this matter. The general viewpoint on implementing forced measures such as a gender quota, was critical, mainly because it labels the person a 'diversity hire.' Instead, some participants thought applicants should be 'hired based on their skills,' and that it is an 'artificial thing' to impose. However, seventeen participants do support a quota, and a big group (n=21) view working in a diverse team as positive. The participants who support the quota mentioned it to be necessary because 'it has been shown that you tend to hire someone with the same profile' and 'the quota could achieve quick results.'

These observations underscore the disparity in opportunities and representation between men and women in the tech sector, with disruptive technologies primarily led and created by men. This leads to a male-dominated power dynamic controlling development and decision-making processes. The limited number of women in leadership positions or as CEOs highlights the perpetuation of existing power structures, particularly when male-dominated networks exclude women from networking and accessing similar opportunities.

7 Discussion

Previous research unveiled the substantial influence of implicit gender bias, which contributes to women's reduced probability of working in male-dominated sectors. This bias manifests through subtle and subconscious disparities in treatment, driven by assumptions and associations (Ellemers, 2014). Implicit biases affect hiring and promotions and limit opportunities for individuals/women who don't fit stereotypes, causing a workplace gender gap (Chadwick & Baruah, 2019). Our findings show that implicit gender bias is not only visible in the work environment, but also in the environment outside of work. Just as previous research shows, our findings regarding the path to technology demonstrate that women's environments, through either parents or teachers, have an influence on them choosing a path in technology (Besecke & Reilly, 2017; Papastergiou, 2008). Peers also have an influence (Zarrett & Malanchuk, 2005). However, we do see that such influence was also unintentional or by chance, while boys receive more encouragement to pursue a tech related career (SEOR & Sterk Techniekonderwijs, 2020). Here the implicit gender bias is present as young girls do not receive this encouragement, perhaps unintentionally. In addition, just as previous research (Beede et al, 2011) indicated, the representation in the form of role models in technology is lacking. Often women replace this missing aspect with family members; in the case of our findings, the father was more than often the role model. This shows that male support and encouragement are important. Female mentors in tech could relate more to the challenges and could provide additional help (Tandrayen-Ragoobur & Gokulsing, 2022). As we see in the findings, the family does not really understand what the participants do.

Unexpectedly, the role of the father emerged across three themes: the path to technology, role models, and family support. This finding contrasts with previous research, which suggested that the father's role was primarily limited to the path to technology. The significance of family, as supported by existing theories, is evident, with approximately half of the participants attributing their entry into the tech field to family influence. However, what stands out is the explicit inclusion of the father in relation to the importance of technology, acting as a role model, and providing family support. Furthermore, another unexpected finding was the influence of peers and acquaintances in shaping individuals' journey into the tech field (Zarrett & Malanchuk, 2005). This highlights the role of social networks and knowledge-sharing in facilitating entry into the tech industry.

The findings highlight a pattern where women are often underestimated and need to prove themselves capable before their abilities are recognized. This tendency may be experienced by women without their awareness, and its impact can vary depending on individual responses and mindset. The bias can be both implicit and explicit, affecting women in small and large proportions. Research reveals that male-dominated industries can be particularly unwelcoming for women (Cauterucci, 2017; Edgar, 2021). Interestingly, when women were asked about bias (specifically, the women who initially denied its existence), they acknowledged its presence upon further discussion. However, it is important to note that the participants did not express strong criticism towards their male colleagues, often mentioning positive interactions and good relationships with them. Working in a diverse environment is considered highly beneficial as it positively impacts the dynamics, atmosphere, and culture, while fostering a more creative approach. Related to this, the general viewpoint on implementing forced measures such as a gender quota to reduce the gap, was critical. While increasing diversity through hiring practices may be a good start, it may not be enough to create lasting change. Nevertheless, this change is needed. The gender gap in the tech sector refers to the disparity in opportunities and representation between men and women within the industry. When we delve deeper into new technologies, we often find that they are predominantly led and created by men. This leads to a power dynamic where men hold control over the development and decision-making processes (Wright, 2016) in fast-paced technology advancements, as the findings from this study also show. This power imbalance is evident in the limited number of women in leadership positions or as CEOs. When there are very few women employed in the sector and they are not given equal opportunities to advance and become CEOs, the chances of having women in top leadership roles in technology remain slim. This perpetuates the existing power structure, particularly when male-dominated networks exclude women from networking and accessing the same opportunities.

To sum up, to combat implicit gender bias it is important to create an inclusive work environment by increasing the visibility of role models and changing workplace culture. When discussing the gender gap and implicit gender bias, workplace culture encompasses the overall environment, and perspectives that influence the experiences and opportunities based on gender such as never being truly part of 'the boys' club' (Cauterucci, 2017; Edgar, 2021). Understanding and addressing workplace culture is crucial for promoting gender equality in the tech industry.

8 Conclusion

Currently, there is a significant underrepresentation of women occupying technical jobs, also known as the gender gap in the technology sector. Previous research has consistently demonstrated the existence of implicit gender bias, whereby assumptions and stereotypes contribute to differential treatment based on gender. This bias has been observed across various factors that are distinct but interconnected, and which could have an influence on the gender gap. This is worrying as technology triggers profound transformation in society, influencing innovations intended to better people's daily lives. For this study, interviews were conducted with women working in Dutch technological jobs. This study delved into their career origins, perspectives on the gender gap in the workplace, experiences with bias, and the impact of role models and the social environment. The findings reveal that the gender gap and implicit gender bias are intricately linked, influencing women's experiences and perspectives in technology. This influence is evident in their work within non-diverse teams and their views on potential solutions and policies to address the gender gap. Additionally, the study highlights that most women encounter a mix of implicit and explicit gender biases, impacting their career choices, experiences throughout their careers, and the level of family support and representation they receive.

In addition to the findings from the initial research, three experts specialising in the gender gap in technology were consulted during this study. These included two representatives from two organisations focused on encouraging women in the tech industry and one expert specialising in the gender gap within the Dutch technology sector. One expert emphasised that the tech culture in the Netherlands is not particularly female-friendly. Her input shed light on the challenges faced by women in this industry. Additionally, she provided valuable recommendations to address and improve the situation, such as the role of male colleagues. Male allies are uniquely helpful in reducing the negative effects of underrepresentation among women through practices such as being a mentor, giving space to women to be visible and express their views, and by giving women recognition (Schinkels, 2021). Specifically, having a male ally who supports gender equality can reduce negative effects on women, such as isolation and hostility, and can empower women to anticipate respect, support, and equality norms in the workplace (Moser & Branscombe, 2021).

Furthermore, experts shared their practical recommendations on how to tackle the gender gap in the technology industry. For women in science, technology and IT, it is not the content of the work that is the cause of their departure. Women leave the work environment because they do not like the work environment. They encounter gender bias, do not feel safe enough, cannot combine work and private life, or face an unpleasant culture. These are causes in which an employer can play a role (VHTO, 2022). Concrete measures that employers can take include, for example, changing the working environment by increasing role model visibility, fostering networks, creating family-friendly and safe working environments, and providing gender bias training. During a gender bias training or workshop, awareness is created about what gender bias is and tools are provided on how to deal with it (VHTO, 2022). Finally, the contributions to practice and policy involve those responsible for the hiring process, those shaping company culture, and parents. Fathers are particularly significant, as they all play essential roles in promoting gender equality and addressing gaps in the workplace.

While this study provides insights into the broad field of technology, the participants represented a diverse range of backgrounds, including consulting, management, and software development. However, as the sample size is relatively small, caution should be exercised when generalising the findings to a larger population. Future research could explore the possibility of scaling up the study to include a larger and more diverse sample. Additionally, it is important to note that the focus of this study was on the broader tech industry. However, it would be valuable for future work to examine potential differences between sub sectors or even narrow the focus to a specific sub sector, such as engineering, given its male-dominated nature. This could provide a deeper understanding of the dynamics within specific areas of the tech industry. Future research could delve deeper into the specific role of fathers, examining their impact across different themes, through for example exploring their influence in relation to role modelling, family support, and other relevant areas. This could provide valuable insights into the unique contributions of fathers in shaping individuals' experiences and pathways within the tech industry. Finally, although cultural background was not examined in this study, since culture affects upbringing, it could be included in future work.

Acknowledgments

We express our sincere gratitude to all the participants and experts who contributed to this study. Your valuable insights and expertise have been instrumental in enhancing our understanding of the subject matter. We would also like to extend our appreciation to those who reviewed the study and provided feedback.

References

Amon, M. J. (2017). Looking through the Glass Ceiling: A Qualitative Study of STEM Women's Career Narratives. *Frontiers in Psychology*, 8, 1–10. https://doi.org/10.3389/fpsyg.2017.00236

Banaji, M. R., & Greenwald, A. G. (2016). *Blindspot: Hidden biases of good people.* Bantam.

Baron, A. S., & Banaji, M. R. (2006). The development of implicit attitudes: Evidence of race evaluations from ages 6 and 10 and adulthood. *Psychological Science*, 17(1), 53–58. https://doi.org/10.1111/j.1467-9280.2005.0166

Beede, D. N., Julian, T. A., Langdon, D., McKittrick, G., Khan, B., & Doms, M. E. (2011). Women in STEM: A Gender Gap to Innovation. SSRN Electronic Journal. https://doi.org/10.2139/ssrn.1964782

Besecke, L. M., & Reilly, A. H. (2017). Factors influencing career choice for women in science, mathematics, and Technology: the importance of a transforming experience. Advancing Women in Leadership Journal, 21. https://doi.org/10.21423/awlj-v21.a258

Boysen, G. A., & Vogel, D. L. (2008). The relationship between level of training, implicit bias, and multicultural competency among counselor trainees. *Training and Education in Professional Psychology*, 2(2), 103–110. https://doi.org/10.1037/1931-3918.2.2.103

Cauterucci, C. (2017, February 21). The sexism described in Uber employee's report is why women leave Tech – Or don't enter at all. *Slate Magazine*. https://slate.com/human-interest/2017/02/the-sexism-in-uber-employees-report-is-why-women-leave-tech-or-dont-enter-at-all.html

CBS. Centraal Bureau voor de Statistiek (2023). AL 23 jaar op rij meer vrouwen dan mannen in hoger onderwijs. Centraal Bureau voor de Statistiek. https://www.cbs.nl/nl-nl/nieuws/2023/10/al-23-jaar-op-rij-meer-vrouwen-dan-mannen-in-hoger-onderwijs

Centraal Bureau voor de Statistiek (2024). Beroepen van werkenden. https://www.cbs.nl/nl-nl/visualisaties/dashboard-arbeidsmarkt/werkenden/beroepen-van-werkenden

Chadwick, A. J., & Baruah, R. (2019). Gender disparity and implicit gender bias amongst doctors in intensive care medicine: A 'disease' we need to recognise and treat. *Journal of the Intensive Care Society*, 21(1), 12–17. https://doi.org/10.1177/1751143719870469

Dixon-Fyle, S., Dolan, K., Hunt, D. V., & Prince, S. (2020). Diversity wins: How inclusion matters. In *McKinsey & Company*. https://www.mckinsey.com/featured-insights/diversity-and-inclusion/diversity-wins-how-inclusion-matters

Duhaime-Ross, A. (2014, September 25). Apple promised an expansive health app, so why can't I track menstruation? *The Verge*. https://www.theverge.com/2014/9/25/6844021/apple-promised-an-expansive-health-app-so-why-cant-i-track

Edgar, G. (2021). 5 actions tech employers can take to retain women. *Women in Technology*. https://www.womenintech.co.uk/5-actions-tech-employers-can-take-to-retain-women

Ellemers, N. (2014). Women at work. *Policy Insights From the Behavioral and Brain Sciences*, 1(1), 46–54. https://doi.org/10.1177/2372732214549327

FME Actieagenda 'Vrouwen in Techniek – Op weg naar 30% in 2030' | FME. (n.d.). FME. https://www.fme.nl/fme-actieagenda-vrouwen-techniek-op-weg-naar-30

-2030#:~:text=Het%20aandeel%20vrouwen%20in%20de,om%20deze%20ambi tie%20te%20behalen.

Friedmann, E., & Efrat-Treister, D. (2022). Gender Bias in STEM hiring: Implicit In-Group gender favoritism among men managers. *Gender & Society*, 37(1), 32–64. https://doi.org/10.1177/08912432221137910

Global Gender Gap Report. (2023). World Economic Forum. https://www.weforum .org/reports/global-gender-gap-report-2023/

Idris, R., Govindasamy, P., Nachiappan, S., & Bacotang, J. (2023). Examining moderator factors influencing students' interest in STEM careers: the role of demographic, family, and gender. *International Journal of Academic Research in Progressive Education and Development*, 12(2). https://doi.org/10.6007/ijarped/v12 -i2/17609

Kenny, E. J., & Donnelly, R. (2019). Navigating the gender structure in information technology: How does this affect the experiences and behaviours of women? Human Relations, 73(3), 326–350. https://doi.org/10.1177/0018726719828449

Kitzinger, J., University, C., Chimba, M., Williams, A., Haran, J., Boyce, T., Cardiff School of Journalism, Media and Cultural Studies Cardiff University, & UK Resource Centre for Women in Science, Engineering and Technology (UKRC). (2008). Gender, stereotypes and expertise in the press: How Newspapers Represent Female and Male Scientists. Gwasg y Bwthyn. https://citeseerx.ist.psu.edu/ document?repid=rep1&type=pdf&doi=39c4660cc9bb56278e881cad1c4e 98b30558afa4

Kuckartz, U. (2014). Qualitative Text analysis: A Guide to Methods, Practice and Using Software (1st ed.). SAGE Publications Limited. https://doi.org/10.1007/978 -3-030-15636-7_8

Makarem, Y., & Wang, J. (2019). Career experiences of women in science, technology, engineering, and mathematics fields: A systematic literature review. *Human Resource Development Quarterly*, 31(1), 91–111. https://doi.org/10.1002/hrdq.21380

Microsoft Diversity & Inclusion Report (2023). Microsoft. https://www.microsoft .com/en-us/diversity/inside-microsoft/annual-report?activetab=innovation-spot lights%3aprimaryr4

Ministry of General Affairs (2022). Ministerie van Algemene Zaken. Emancipatie een opdracht voor ons allemaal. Nieuwsbericht | Rijksoverheid.nl. https://www .rijksoverheid.nl/actueel/nieuws/2022/11/18/emancipatie-een-opdracht-voor-ons -allemaal

Moser, C. E., & Branscombe, N. R. (2021). Male Allies at Work: Gender-Equality Supportive Men Reduce Negative Underrepresentation Effects among women. Social Psychological and Personality Science, 13(2), 372–381. https://doi.org/10.1177/ 19485506211033748

Nguyen, D., & Beijnon, B. (2023). The data subject and the myth of the 'black box' data communication and critical data literacy as a resistant practice to platform exploitation. *Information Communication & Society*, 27(2), 1–17. https://doi.org/10.1080/1369118x.2023.2205504

NOS (2023). Deeltijdwerk al snel na diploma in trek onder vrouwen, 'niet per se slecht'. NOS. https://nos.nl/artikel/2471090-deeltijdwerk-al-snel-na-diploma-in-trek-onder-vrouwen-niet-per-se-slecht

Papastergiou, M. (2008). Are Computer Science and Information Technology still masculine fields? High school students' perceptions and career choices. *Computers & Education*, 51(2), 594–608. https://doi.org/10.1016/j.compedu.2007.06.009

Perez, C. C. (2019). *Invisible women: Data bias in a world designed for men*. Abrams Press.

Sassler, S., Glass, J., Levitte, Y., & Michelmore, K. M. (2017). The missing women in STEM? Assessing gender differentials in the factors associated with transition to first jobs. *Social Science Research*, 63, 192–208. https://doi.org/10.1016/j.ssresearch.2016.09.014

Schinkels, C. (2021). *De IT girl: Hoe overleef je een door mannen gedomineerde werkvloer?* Culemborg: van Duuren Management.

Schorr, A. (2019). Pipped at the post: Knowledge gaps and expected low parental IT competence ratings affect young women's awakening interest in professional careers in information science. *Frontiers in Psychology*, 10, 1–18. https://doi.org/10.3389/fpsyg.2019.00968

Statista (2023). Google: Global Corporate Demography 2023, by gender and department. https://www.statista.com/statistics/311805/google-employee-gender-department-global/

Tandrayen-Ragoobur, V., & Gokulsing, D. (2021). Gender gap in STEM education and career choices: what matters? *Journal of Applied Research in Higher Education*, 14(3), 1021–1040. https://doi.org/10.1108/jarhe-09-2019-0235

VHTO (2022).Vrouwen in bèta, techniek en IT Hoe behoud je ze als organisatie? VHTO. 1–27. https://www.vhto.nl/wp-content/uploads/2022/09/Whitepaper_Vrouwen_in_beta__techniek_en_IT__hoe_behoud_je_ze_als_organisatie.pdf

De Vleeschouwer, E., Wiemers, S., Zandvliet, K., & SEOR BV. (2020). *Kiezen voor technisch vmbo: de rol van ouders en hun beeld van techniek*. In SEOR. https://www.seor.nl/Cms_Media/S1311KiezenvoortechnischvmboDerolvanoudersenhunbeeld-van-techniek.pdf

Wang, M., & Degol, J. L. (2016). Gender gap in Science, Technology, Engineering, and Mathematics (STEM): current knowledge, implications for practice, policy, and future directions. *Educational Psychology Review*, 29(1), 119–140. https://doi.org/10.1007/s10648-015-9355-x

Yi Wang and David Redmiles 2019. Implicit gender biases in professional software development: an empirical study. In Proceedings of the 41st International Conference on Software Engineering: Software Engineering in Society (ICSE-SEIS '19). IEEE Press, 1–10. https://doi.org/10.1109/ICSE-SEIS.2019.00009

Wentling, R., & Thomas, S. (2009). Workplace culture that hinders and assists the career development of women in information technology. *Information Technology, Learning & Performance Journal*, 25(1). 25–42.

De Wit, J., & Kalkhoven, F. (2019). *Tekorten ict-beroepen nemen verder toe*. https://www.uwv.nl/imagesdxa/factsheet-ict-beroepen_tcm94-447183.pdf

Wright, T. (2015). Women's experience of Workplace Interactions in Male-Dominated Work: the intersections of gender, sexuality and occupational group. *Gender Work and Organization*, 23(3), 348–362. https://doi.org/10.1111/gwao.12074

Yerkes, Mara A., and Belinda Hewitt, 'Part-time strategies of women and men of childbearing age in the Netherlands and Australia', in Heidi Nicolaisen, Hanne Kavli, and Ragnhild Steen Jensen (eds), *Dualisation of Part-Time Work: The Development of Labour Market Insiders and Outsiders* (Bristol, 2019; online edn, Policy Press Scholarship Online, 23 Jan. 2020), https://doi.org/10.1332/policypress/9781447348603.003.0011

Zacharia, Z. C., Hovardas, T., Xenofontos, N., Pavlou, I., & Irakleous, M. (2020). Education and employment of women in science, technology and the digital economy, including AI and its influence on gender equality. In European Parliament, *European Parliament* (Study PE 651.042). https://www.europarl.europa.eu/RegData/etudes/STUD/2020/651042/IPOL_STU(2020)651042_EN.pdf

Zarrett, N. R., & Malanchuk, O. (2005). Who's computing? Gender and race differences in young adults' decisions to pursue an information technology career. *New Directions for Child and Adolescent Development*, 2005(110), 65–84. https://doi.org/10.1002/cd.150

Zollman, A. (2012). Learning for STEM literacy: STEM Literacy for Learning. *School Science and Mathematics*, 112(1), 12–19. https://doi.org/10.1111/j.1949-8594.2012.00101.x

CHAPTER 13

Beyond the Strictest Computation of the General Proportion

Language Automation, Machine Translation and Their Complicated Possibilities for Cultural Exchange

Christophe Declercq and Gys-Walt Van Egdom

1 Introduction[1]

This chapter offers a critical assessment of technology's potential to empower democracy within the global book market, in particularly with regard to lesser-resourced languages and languages of lesser diffusion.

As language and translation technologies have limitations, challenges, and biases, this chapter also provides an analysis of the benefits and risks of extensive use of language technology in intercultural exchange. Central to this discussion is empirical research that involves gathering and scrutinising literature-derived data as well as building on relevant historical perspectives in order to address the research question: *To what extent can language automation and machine translation contribute to the literary translation book market and the democratisation of literatures from lesser-resourced languages?* By amalgamating existing research and drawing from personal experience and observations of the last 25 years, the authors present a comprehensive overview of the existing body of knowledge regarding the impact of language automation and machine translation on the translation market, particularly in the literary domain. Throughout the chapter, the broader concept of automated translation (AT) is used to refer to translation facilitated by machine technology, which includes both machine translation (MT) and artificial intelligence (AI).[2] This AT umbrella term acknowledges the interconnectedness of these concepts, despite conceptual distinctions.

1 The first part of the title is taken from Jonathan Swift's *Gulliver's Travels*, which features in the chapter.

2 The development of language and translation technology for automated translation purposes is increasingly accelerating. The differentiation between MT and AI (generative AI / large language models) was still relevant at the time the abstract for this contribution was

 | DOI:10.1163/9789004711396_015

Sociological research on intercultural exchange has shown that intercultural relations are strongly driven by two factors: hierarchy and proximity – including kinship (see Heilbron & Sapiro, 2007; Collet, 2015; Elias & Mansouri, 2020). The opportunity to come into close contact with people from different linguistic and cultural backgrounds is crucial in the formation of intercultural relations, in the erosion of stereotypes, and in establishing affective ties and positive emotions such as empathy and solidarity (Pettigrew, 1998, pp. 71–72; Visbal 2009, p. 1). Societies, communities, audiences, and users seek cultural proximity with cognate cultures and their languages in goods as widely varying as trade, bilateral agreements, membership of international organisations, tourism and trade on the one hand, and information dissemination, cultural production, and popular entertainment on the other – all through and by means of intercultural exchange and interlingual translation. Here translation studies intersect with sociology in that for instance Even-Zohar's polysystem theory explores the influence of different cultures on literary systems, while proximity guides decisions for entire translation projects based on cultural closeness (Even-Zohar, 1979; Shuttleworth, 2019).

Together, these theories offer insights into translation dynamics. Polysystem theory helps explain why certain source cultures or literary traditions are more 'proximate' than others in a given target culture. It considers historical factors, cultural trends, and the relationships between different literary systems. Even-Zohar's concept of 'center' and 'periphery' within the polysystem reflects this idea, where certain cultures or literatures are considered central and influential, while others are marginalized or peripheral. Proximity is a practical consideration in the translation process, but it is also a reflection of the power dynamics and the positioning of literary traditions within the polysystem. Motives of cultural exchange are driven by the idea of wanting to relate to another entity: translating either the other more powerful, deemed more prestigious or better resourced language to one's own – such as the uptake of American televised series, subtitled or dubbed; the desire to cascade one's own cultural conceptualisations of reality – cultural products such as Danish or Swedish detective series; or a cross-fertilisation of both.[3]

Mediated through technology, direct access to locations not previously deemed as being in the proximity of one's own cultural space has been

---

accepted, but the distinction was increasingly much less obvious by the time of finalising the text nearly a year later.

3 A possible example of this could be the Eurovision song contest and the role of cultural proximity in its scoring performances.

allocated a higher degree of potential proximity through media – traditional and social. The instant publication of news content as well as the possibilities of accessing content distributed for dissemination or publication purposes, make this proximity possible. In the dissemination of information through machine translated content, received as partial or superficial by the target audience, *gisting* relates to the broad idea of what is expressed in a language different from the original language. With content that is properly and fully translated, publication supports increased productivity and capacity for more varied reception or cascading of cultural production. To that end, machine translation (MT) for gisting and for disseminating the main idea of a verbal product, is similar to the entertainment function of media. Both provide the opportunity for one culture to broadly penetrate the other, regardless of possible quality issues beyond a *fit for purpose* approach. This can also be regarded as a form of *lean localisation*, where the target locale is not fully but broadly addressed, or as *constrained cultural transduction* where a certain genetic essence of the information is kept but the content container itself is appropriated by the user only in a superficial and partial manner.

Following these approaches to asymmetrical interdependence in media and the reconceptualisation of an active target audience (Straubhaar, 1991; Straubhaar, 2003) possibly settling for partial information appropriation, lesser-known languages[4] will have difficulty in forking out a stake in the linguistic landscape of a better-known language or of a language used as a primary means of communication across different cultural barriers such as English, Spanish, or Arabic. In the context of intercultural exchange, literature plays a central role as a rich source of cultural resources, typically aiming at complete assimilation through (proper, human) translation and publication. Literature encompasses the production, distribution, and reception of literary texts, as well as the complex realm of translation, which is conceptually universal but executed in dedicated target languages. Cultural exchange and the translation of literature are closely intertwined. Translations not only facilitate the study of literature across different languages but also serve as vehicles for understanding the workings of cultural systems elsewhere (Kloepfer & Shaw,1981, p. 36; Broomans et al., 2022), in part through a mirroring effect with respect the first language.

4 Terminology used in contemporary digitised and automated societies often refers to languages that are 'less resourced' or have 'lower diffusion.'

By the late 19th century and the 20th century, the literary field emerged as an increasingly autonomous sphere and the subsequent liberalisation of the book market has created a relative liberation from state control (e.g. centralised regulation of production) in Western countries (Bourdieu, 1983 in Sapiro, 2003). The reduced influence of politics (Van Voorst, 1997) impacted hierarchical relationships quite heavily. Along with the Anglicisation of post-World War II audiovisual popular culture, the emergence of English as a lingua franca appeared in the world of higher education and in international organisations such as the UN. English acted as a prime container and carrier in the rapidly developing domain of information and communication technology (ICT) and in the developments of globalisation and localisation (House, 2003; Nickerson, 2005; Phillipson, 2008; Dewey, 2010; Blommaert, 2010). These socio-economic trends further diminished politically driven factors while enhancing greater flexibility for emerging market dynamics. In addition, the influential and mainly Anglo-Saxon mainstream culture that expanded following World War II fostered a growing prevalence of the English language and its literature, not just in specific fields like business, marketing, healthcare, and technical documentation, but also in literature, including translated works.When interpreting these developments as a progression within an international field of power, as is customary in sociology, it can therefore be assumed that the dominance of Anglo-American influences has come to weaken the position of other cultures. This is – as the chapter will show – only exacerbated by the dominance of English data in the most widely used AI applications.

In publishing, diversity is constrained by the predominance of translations from central languages (Franssen, 2015). Despite this persistently high share of English, there is relatively increasing room for a broader array of source languages – though initiatives often appear scattered and seem to stem more from individual passion than systematic approaches. One rather unexpected solution to address the increasingly unequal distribution of cultural exchanges and to promote cultural diversification in the international book market is the use of language technology, more specifically automated translation (AT). Automated translation is best known for its two main forms: machine translation (MT) and artificial intelligence (AI). The two terms become increasingly more intimately related conceptually but have a different background in terms of origins, professional use, and possible advantages and disadvantages. At the time of writing, the latest wave of machine translation technology, neural machine translation (NMT), not only contains an artificially intelligent component – machine learning – but is also renowned for potentially being the most useful of all iterations of different MT technologies up to now (Bentivogli, et al., 2016; Benkova, 2021).

2 Early Attempts at Language and Automation

Ever since the Tower of Babel, humankind has always dreamed of being able to effortlessly convert unknown languages into comprehensible messages. Ideally this would occur through a utopian intermediary carrier that stores all known knowledge and is capable of generating new knowledge based on existing data across language barriers, or as an intralingual tool of text production. Jonathan Swift's 'The Engine' in *Gulliver's Travels* (1726) exemplifies such a concept, generating word permutations and rearrangements within a seemingly closed set of possibilities. Though a fictional creation, Swift's portrayal satirizes philosophers like Ramon Llull, Blaise Pascal, John Wilkins, and Gottfried Leibniz (Booth & Booth, 1956; Daylight, 2015), who sought to harness thought and establish a universal language, often through numerical codes (Bonner, 1994; Rodgers, 2017). The story also serves as a critique of societies risking the overshadowing of human concerns by technology: 'tyranny becomes devastating when a ruler has technology as an assistant. Moral decadence is the product where manners overpower morals in value' (Barnes, 1972, p. 54).

Core to the supposed activation of the participation of technology that eventually lulls morals into a deep sleep is the idea that the Swiftian device was designed for a purpose. This early alignment of technology use with intended purpose (fit for purpose) occurred alongside data collection, quality assurance, and the machine's intended capacity for self-learning. To even remotely suggest that Jonathan Swift ought to be seen as one of the pioneers of language and automation is an exaggeration. However, over subsequent centuries, non-fictional devices emerged, products of inventive minds aiming to decipher the codes of language. Figures like Thomas Bayes (1701–1761) and Charles Babbage (1791–1871) contributed significantly. Bayes's statistical inference became integral to machine learning, while Babbage's Analytical Engine marked a milestone in computing history as the first general-purpose automatic computer (Haigh & Priestley, 2016).[5] Shortly after Babbage's work, George Boole (1815–1864) introduced Boolean logic, facilitating the creation of binary operating languages and count-based models (Press, 2016). Computing logic can often be expressed in Boolean terms (Christensson, 2011).

Initiated by Alan Turing's 1936 concept of a universal machine, by 1945, multiple individuals were endeavoring to realize such a device (Robinson,

5 The proper first general-purpose computer, the German Z3, was not operational until 1941.

1994). In the UK, several groups aimed to construct practical stored-program computers by 1948 (Lavington, 2012, p. XIII). Aligned with computing ideas of the 1930s and 1940s and the concept of NLP, which applies computational techniques to learning, understanding, and generating human language content (Hirschberg & Manning 2016, 261), this era also witnessed pioneering advancements in automated translation. Despite earlier efforts in the 1930s by George Artsrouni to create a translating mechanical brain (Hutchins, 2014), it was not until Warren Weaver's 1949 Memorandum that the notion of machine translation and language technology gained proper recognition. Weaver proposed resolving the simplistic word-for-word approach of automated translation through statistical methods and ideas from information theory (Brown et al. 2003, p. 335), anticipating future statistical machine translation. He suggested that multiple meanings could be addressed by examining immediate context, and drew inspiration from early mathematical models of neural networks (McCulloch & Pitts, 1943), envisioning computers deducing logical outcomes automatically (Hutchins, 1999, p. 5). Although Weaver presented his ideas modestly, the memorandum sparked significant research into machine translation, particularly during the Cold War era. Research teams emerged in the US, UK, and Russia, with the mutual desire to automatically translate messages between English and Russian seen as pivotal for gathering information from both sides. The 1954 Georgetown-IBM experiment, aiming to translate political and military messages from Russian into English, is one such example of these endeavors, which continued into the 1960s.

However, in 1966, a report by the Automatic Language Processing Advisory Committee (ALPAC) curtailed all optimism surrounding computers' potential in translation and linguistics. The ALPAC report led to a significant reduction in funding for research into machine translation, particularly in the United States (Hutchins, 2013, p. 13). Despite this setback, it also spurred more creative endeavours, including the development of the Canadian METEO system, operational since 1976. Designed for ENG<>FRE translation, METEO was tailored for the specialised vocabulary and syntax of meteorological forecasts (Hutchins, 1994). This marked the first practical application of Swift's finite set of language tokens and the purpose-driven concept of The Engine in real-life translation tasks. While machine translation began to regain traction, focusing on more limited yet relevant applications, another significant development emerged in the late 1970s: computer-aided translation (CAT, or Computer-Assisted Translation). This idea would become pivotal for the language technology industry that evolved from machine translation.

3 Computer-Aided Translation

If METEO's controlled language, limited to specific vocabulary and syntax within a language pair, addressed the need for robust machine translation solutions, then computer-aided translation (CAT) represented another step in automation, mitigating the limitations of machine translation in a different manner. In the late 1970s, Peter Arthern, associated with the English Translation Division of the Council of European Communities in Brussels, proposed a different approach. Instead of relying on computers to generate translations autonomously, he suggested leveraging previously translated materials stored in a memory system. These translations could be easily retrieved, edited, and reused for similar or repetitive texts – a concept that gave birth to translation memory, with added terminology management for quality assurance (Snell, 1978, p. VI; Hutchins, 1998, pp. 294–5). Arthern recognised the broader potential of this approach beyond European institutions and emphasised its relevance to users at smaller scales too (Arthern, 1981, p. 309). However, such widespread grassroots adoption was hindered by limited access to computing power and the lack of available hardware and software for the general user.[6]

Arthern's idea of a memory for translations was a way to overcome shortcomings of language automation and machine translation. Following this, in the 1980s bi-text and the parallel corpus emerged as the basis of the concept of the translation unit. This unit represents a segment in the source language that corresponds to a segment in the target language, potentially available for reuse if a future source language sentence matches an already existing sentence, clause, phrase, or word(s). The concept of creating a database with previously translated material for potential reuse, provided the match is substantial, significantly expanded the translation market. This breakthrough facilitated easier movement across previously stagnant spheres characterised by limited translator capacity, project durations, client diversity, and specific domains. Consequently, it became a logical progression for a new wave of translation agencies to capitalise on the

6 The 1990s witnessed a significant acceleration in access to computing power, leading to the democratisation of information technology (IT). Over the last few years of the 20th century and the first decade of the 21st century, access to computing power saw a remarkable shift, with the cost of a decent PC dropping from $6000 (adjusted for inflation) in 1997 to one-tenth of that amount by 2009 (figures from Comen (2018)).

opportunities offered, emphasising benefits such as consistency, productivity, and reductions in production cycle time.[7]

In the same broader period, pioneers in translation technology like Trados and INK were established, in Germany and the Netherlands respectively, focusing on bilingual database segments to boost productivity and ensure quality consistency. While tools such as Trados Workbench, Trados Multiterm, IBM Translation Manager, Star's Transit, and Atril's Déjà Vu emerged in the early 1990s, the language and translation industry awaited the democratisation of computing power and the declining costs in the late 1990s for further realisation of translation automation, albeit the CAT way. By the turn of the century, computer-aided translation became a driving force behind the language and translation technology sector's expansion in the globalised market.[8] In the late 1990s and early 2000s, the emphasis on accessing local markets was central to globalisation ('think global, act local'), with localisation primarily focused on software development and technical documentation. However, localisation soon expanded to encompass websites, as well as help files, video games, and other forms of audiovisual media. Localisation serves a distinct purpose: it aims to appeal to a specific target audience, adapt the source product to fit local norms, and minimise the otherness of the source product (Declercq, 2012, p. 264).

The localisation expansion propelled the translation industry to grow at an annual rate of five to ten percent since the mid-2000s (Kotzsch, 2021). However, in that emerging era of globalisation and localisation and with the advent of computer-aided translation, translation became a tradable commodity. These models commodified the translation process as a symbol of trade and consumption targeted at specific audiences; firms had to 'adapt their products, services, and practices to the host country's (local) way of doing business' (Kofi Adarkwah, 2022, p. 1). With the rise of translation technology such as computer-aided translation tools, cultural exchange can benefit from increased access to information previously unavailable because it was too costly. Language technology available globally can be utilized by local users for their benefit, relying on mechanisms that emphasise a local perspective. The idea of think global, act local can, however, also be seen as cultural imperialism, where external parties dictate what is presented abroad and tailor target audiences based on perceived majority preferences.

7 For more on early concepts and the translation market increasingly building on translation memories see Melby (1996) and Hartley & Paris (1997).

8 See also Garcia (2014); Zhang & Cai (2015).

Probabilities are limited to assumptions relating to marketing but not yet to translation re-use, which still rely nearly entirely on human input and assumptions about the locale, a country or region and/or a language. This local perspective can be geographical but also geo-localised. Whereas the former relates to physical proximity but not necessarily critical proximity (or cultural cognateness), digital access to textual material can be the other way around: provision of information is in place nonetheless, but selection of it is related more to critical proximity than physical location. Local users can be detached from their location as part of the target audience but still be connected to the target locale through cloud applications, i.e. with potential accessibility anywhere in the world, and yet remain part of a specific locale.

As digital access to textual material prioritises critical proximity over physical location, any technology in support of language content and/or translation purposes facilitates the fragmentation of the target audience. In more practical terms: language and translation technology can be used by those requiring access to other lingual information while based in the locale of the target language, but that target language can equally address fragmented audiences elsewhere in need of access to information in the same language. This divergent fragmentation was an a priori condition for the global uptake of machine translation. Along with the democratisation of computing power of the 2000s, language automation for translation purposes rose to the occasion.

4 Machine Translation

The role of machine translation in cultural production has been a much-contested domain for decades. With the increased adoption of machine translation – in its many current or previous forms – for dissemination of communication (access to information with lack of an accuracy guarantee and very likely lacking proper fluency) and publication/assimilation of information purposes (Hutchins, 2001, p. 8), MT has become an established nexus of contact. Access to other lingual content has long surpassed the spheres of professional translation, interpretation, or mediation. While translating into the (preferably native) target language, human translators appeal to their competencies and skills in the language and culture of both source text and target text. It is inevitable that the source language has an impact on the target text. This interference in traditional translation paradigms (Toury, 1995) can increase a 'shining through' phenomenon (a mirroring effect of the source language) in any target language produced by that translator, and ultimately by the readership of the target text. Access to information is guaranteed; accuracy is not, nor is fluency. If translation is therefore a language contact event with the capacity to result in target language products influenced by that contact, then the addition of machine translation takes this form of contact to an entirely new dimension. This is not least because in the case of language contact through target language influenced by human translation, processes involved are constrained by a range of cognitive, linguistic, and social factors (Muysken, 2013; Kotze, 2020), whereas MT lacks acknowledgement of these factors.

Acknowledging the history of machine translation in terms of the development of architecture is beyond the scope of this chapter. However, roughly three periods need to be presented, even though the boundaries between them are not always obvious. The first period of machine translation concerns the broad organisation of any type of MT prior to statistical machine translation (SMT). When exactly SMT actually emerged is a debate in its own right, but it definitely was present in the background and in research and development (R&D) during the 1980s and 1990s, and foregrounded in the 2000s, not least with the leap that was Google Translate's SMT move in 2006. Prior to SMT, MT systems were characterised by rule-based systems (dictionary-based, transfer-based, or interlingual), basically allowing machine translation platforms to perform only when the requested information is available as dictionaries and grammars in both source and target language. Typically, output was very source-language driven and the implied one-on-one source term–target term relationship did not allow for nuance at all. SMT, however, sought to retrieve information as relevant as possible from statistical models that derived their decisions on the basis of natural language presented in large bilingual corpora. Still, it can be argued that SMT still relates to statistical learning, a subfield of statistics, and has its statistical models as a formal representation of a theory. SMT learns from samples and populations, using relatively small data sets to provide the best estimates and inferences, and to establish a mathematical relationship between one or more random variables and other non-random variables (Ramzai, 2020, n.p.). Neural MT (NMT), however, learns from data, has strong predictive abilities, and discovers patterns (Verma et al., 2020).

5 Neural Networks, Machine Learning and AI

If statistical machine translation (SMT) relies on a language model for the source language (SL) and a translation model for the target language (TL), using relatively small datasets, it can be expected that most assumptions made during the model design phase will be reasonably accurate. At the same time, it is equally expected that there will be uncertainties in the predictions. With NMT, however, prediction is not based on other independent variables (Masera, 2022) but on networks and graphs, on allocating weight to learning of large data sets, an idea that floated conceptually already in the 1940s (Lebedev & Nicolelis, 2017). Ultimately, in comparison to SMT, machine learning algorithms of NMT need to deal with far fewer assumptions, if at all, 'as they don't have rigid requirements about collinearity, normal distribution of residuals, etc.' (Ramzai, 2020, n.p.). As NMT's machine learning is more flexible than statistical models, NMT systems allow for higher uncertainty tolerance (Ramzai, 2020) than SMT. As well, the flexibility of NMT can very likely produce seemingly more fluent (fake/false fluency) and seemingly more accurate translation results. But ultimately the built-in tolerance can lead at times to more awkward solutions than SMT.

Yet, as the popular uptake of NMT signalled a next step in the development of machine translation, it was adopted rapidly. The hype went as far as to elicit explanations because human parity had been reached, for Chinese to English no less (Hassan et al. 2018). Yet that mantra was quickly dismantled and brought back to a more reasonable level of limited usefulness because several seminal variables had not been considered previously (Toral et al., 2018). Whether this new dimension of automated translation was in fact suitable for literary texts or creative translation remained to be seen. Yet, the discussion about neural machine translation's capacity for creative purposes has been shaken to the core with the advent of artificial intelligent chatbots, the most renowned proponent of which at the time of writing is ChatGPT.

Neural networks, machine learning and AI are not exclusive to translation. They have been assumed in many domains, but NMT has been at the forefront and even the core of the wave of AI that emerged in the second half of 2022 and first half of 2023. Indeed, being an end-to-end approach to automated translation, 'taking advantage of deep learning methods and neural networks closely based on the human brain' to predict the likelihood of a sequence of words – the broken sentences from Swift's Engine – NMT has been considered by many as artificial intelligence already (Ramzai, 2020, n.p.). Neural networks called transformers, like GPT, learn how to track relationships in sequential data and therefore have the potential to learn context, something SMT was never capable of.

Deep learning algorithms that gain knowledge from large datasets are called Large Language Models (LLMs). Their capabilities in relation to working with text include summarisation, creation, and translation, but their application extends far beyond the worlds of text proper and concerns any form of communication, including in fields as diverse as software code or protein or molecular sequences (Lee, 2023). Being able to handle a broad range of tasks makes LLMs a successful example of generative foundation models, which is why, for instance, OpenAI's GPT-4 is actually an acronym that stands for Generative Pre-Trained Transformer 'designed to understand and generate human-like text, making them powerful tools for a wide range of applications' (ML6.EU, 2023).

Whether through an increase in productivity through the use of previously translated material contained in translation memory or improved consistency through the strict appliance of specific terminology, translation technology has liberated the translation industry from its traditional chains of reliance on the expertise of singular translators. Especially for source texts with levels of repetition and needs of strict terminological control, translation technology has been a blessing for increasing productivity at all levels of the translation pipeline: translator, company/client, intermediaries, and by extension, even global actors. This has allowed for more flexibility across translation projects as well as increased cost cutting. By foregrounding the technology in both process and as a frame for the end product for specialised domains, translators increasingly lost their own voice to the rapid spinning mill of profit, productivity, quantity, and usability. Language technology such as automated translation by either machine translation or artificial intelligence has only added to the industry's 'increasingly pragmatic view of product and process quality' (Moorkens, 2020, p. 323).

When it comes to using the output of translation technology such as translation memories or automated translation such as MT or AI, further adaptation or editing is needed. In relation to machine translation this is called post-editing: editing the machine translation results after the automated process. While within the confinements of an AI model automated results can also be amended, such editing happens through using engineering. In AI, this involves the design, development, and optimisation of systems and algorithms that enable natural language processing, machine learning, and neural networks to generate coherent and contextually relevant responses to user prompts. This includes techniques like few-shot prompting, where models are fine-tuned when demonstrations are included in the prompt to steer the model to better performance with limited examples. It also extends to multiple-shot prompting, where concepts and ideas can be combined, and by assigning specific weight to each the broader range of data enhances AI performance and output. At that stage, information is presented which the small user still needs to adapt to the purpose of the message.

Another option for potentially improving the quality of machine translation, especially literary texts, concerns transformer models. This is largely due to their attention mechanism, which enables the model to focus on parts of the input text and aims to capture intricate relationships between words and phrases essential for preserving stylistic subtleties and nuances in translation. By effectively aiming to handle long-range dependencies, transformer models should be able to maintain coherence and fidelity in translations and to convey not only literal meaning but also the intended tone, mood, and style of the original work (see for instance Van de Cruys (2022) on machine translations of poetry).

6 Automated Translation and Creativity

Within the translation production pipeline, post-editing automated results does not yet match human translation quality. One study involving the translation of a fictional story from English into Catalan covered each of these modes: human translation not only presented a higher creativity score compared to post-editing and direct MT output, it also ranked higher in narrative engagement and translation reception (Guerberof-Arenas & Toral, 2020). Considering the economic pressures on translation production, particularly in the aftermath of the 2007/8 economic downturn (see Moorkens, 2020), it may be contended that integrating two facets of automated translation (MT or AI) could be beneficial. This integration involves utilizing direct output for information dissemination to grasp the source text substantially while applying this 'good enough' approach to language pairs where one language (either the source or target) lacks resources in terms of technology, data, or proximity. In fact, this approach for lesser-resourced languages not only mirrors application of automated translation between major resourced languages such as English, French, German, Spanish, and Italian (to name but a few). It also concerns translations between distant languages that have distinct formal differences such as English and Chinese, as well as between languages that require building on much fewer resources than others.[9]

9 Because information is not generally available in minority languages, Meta AI launched the No Language Left Behind project (https://ai.meta.com/research/no-language-left-behind/), consisting of open-sourced models capable of delivering translations that are useful enough for the purpose they would serve with respect to 200 languages.

In an attempt to gauge the performance of the neural machine translation system DeepL in relation to the strengths and limitations of English-Chinese translation of literary texts, Kaibao and Xiaoqian (2023) investigated the accuracy and fluency rates as well as linguistic features of translating Shakespeare (*Coriolanus* and *The Merchant of Venice*) into Chinese and found that DeepL reached accuracy and fluency rates of above 80%. This was not yet possible with statistical machine translation for Chinese into English (Hoover & Sommer, 2010). Although creativity was still limited, the 'use of translation methods such as addition, explicitation, conversion and shift of perspective, and in the use of Chinese sentence-final modal particles' (Hu & Li, 2023, p. 546) was present to a degree nonetheless. This resonates with assessments that appeared with SMT of the mid-2010s, when post-edited machine translation of a Richard Powers essay from English into French was deemed acceptable by readers but less so by professional translators (Besacier & Schwartz, 2015). Although it has been argued that the features of literary translation have been an insurmountable obstacle for neural networks (van Egdom et al., 2023), the English into Chinese study supports the belief that beyond the difficulty of accurately interpreting figures of speech and puns, imagery, humour, and irony, recent NMT capabilities actually offer possibilities for readership.

Existing tools exhibit limited user-friendliness, posing a risk of linguistic impoverishment in the case of automatic translation. Furthermore, technology constrains autonomy and creativity. This assertion is supported by Ehrensberger-Dow and Heebs (2016), who focused on cognitive-ergonomic research. They characterised existing tools as lacking user-friendliness and identified the inadequate alignment of interfaces and functionalities with translators' intuitive working methods, leading to cognitive overload (van Egdom, 2022). Similarly, Van Massenhove et al. (2019) conducted research on machine translation and post-editing, revealing a decrease in lexical richness in translations, including literary translations, due to the influence of automatic translation. Additionally, Guerberof-Arenas and Toral (2022) observed a dearth of creative solutions in (semi-)automated processes, as translators tend to opt for pre-digested solutions. It should therefore be noted that creative solutions for literary translation using automation often have been operationalized in ways that appear far from creative themselves, i.e. they are still far removed from the creative hub and buzz that is human translation.

7 Automated Translation in Cultural Production

Against a background of selected historical developments in technology in the field of translation production introduced in the previous section, this next section outlines the benefits and advantages of the use of automated translation in cultural production. As these characteristics also include the commodification of language, seemingly a very democratic development, the authors posit that through the use of English as a relay or pivot language, evident in the added and renewed dominance of English in AI, recent technological advances are tantamount to stealth cultural imperialism.

In the winter of the academic year 2002–2003, the translation studies programme MScTrans at Imperial College hosted Yves Champollion, the CEO of Wordfast (both a company and CAT tool), as a visiting lecturer. In an inspiringly different take on technological developments, Champollion developed his vision of the future of translation and language technology by using an analogy. In the 1987 film *No Way Out* (1987), Kevin Costner plays the role of a double agent, being an American officer as well as a Russian spy. A Polaroid picture had been taken during a romantic encounter, but the photo had been pulled out too fast. Although it was a fictional prospect at the time, the story continued with the intelligence services developing a way in which the incomplete Polaroid could be improved to such an extent that the real picture would eventually show: algorithms read all the pixels and dots of the picture deciding on colour, shape, and content of each depending on the surrounding pixels. By reading line after line, these became clearer and started to become a proper picture, on the basis of which the identity of the double agent would be revealed. According to Champollion, this technique pointed to the future of machine translation: semi- to nearly fully independent reasoning and analysis of data to determine the most probable outcome.

Neural machine translation mimics the neural networks of the brain, processing lots of data points and working in a very similar way. Moreover, given the recent boost of artificial intelligence via inventions such as ChatGPT, the allure of technology capable of producing (synthetic) text or of seemingly seamless translation between two different language pairs is evident from fictitious examples over the centuries. These capabilities are now at the fingertips of many users and digital consumers. How-ever, while language technology such as neural machine translation and translation by AI platforms appear to be digital manna that offers promising opportunities, limitations and challenges emerge as well.

Among the main challenges of both automated translation approaches (MT and/or AI) is accuracy and nuance in translations. Machine translation systems struggle with rendering source text specificities such as cultural nuances, idiomatic expressions, and context-specific meanings (see also Hutchins, 1997; Krings, 2001). This often results in loss or distortion of the original intent. Additionally, the quality and accuracy of translations can vary depending on the complexity of the source text. Highly technical, literary, or culturally specific works pose significantly proprietary challenges for machine translation algorithms and large language models. Typically, automated translations of literary works are not deemed fit for purpose, if feasible at all, because of their complex nature, anticipated distortion of the original intent, and renderings of source text specificities that make up the quality of a literary source text (see above). While translations that are the result of automation might have relevance in specific domains and for specific uses (as explored above), they struggle to capture the richness of literary works, including their stylistic subtleties and cultural nuances. This deficiency can lead to distorted translations that lack the depth and authenticity of the original. Furthermore, over-reliance on machine translation may result in a homogenization of literary styles (see Kenny & Winters, 2020), erasing the unique voices of authors, reducing diversity in the literary landscape, and flattening the much-coveted rich exuberances of (translated) literature. Studies suggest that even post-editing by human translators may not fully mitigate this issue (Macken et al., 2022), as automated translation creates a priming effect, allowing the mechanical and source-driven voice of the MT to affect and permeate post-edited texts (Van Egdom, 2023; Castilho & Resende, 2022). However, in a scenario where no financial resources or human translator capacity is available for a language pair in which one is of a lesser diffusion composure, automated translation can offer a first glimpse of the literature of the source language. It does so in terms of new output, more historical output, ideas for a retranslation (either by assumption or by critical rejection), detecting societal attitudes in previously untranslated works, and proposing a new human translation on the basis of a (partially) automated translation.

Critical considerations of the use of language technology for increased up front dispersal of cultural production not only relate to the features of the source text. The target text has issues, too, not least because it is the result of a specifically designed architecture and the assorted dataset on which it was trained initially. Therefore, the whole automated aspect of the language and cultural production does not correlate well with the intention of a literary text, let alone the anticipated fit for the purposes of its translation. A first consideration here relates to the presence of bias. Both neural machine translation models and large language models for AI text generators are developed using vast amounts of training data. The data used to train LLMs can reflect societal biases and prejudices present in the source material, which can then be perpetuated or even amplified by the model's responses (see for instance Fitria (2021) and Li (2023)). If training data predominantly represent certain demographics (85% of data from English and main Western languages) and thus include its assorted biased content, the resulting model will rely on that bias as well to produce text. Arguably, the hyped craze about AI capabilities is still very much limited to those on the beneficiary side of the resourcedness gap: most benefits befall English-speaking users (Vashee, 2023). This dominant position of English in AI is a given, whether within or outside a translation context.

If training data predominantly consists of texts written by or about specific racial or gender groups, systemically biased behaviour will result (Scott et al., 2023). Unequal representation in terms of cultural, gender, or racial biases in translated texts is not easily remedied technically. Such bias requires measurjes throughout the development process: pre-training and fine-tuning techniques can be employed to mitigate bias, including careful curation of training data, i.e. conscious diversification of data sources instead of 'blind' harvesting of data. Also, ongoing evaluation can help identify and rectify biases present in the model's outputs. Regardless of the measures in place to overcome possible bias, upfront content warnings are initially required.

Warnings in terms of using automated translation platforms, whether they are MT and/or AI, are also needed for the privacy and security of data, whether the ones used for training, or the ones being used for input to generate results (see also Ranade et al., 2018; Federici et al., 2023; Li, 2023). Ideally, initiatives to warn users of the dangers of relying on AI should coincide with campaigns to promote the visibility of translators and their copyrighted work to be excluded from data harvesting prior to training AI models (Lison et al., 2021). In order to avoid potential privacy implications, measures need to be put in place to anonymise or aggregate the data (see for instance Senavirathne & Torra, 2020) and therefore to mitigate the risk of re-identification.

The privacy of data that is added after the initial creation of the LLM must be considered: if NMT or LLM is designed to interact with or generate responses based on user input, there is a need to handle user data with utmost care so that there is no breach of trust (as this leads to failure to attract or keep users). In the absence of a system in which informed consent is needed upfront, transparency becomes crucial in ensuring that users understand how their data will be used and that the safeguards in place to protect their privacy are adequate and airtight. Strict encryption techniques should be implemented to safeguard the data during transmission and secure communication protocols need to be in place for the data that are being kept. There are measures in place to address these concerns. Adherence to privacy regulations and standards is a noble intention but given possible national differences, virtual and critical proximity but physical distance might easily mean that different data handling protocols are in place for the location of input, the location of data storage, and the main location of the service providing entity, respectively. This is even though data protection laws such as the General Data Protection Regulation (GDPR) provide guidelines on handling personal data and require organisations to implement necessary safeguards. Additionally, privacy impact assessments and audits can help identify and mitigate potential privacy risks associated with LLMs.

8 Discussion

In recent years, there has been a growing call for diversification in the literary landscape, both in terms of representation of individuals in society and the broad palette of cultural influences introduced through translation. This call aims to break the dominance of English and promote a more inclusive and diverse literary scene.

Representation and inclusivity are crucial aspects of any creative field, including literature. The lack of diversity in the translation sector is a significant concern, particularly regarding ethnicity, gender, sexual orientation, ability, age, and religious and cultural background. Efforts are being made to address this issue and bring more diversity into the profession of literary translation such as the EU's *Translators on the Cover*, American PEN's *The 2023 Manifesto on Literary Translation*, and the Dutch Language Union's *Translating for the Future*. One key aspect of promoting diversity is improving transparency, accountability, and inclusivity in publishing processes. This involves hiring a diverse range of editorial staff, freelance translators, and editors. However, addressing diversity goes beyond recruitment. It requires interventions in the training of translators starting at an early stage, outreach in schools to break down implicit barriers to language learning and literature, and offering a much more prolific choice in terms of source languages of literary cultures. By creating a more inclusive and accessible environment for aspiring translators, we can foster a diverse pool of talent in the translation profession.

As mentioned earlier, the global literature market is often criticised for its lack of multilingualism. Instead of embracing linguistic and cultural diversity, national markets are often dominated by domestic literature and translated Anglo-American books. This trend limits the exposure of readers to diverse voices and perspectives. To combat this lack of diversity, a good many publishers are taking steps. However, it has been reported that these actors face financial risks due to the uncertain market for literary works, particularly translated works. Public funding at national and European levels has proven to be indispensable in the efforts to mitigate these risks and enable the circulation of diverse literary works across borders. Adequate cultural funding brings diverse voices to readers across the world, as is the case for instance with poetry in the Netherlands, where English does not retain the dominant presence it has as in fiction such as romance literature, Young Adult literature, and crime fiction. Still, the dominance of English in the literary landscape is not an isolated issue; it reflects a broader trend in cultural content sectors. Music, films, and television series also face similar challenges, with English content dominating the global market. In order to break free from this cultural hegemony, concerted efforts are required to promote content in other languages and facilitate cross-border circulation.

It should be noted that isolated attempts to bring about change will probably have little effect: attaining diversity and inclusion in the literary landscape is a goal that requires collaboration and commitment from all key players in the cultural industry, including publishers, translators, educators, and policymakers. By actively promoting representation, providing adequate funding, and encouraging the circulation of diverse literature, a vibrant and inclusive literary scene that reflects the richness of global society is within reach, and always with the name of real translators on the cover. Publishers can therefore play a crucial role as cultural gatekeepers. While there are actors such as scouts and agents who play a leading role in the selection of works (see Van Voorst, 1997, for instance), it is undisputed that publishers make the actual choice of works. This means they can also play a key role in promoting diversity in terms of the origin of works in the target culture.

Machine translation can serve as a cultural-political tool to achieve the emancipatory ideal of a global literary revival on a larger scale ('Weltliteratur'). This entails a profound emancipation of world literature, representing peripheral and hyper-peripheral cultures beyond the boundaries of their own setting. Placing machine translation within a cultural-political framework and connecting it with the notion of 'Weltliteratur' opens up perspectives. We argue that machine translation can be a key to realising a 'Weltliteratur.' In terms of cultural imperialism, machine translation can be employed as a cultural-political instrument to enrich Dutch culture for instance. The digital book market plays a crucial role in this regard. It is known that the Dutch book market heavily relies on original publications and translations from English. Due to the cultural dominance of Anglo-Saxon literature, the economic risks associated with publishing translations from this literature are considered smaller. With the digitalisation of the book market, we observe a decrease in the costs associated with publishing a book (including translations). Consequently, it may become more attractive to publish works from less 'central' languages. By strategically utilizing machine translation, publishers could make titles from other language areas available without significant risks, thereby contributing to the emancipation of world literature. Consequently, the Netherlands can enhance its global reputation by becoming a shining example of (culturally inclusive) literature. Furthermore, this creates a fertile ground for the local culture in question, as importing foreign literature is a proven means of enriching a 'peripheral' culture. Finally, the pioneering (machine) translation can serve as a catalyst for retranslation by skilled translators from the Dutch literary field.

9 Conclusion: Implications and Future Directions

This chapter establishes the use of language technology, in particular automated translation (MT and/or AI), as a potential solution to address unequal cultural exchanges and promote diversification of cultural production and reception, especially the translation of literary works when no human translation is available (just yet). Regardless of physical proximity, virtual proximity aligned with critical proximity allows cultural production in a language of lesser diffusion or a lesser-resourced language to open new pathways of cultural exchange through automated translation. This happens along two lines: first, cultural production in a more widely available language seeking to increase its uptake of lesser-known languages for which no sufficient human translator capacity is available most certainly has use for automated translation. Second, AT is even more crucial for inbound inclusion – the more resourced receptor culture extending to lesser-resourced languages through the dissemination/gisting prism. Questions remain, however, whether or not natural voices and cultures are affected by the automation of (likely biased and limited) language models. Such automation possibly adds more to stealth imperialism of the core language of the data AT relies on in the more dominant culture than actually providing opportunities for lesser-resourced languages and their assorted cultures. If we settle for accepting a decent gist of a new literary work in a language our human translation capacity has no means of accommodating, will literary nuances and cultural intricacies not be lost upon the readership of automated translation? Reception studies are very much needed so that those who mull over this right now will not be considered old bores in the near future because they keep referring to how things used to be without being able to properly back up claims.

However difficult if nigh impossible ensuring fairness is from the user's perspective, there is one major opportunity that automated translation offers which can to some extent help and compensate for the challenges. By opening up lesser-resourced languages to more widely used ones through automated translation of literary works of those smaller languages, a more accurate representation of diverse voices and cultures in the literary field is possible. Therefore, language technology, particularly automated translation (MT and/or AI) holds promising potential as an emancipatory facilitator in the democratisation of the global book market, to help overcome linguistic divides that are still in place. Therefore, MT/AI allows for increased cultural proximity through the liberalisation of access to literatures of lesser-resourced cultures.

Future research should investigate the responses and preferences of readers and consumers towards machine-translated literature from lesser-resourced languages. It is important to examine how local literary traditions and cultural specificities influence the adoption and adaptation of machine translation technologies for translating literature. Understanding these aspects would contribute to enhancing the effectiveness and acceptance of machine-translated literary texts and promoting linguistic and cultural diversity in translated literature.

While acknowledging reservations, critical considerations, and potential objections, the authors do not aim to anticipate widespread support for taking the chapter's proposal of using automated translation as a key to unlock cultural transfer for lesser-resourced languages. Determining who should take the lead in implementing machine translation, voicing doubts about the impact of automatic translation quality on foreign authors and cultures, determining the potential loss of literary qualities, and assessing the risk of undermining the societal image of translators are all themselves ethical dilemmas, not to mention copyright protection. Does the use of NMT or LLMS for this lean localisation of cultures by means of English and Western languages dominating data sets and architectures not unintentionally reinforce the status and dominance of said languages? Despite these barriers, the authors believe that publishers, driven by idealism and the pursuit of opening up to global literatures, will eventually consider automatic alternatives to non-translation.[10]

References

Arthern, P. J. (1981). Aids unlimited: the scope for machine aids in a large organization. *Aslib Proceedings*, 33(7), 309–319.

Barnes, L. W. (1972). *The language of incentive in Jonathan Swift's* Gullivers Travels. Morehead State University.

Benkova, L., Munkova, D., Benko, L., & Munk, M. (2021). Evaluation of English–Slovak neural and statistical machine translation. *Applied Sciences*, 11(7), 2948.

Bentivogli, L., Bisazza, A., Cettolo, M., & Federico, M. (2016). Neural versus phrase-based machine translation quality: A case study. In J. Su, K. Duh, & X. Carreras (Eds.), *Proceedings of the 2016 Conference on Empirical Methods in Natural Language Processing* (pp. 257–267). Association for Computational Linguistics. Austin, Texas, USA. https://doi.org/10.18653/v1/D16-1025.

Besacier, L. & Schwartz, L. (2015). Automated translation of a literary work: a pilot study. *Proceedings of NAACL-HLT Fourth Workshop on Computational Linguistics for Literature* [Denver, Colorado, June 4, 2015], 114–122.

Blommaert, J. (2010). *The sociolinguistics of globalisation*. Cambridge: Cambridge University Press.

Bonner, A. (1994). *Doctor illuminatus: A Ramon Llull reader*. Princeton: Princeton University Press.

Booth, A. D. & Booth, K. H. V. (1956). *Automatic digital calculators* (2nd ed.). London: Butterworths Scientific Publications.

Broomans, P., F. Duan, & Hedberg, A. (2022). Views on World Literature: Cultural transfer and translation in the context of an Online International Exchange (OIE) project – A case study of China, the Netherlands and Sweden. In T. Clark & P. Gordon (Eds.), *Beyond Babel. Scholarly organizations and the study of languages and literatures* (pp. 77–95). Amsterdam/Philadelphia: John Benjamins Publishing Company.

10 For more on machine translation as a cultural-political means, see van Egdom (2022).

Brown, P. F., J. Cocke, S. A. Della Pietra, V. J. Della Pietra, F. Jelinek, J. D. Lafferty, R. L. Mercer, & Roossin, P. S. (2003). A Statistical Approach to Machine Translation. In S. Nirenbrug, H. L. Somers & Y. Wilks (Eds.), *Reading in Machine Translation,* pp. 355–362. Cambridge/London: MIT Press.

Castilho, S., & Resende, N. (2022). Post-editese in literary translations. *Information,* 13(2), 66.

Christensson, P. (2011, March 12). Boolean Definition. *TechTerms,* https://techterms.com/definition/boolean.

Collet, B. (2015). From intermarriage to conjugal mixedness: Theoretical considerations illustrated by empirical data in France. *The ANNALS of the American Academy of Political and Social Science,* 662(1), 129–147.

Comen, E. (2018). Check out how much a computer cost the year you were born. *USA Today,* https://eu.usatoday.com/story/tech/2018/06/22/cost-of-a-computer-the-year-you-were-born/36156373/.

Daylight, E. G. (2015). Towards a historical notion of Turing – the father of computer science. *History and Philosophy of Logic,* 36(3), 205–228.

Declercq, C. (2012). Advertising and localization. In K. Malmkjær & K. Windle (Eds.), *The Oxford Handbook of Translation Studies,* pp. 262–272. Oxford: Oxford University Press.

Dewey, M. (2010). English as a lingua franca and globalization: An interconnected perspective. *International Journal of Applied Linguistics,* 17(3), 332–354.

Ehrensberger-Dow, M., & Heeb, A. H. (2016). Investigating the ergonomics of a technologized translation workplace. In R. Muñoz Martín (Ed.), *Reembedding translation process research,* pp. 69–88. Amsterdam: John Benjamins.

Elias, A., & Mansouri, F. (2020). A systematic review of studies on interculturalism and intercultural dialogue. *Journal of Intercultural Studies,* 41(4), 490–523.

Even-Zohar, I. (1979). Polysystem theory. *Poetics Today,* 1(1/2), 287–310.

Federici, F. M., Declercq, C., Cintas, J. D., & Piñero, R. B. (2023). Ethics, automated processes, machine translation, and crises. In H. Moniz & C. Parra Escartín (Eds.). *Towards responsible machine translation: Ethical and legal considerations in machine translation* (pp. 135–156). Cham: Springer International Publishing.

Fitria, T. N. (2021). Gender bias in translation using google translate: Problems and solution. *Language Circle: Journal of Language and Literature,* 15(2), 285–292.

Franssen, T. (2015). Diversity in the large-scale pole of literary production: An analysis of publishers' lists and the Dutch literary space, 2000–2009. *Cultural Sociology,* 9(3), 382–400.

Garcia, I. (2014). Computer-aided translation. In Chan Sin-wai (Ed.), *The Routledge Encyclopedia of Translation Technology,* pp. 68–87. Abingdon: Routledge.

Guerberof-Arenas, A. & Toral, A. (2020). The impact of post-editing and machine translation on creativity and reading experience. *Translation Spaces,* 9(2), 255–282.

Guerberof-Arenas, A. & Toral, A. (2022). Creativity in translation: Machine translation as a constraint for literary texts. *Translation Spaces* 11(2), 184–212.

Haigh, T. & Priestley, M. (2016). Historical reflections: Where code comes from: architectures of automatic control from Babbage to Algol. *Communications of the ACM*, 59(1), 39–44.

Hartley, A., & Paris, C. (1997). Multilingual document production from support for translating to support for authoring. *Machine Translation*, 12(1–2), 109–129.

Hassan, H., Aue, A., Chen, C., Chowdhary, V., Clark, J., Federmann, C., Huang, X., Junczys-Dowmunt, M., Lewis, W., Li, M., & Liu, S. (2018). Achieving human parity on automatic chinese to english news translation. *arXiv preprint arXiv:1803.05567*.

Heilbron, J., Sapiro, G. (2007). Outline for a sociology of translation: Current issues and future prospects. In M. Wolf & A. Fukari (Eds.), *Constructing a Sociology of Translation*, pp. 93–107. Amsterdam: John Benjamin Publishing.

Hirschberg, J. & Manning, C. D. (2015). Advances in natural language processing. *Science* 349(6245), 261–266.

Hoover, C. & Sommer, H. (2010). Automated translation of Chinese-to-English creative literature. *Minds@UW*, University of Wisconsin, Student Research Day. https://minds.wisconsin.edu/handle/1793/47395

House, J. (2003). English as a lingua franca: A threat to multilingualism? *Journal of Sociolinguistics*, 7(4), 556–578.

Hu, K. & Li, X. (2023). The creativity and limitations of AI neural machine translation. A corpus-based study of DeepL's English-to-Chinese translation of Shakespeare's plays. *Babel* 69(4), 546–563.

Hutchins, J. (1994). Machine translation: History and general principles. In R.E. Asher (Ed.), *The Encyclopaedia of Languages and Linguistics* (vol. 5), pp. 2322–2332. Oxford: Pergamon Press.

Hutchins, J. (1997). From first conception to first demonstration: The nascent years of machine translation, 1947–1954. a chronology. *Machine Translation*, 12, 195–252.

Hutchins, J. (1998). The origins of the translators workstation. *Machine Translation*, 13, 287–307.

Hutchins, J. (1999). From the archives … Warren Weaver Memorandum, July 1949. *MT News International*, 22, 5–6.

Hutchins, J. (2001). Machine translation and human translation: In competition or in complementation? *International Journal of Translation*, 13(1–2), 1–20.

Hutchins, J. (2013). ALPAC: The (In)Famous Report. In S. Nirenbrug, H.L. Somers & Y. Wilks (Eds.), *Reading in Machine Translation*, pp. 131–136. Cambridge/London: MIT Press.

Hutchins, J. (2014). The history of machine translation in a nutshell. *Technical Report*. University of East Anglia.

Kaibao, H. & Xiaoqian, L. (2023). The creativity and limitations of AI neural machine translation. *Babel*, 69(4), 546–563.

Kenny, D., & Winters, M. (2020). Machine translation, ethics and the literary translator's voice. *Translation Spaces*, 9(1), 123–149.

Kloepfer, R. & Shaw, P. (1981). Intra- and Intercultural Translation. *Poetics Today* 2 (4), 29–37.

Kofi Adarkwah, G. (2022). Why "Think Globally, Act Locally" is a dangerous strategy for emerging markets. *California Management Review*. Online, https://cmr.berkeley.edu/2022/07/why-think-globally-act-locally-is-a-dangerous-strategy-for-emerging-markets/

Kotze, H. (2020). Translation, contact linguistics and cognition. In F. Alves & A. Jakobsen (Eds.) *The Routledge Handbook of Translation and Cognition*, pp. 113–132. Abingdon: Routledge.

Kotzsch, R. (2021). 5 CEO Insights for the Localization Industry in 25 Years. *LinkedIn post*, https://www.linkedin.com/pulse/7-ceo-insights-localization-industry-25-years-roman-kotzsch/

Krings, H. P. (2001). *Repairing texts: Empirical investigations of machine translation post-editing processes* (Vol. 5). Kent Ohio & London: Kent State University Press.

Lavington S. (Ed.) (2012). *Alan Turing and his contemporaries: Building the world's first computers*. Swindon: BCS.

Lebedev, M. A., & Nicolelis, M. A. (2017). Brain-machine interfaces: From basic science to neuroprostheses and neurorehabilitation. *Physiological Reviews*, 97(2), 767–837.

Lee, A. (2023, January 26). 'What Are Large Language Models Used For?'. *Nvidia.com* https://blogs.nvidia.com/blog/2023/01/26/what-are-large-language-models-used-for/.

Li, B. (2023). Ethical issues for literary translation in the Era of artificial intelligence. *Babel*, 69(4), 529–545.

Lison, P., Pilán, I., Sánchez, D., Batet, M., & Øvrelid, L. (2021, August). Anonymisation models for text data: State of the art, challenges and future directions. In *Proceedings of the 59th Annual Meeting of the Association for Computational Linguistics and the 11th International Joint Conference on Natural Language Processing*, Vol.1., 4188–4203. Online. Association for Computational Linguistics.

Macken, L., Vanroy, B., Desmet, L., & Tezcan, A. (2022). Literary translation as a three-stage process: Machine translation, post-editing and revision. In *23rd Annual Conference of the European Association for Machine Translation*. European Association for Machine Translation, 101–110. Ghent, Belgium. European Association for Machine Translation.

Masera, G. (2022). Statistics vs. Machine Learning – and When to Use Either One? Levity.ai, https://levity.ai/blog/statistics-vs-machine-learning

McCulloch, W. & Pitts, W. (1943). A logical calculus of the ideas imminent in nervous activity. *Bulletin of Mathematical Biology*, 52(1/2), 99–115.

Melby, A. (1996). Machine translation and other translation technologies. *Annual Review of Applied Linguistics*, 16, 86–98.

ML6.EU. (2023). Large Language Models, https://www.ml6.eu/resources/large-language-models

Moorkens, J. (2020). Translation in the neoliberal era. In E. Bielsa & D. Kapsaskis (Eds.), *The Routledge Handbook of Translation and Globalization*, pp. 323–336. Abingdon: Routledge.

Muysken, P. (2013). Language contact outcomes as the result of bilingual optimization strategies. *Bilingualism: Language and Cognition*, 16(4), 709–730.

Nickerson, C. (2005). English as a lingua franca in international business contexts. *English for Specific Purposes*, 24(4), 367–380.

Pettigrew, T. F. (1998). Intergroup contact theory. *Annual Review of Psychology*, 49 (1), 65–85.

Phillipson, R. (2008). Lingua franca or lingua frankensteinia? English in European integration and globalisation. *World Englishes*, 27(2), 250–267.

Press, G. (2016). A very short history of Artificial Intelligence (AI). *Forbes*, https://www.forbes.com/sites/gilpress/2016/12/30/a-very-short-history-of-artificial-intelligence-ai/?sh=33f261376fba

Ramzai, J. (2020). Clearly explained: How machine learning differs from statistical modeling. *Towards Data Science*, https://towardsdatascience.com/clearly-explained-how-machine-learning-differs-from-statistical-modeling-967f2c5a9cfd.

Ranade, P., Mittal, S., Joshi, A., & Joshi, K. (2018, November). Using deep neural networks to translate multi-lingual threat intelligence. In *2018 IEEE International Conference on Intelligence and Security Informatics (ISI)*. IEEE, 238–243. Miami, FL, USA.

Robinson, J. A. (1994). Logic, computers, Turing, and von Neumann, in K. Furukawa, D. Michie, and S. Muggleton, *Machine Intelligence 13: Machine Intelligence and Inductive Learning*, pp. 1–35. Oxford: Clarendon Press.

Rodgers, J. (2017). The Genealogy of an Image, or, What Does Literature (Not) Have To Do with the History of Computing?: Tracing the Sources and Reception of Gulliver's Knowledge Engine. *Humanities*, 6(4), 85–93.

Sapiro, G. (2003). The literary field between the state and the market. *Poetics*, 31(5–6), 441–464.

Scott, B., J. Woods & Chang, A. (2023). How AI could perpetuate racism, sexism and other biases in society. *NPR*, https://www.npr.org/2023/07/19/1188739764/how-ai-could-perpetuate-racism-sexism-and-other-biases-in-society.

Senavirathne, N., & Torra, V. (2020, December). On the role of data anonymization in machine learning privacy. In *2020 IEEE 19th International conference on trust, security and privacy in computing and communications (TrustCom)*. IEEE, 664–675. Guangzhou, China.

Shuttleworth, M. (2019). Polysystem theory. In M. Baker & G. Saldanha (Eds.), *Routledge Encyclopedia of Translation Studies*, pp. 419–423. London: Routledge.

Snell, B. (1978). Introduction [to Translating and the Computer 1978]. https://aclanthology.org/1978.tc1.0.pdf,vi.

Straubhaar, J. D. (1991) Beyond media imperialism: Asymmetrical interdependence and cultural proximity. *Critical Studies in Mass Communication*, 8(1), 39–59.

Straubhaar, J. D. (2003). Choosing National TV: Cultural capital, language, and cultural proximity in Brazil. In M.G. Elasmar (Ed.), *The Impact of International Television: A Paradigm Shift*, pp. 77–110. Abingdon/New York: Routledge.

Swift, Jonathan. (1726/1977). *Gulliver's Travels*. New York/ Toronto: Oxford University Press.

Toral, A., S. Castilho, K. Hu & Way, A. (2018). Attaining the unattainable? Reassessing claims of human parity in neural machine translation. *Proceedings of the Third Conference on Machine Translation: Research Papers* [Brussels, Belgium. Association for Computational Linguistics], 113–123.

Toury, G. (1995). *Descriptive Translation Studies and Beyond*. Amsterdam/Philadelphia, John Benjamins.

Van de Cruys, T. (2022). Constraint-based neural architectures for the translation of literary texts. Paper presented at *Network of Interdisciplinary Translation Studies in the Netherlands and Flanders (NITS) Conference*. Groningen, The Netherlands.

Van Egdom, G.-W. (2022) Machinevertaling als cultuurpolitiek instrument. *Webfilter* dossier. https://www.tijdschrift-filter.nl/webfilter/dossier/literair-vertalen-en-technologie/september-2022/machinevertaling-als-cultuurpolitiek-instrument/.

Van Egdom, G. M. W. (2023). Translation criticism in the digital literary sphere: Reader responses to Portuguese literature across the globe and translated literature in Lusophone countries. *Qorpus*, 13(1), 39–60.

Van Egdom, G.-W., O. Kosters & Declercq, C. (2023). The riddle of (literary) machine translation quality: Assessing automated quality evaluation metrics in a literary context. *Tradumatica*, 21, 129–159.

Van Massenhove, E., D. Shterionov & Way, A. (2019). Lost in Translation: Loss and Decay of Linguistic Richness in Machine Translation. *Proceedings of the 17th Machine Translation Summit* [MTSummit2019, Dublin, Ireland, August 2019].

Van Voorst, S. (1997). *Weten wat er in de wereld te koop is. Goed beeld van liberalisering in NL*. The Hague: SDU.

Vashee, K. (2023). Making generative AI effectively multilingual at scale. *ModernMT.com* blog, https://blog.modernmt.com/making-generative-ai-multilingual-at-scale/.

Verma, C., S. Arora & Jain, A. (2020). *Machine learning in healthcare: Separating myth from reality. International Journal of Advanced Science and Technology*, 29(8), 1348–1355.

Visbal, O. (2009). *The erosion of stereotypes through intercultural exchange programs: Testing Pettigrew's contact theory*. University of Hamburg: PhD Dissertation.

Zhang, C. & Cai, H. (2015). On technological turn of translation studies: Evidences and influences. *Journal of Language Teaching and Research*, 6(2), 429–434.

Index

www.ingramcontent.com/pod-product-compliance
Lightning Source LLC
Chambersburg PA
CBHW071333230926
51849CB00002B/390
* 9 7 8 9 0 0 4 7 1 1 3 8 9 *